Strengthening and rehabilitation of civil infrastructures using fibre-reinforced polymer (FRP) composites

Related titles:

Durability of composites for civil structural applications
(ISBN 978-1-84569-035-9)
This comprehensive book addresses the current lack, or inaccessibility, of data related to the durability of fibre-reinforced polymer composites which is proving to be one of the major challenges to the widespread acceptance and implementation of these materials in civil infrastructure. Part I discusses general aspects of composite durability. Chapters examine mechanisms of degradation such as moisture, aqueous solutions, UV radiation, temperature, fatigue and wear. Part II then discusses ways of using fibre-reinforced polymer composites, including the strengthening and rehabilitation of existing structures. This section also covers monitoring techniques such as structural health monitoring.

Advanced civil infrastructure materials: science, mechanics and applications
(ISBN 978-1-85573-943-7)
In recent decades, material development in response to the call for more durable infrastructures has led to many exciting advancements. Fibre-reinforced composites, with very unique properties, are now being considered for use in many infrastructural applications. Even concrete and steel are being steadily improved to have better properties and durability. This book provides an up-to-date review of several emerging infrastructure materials that may have a significant impact on repair and/or new construction. Chapters include examples of real world applications using such advanced materials.

Durability of engineering structures: design, repair and maintenance
(ISBN 978-1-85573-695-5)
Structures often deteriorate because not enough attention is given to durability during the design stage. Since failure incurs high maintenance and repair costs, knowledge of the long-term behaviour of materials is the basis for avoiding these costs. *Durability of engineering structures* looks at the design of buildings for service life, and effective maintenance and repair techniques in order to reduce the likelihood of failure. It describes the *in-situ* performance of all the major man-made materials used in civil engineering construction. In addition, some relatively new high-performance materials are discussed. Deterioration mechanisms and the measures to counteract these, as well as subsequent maintenance and repair techniques are also considered.

Details of these and other Woodhead Publishing materials books, as well as materials books from Maney Publishing, can be obtained by:

- visiting our web site at www.woodheadpublishing.com
- contacting Customer Services (e-mail: sales@woodhead-publishing.com; fax: +44 (0) 1223 893694; tel: +44 (0) 1223 891358 ext. 130; address: Woodhead Publishing Ltd, Abington Hall, Granta Park, Great Abington, Cambridge CB21 6AH, England)

If you would like to receive information on forthcoming titles, please send your address details to: Francis Dodds (address, tel. and fax as above; e-mail: francisd@woodhead-publishing.com). Please confirm which subject areas you are interested in.

Maney currently publishes 16 peer-reviewed materials science and engineering journals. For further information visit www.maney.co.uk/journals

Strengthening and rehabilitation of civil infrastructures using fibre-reinforced polymer (FRP) composites

Edited by
L.C. Hollaway and J.G. Teng

Woodhead Publishing and Maney Publishing
on behalf of
The Institute of Materials, Minerals & Mining

CRC Press
Boca Raton Boston New York Washington, DC

WOODHEAD PUBLISHING LIMITED
Cambridge England

Woodhead Publishing Limited and Maney Publishing Limited on behalf of
The Institute of Materials, Minerals & Mining

Woodhead Publishing Limited, Abington Hall, Granta Park, Great Abington
Cambridge CB21 6AH, England
www.woodheadpublishing.com

Published in North America by CRC Press LLC, 6000 Broken Sound Parkway, NW,
Suite 300, Boca Raton, FL 33487, USA

First published 2008, Woodhead Publishing Limited and CRC Press LLC
© 2008, Woodhead Publishing Limited
The authors have asserted their moral rights.

This book contains information obtained from authentic and highly regarded sources. Reprinted material is quoted with permission, and sources are indicated. Reasonable efforts have been made to publish reliable data and information, but the authors and the publishers cannot assume responsibility for the validity of all materials. Neither the authors nor the publishers, nor anyone else associated with this publication, shall be liable for any loss, damage or liability directly or indirectly caused or alleged to be caused by this book.

Neither this book nor any part may be reproduced or transmitted in any form or by any means, electronic or mechanical, including photocopying, microfilming and recording, or by any information storage or retrieval system, without permission in writing from Woodhead Publishing Limited.

The consent of Woodhead Publishing Limited does not extend to copying for general distribution, for promotion, for creating new works, or for resale. Specific permission must be obtained in writing from Woodhead Publishing Limited for such copying.

Trademark notice: Product or corporate names may be trademarks or registered trademarks, and are used only for identification and explanation, without intent to infringe.

British Library Cataloguing in Publication Data
A catalogue record for this book is available from the British Library.

Library of Congress Cataloging in Publication Data
A catalog record for this book is available from the Library of Congress.

Woodhead Publishing ISBN 978-1-84569-448-7 (book)
Woodhead Publishing ISBN 978-1-84569-489-0 (e-book)
CRC Press ISBN 978-1-4200-8774-1
CRC Press order number: WP8774

The publishers' policy is to use permanent paper from mills that operate a sustainable forestry policy, and which has been manufactured from pulp which is processed using acid-free and elementary chlorine-free practices. Furthermore, the publishers ensure that the text paper and cover board used have met acceptable environmental accreditation standards.

Typeset by SNP Best-set Typesetter Ltd., Hong Kong
Printed by TJ International Ltd, Padstow, Cornwall, England

Contents

Contributor contact details   xi
Preface   xv

1 Structurally deficient civil engineering infrastructure: concrete, metallic, masonry and timber structures   1
L. De Lorenzis, University of Salento, Italy;
T. J. Stratford, University of Edinburgh, UK;
L. C. Hollaway, University of Surrey, UK

1.1	Introduction	1
1.2	Structural deficiencies	2
1.3	Structural deficiencies in concrete structures	4
1.4	Strengthening concrete structures using FRP composites	5
1.5	Metallic materials used in civil infrastructure	16
1.6	Structural deficiencies in metallic structures	17
1.7	Strengthening metallic structures using FRP composites	18
1.8	Overview of masonry structures	23
1.9	Structural deficiencies in masonry structures	23
1.10	Strengthening masonry structures using FRP composites	24
1.11	Overview of timber structures	31
1.12	Structural deficiencies in timber structures	31
1.13	Strengthening timber structures using FRP composites	32
1.14	Drivers for using FRP composites rather than conventional strengthening options	35
1.15	References	37

2 Fibre-reinforced polymer (FRP) composites used in rehabilitation   45
L. C. Hollaway, University of Surrey, UK

2.1	Introduction	45
2.2	Component parts of composite materials	47
2.3	Properties of matrices	47

2.4	Filaments and fibres	56
2.5	The advanced polymer composite	64
2.6	The mechanical properties of advanced polymer composites	67
2.7	The in-service properties of the advanced polymer composites	71
2.8	Conclusion	78
2.9	Sources of further information and advice	78
2.10	References	79
3	**Surface preparation of component materials**	**83**
	A. R. HUTCHINSON, Oxford Brookes University, UK	
3.1	Introduction	83
3.2	Adhesion and interfacial contact	85
3.3	Primers and coupling agents	86
3.4	Surface preparation	88
3.5	Surface preparation of concrete	90
3.6	Surface preparation of metallic materials	92
3.7	Surface preparation of timber	96
3.8	Surface preparation of FRP materials	98
3.9	Assessment of surface condition prior to bonding	102
3.10	The bonding operation	105
3.11	Quality assurance testing and acceptance criteria	107
3.12	Summary	108
3.13	References	108
4	**Flexural strengthening of reinforced concrete (RC) beams with fibre-reinforced polymer (FRP) composites**	**112**
	J. G. TENG, The Hong Kong Polytechnic University, China; S. T. SMITH, The University of Hong Kong, China; J. F. CHEN, The University of Edinburgh, UK	
4.1	Introduction	112
4.2	Failure modes	113
4.3	Flexural strength of an FRP-plated section	120
4.4	Interfacial stresses	126
4.5	Bond behaviour	127
4.6	Strength models for debonding failures	131
4.7	Design procedure	134
4.8	Application of the design procedure to indeterminate beams	138
4.9	Acknowledgement	138
4.10	References	138

5	Shear strengthening of reinforced concrete (RC) beams with fibre-reinforced polymer (FRP) composites	141
	J. F. CHEN, The University of Edinburgh, UK; J. G. TENG, The Hong Kong Polytechnic University, China	
5.1	Introduction	141
5.2	Strengthening schemes	142
5.3	Failure modes	143
5.4	Shear strengths of FRP-strengthened reinforced concrete beams	144
5.5	Design procedure	152
5.6	Acknowledgement	154
5.7	References	154
6	Strengthening of reinforced concrete (RC) columns with fibre-reinforced polymer (FRP) composites	158
	J. G. TENG and T. JIANG, The Hong Kong Polytechnic University, China	
6.1	Introduction	158
6.2	Behavior of FRP-confined concrete	159
6.3	Design-oriented stress–strain models for FRP-confined concrete	166
6.4	Analysis-oriented stress–strain models for FRP-confined concrete	172
6.5	Section analysis	176
6.6	Design of FRP-confined reinforced concrete columns	180
6.7	Acknowledgement	190
6.8	References	190
7	Design guidelines for fibre-reinforced polymer (FRP)-strengthened reinforced concrete (RC) structures	195
	J. G. TENG, The Hong Kong Polytechnic University, China; S. T. SMITH, The University of Hong Kong, China; J. F. CHEN, The University of Edinburgh, UK	
7.1	Introduction	195
7.2	General assumptions	196
7.3	Limit states and reinforced concrete design	196
7.4	Material properties: characteristic and design values	197
7.5	Flexural strengthening	198
7.6	Shear strengthening	205
7.7	Strengthening of columns with FRP wraps	207

7.8	Acknowledgement	211
7.9	References	211

8	**Strengthening of metallic structures with fibre-reinforced polymer (FRP) composites**	**215**
	T. J. STRATFORD, University of Edinburgh, UK	
8.1	Introduction	215
8.2	Critical issues in the design of FRP strengthening for metallic structures	215
8.3	Selection of strengthening materials	218
8.4	Design of flexural strengthening	221
8.5	Design in cases other than flexural strengthening	230
8.6	References	232

9	**Strengthening of masonry structures with fibre-reinforced polymer (FRP) composites**	**235**
	L. DE LORENZIS, University of Salento, Italy	
9.1	Introduction	235
9.2	General aspects of FRP strengthening for masonry structures	237
9.3	Bond of FRP systems to masonry	239
9.4	Strengthening of masonry panels under out-of-plane loads	243
9.5	Strengthening of masonry panels under in-plane loads	251
9.6	Strengthening of lintels and floor belts	254
9.7	Strengthening of arches and vaults	255
9.8	Confinement of masonry columns	259
9.9	Other applications	261
9.10	Detailing issues	262
9.11	References	264

10	**Flexural strengthening application of fibre-reinforced polymer (FRP) plates**	**267**
	L. CANNING and S. LUKE, Mouchel, UK	
10.1	Introduction	267
10.2	Unstressed FRP composite systems	268
10.3	Prestressed FRP composite systems	272
10.4	Stages within the FRP strengthening installation process	278
10.5	FRP strengthening site activities	286
10.6	Inspection, maintenance and monitoring	288
10.7	Relationship between application and design	289
10.8	Conclusions	290
10.9	References	290

11	Durability of externally bonded fiber-reinforced polymer (FRP) composite systems	292
	T. E. BOOTHBY and C. E. BAKIS, The Pennsylvania State University, USA	
11.1	Introduction	292
11.2	FRP composites	294
11.3	Externally bonded concrete–FRP hybrid systems	298
11.4	Externally bonded wood–FRP hybrid systems	303
11.5	Externally bonded masonry–FRP hybrid systems	306
11.6	Externally bonded steel–FRP hybrid systems	312
11.7	Conclusion	315
11.8	References	316
12	Quality assurance/quality control, maintenance and repair	323
	J. L. CLARKE, Concrete Society, UK	
12.1	Introduction	323
12.2	Deterioration and damage	324
12.3	The need for an inspection regime	326
12.4	Inspection during strengthening	329
12.5	Instrumentation and load testing	339
12.6	Inspection of strengthened structures	342
12.7	Routine maintenance and repair	346
12.8	Summary and conclusions	348
12.9	Acknowledgements	348
12.10	References	349
13	Case studies	352
	L. C. HOLLAWAY, University of Surrey, UK	
13.1	Introduction	352
13.2	The manufacturing systems used in the case studies considered	353
13.3	Case study 1: Reinforced concrete beams strengthening with unstressed FRP composites	355
13.4	Case study 2: Repair and strengthening the infrastructure utilising FRP composites in cold climates	362
13.5	Case study 3: Prestressed concrete bridges strengthening with prestressed FRP composites	366
13.6	Case study 4: Rehabilitation of aluminium structural system	367
13.7	Case study 5: Metallic structures strengthening with unstressed FRP composites	369

13.8	Case study 6: Metallic structures strengthening with prestressed FRP composites	373
13.9	Case study 7: Preservation of historical structures with FRP composites	376
13.10	Acknowledgements	381
13.11	Sources of further information and advice: design codes	381
13.12	References	383

Index   *387*

Contributor contact details

(* = main contact)

Editors

L. C. Hollaway
Faculty of Engineering and
  Physical Sciences
University of Surrey
Guildford
Surrey
GU2 7XH
UK
e-mail: l.hollaway@surrey.ac.uk

J. G. Teng
Department of Civil and Structural
  Engineering
The Hong Kong Polytechnic
  University
Hong Kong
China
e-mail: cejgteng@polyu.edu.hk

Chapter 1

L. De Lorenzis*
Department of Innovation
  Engineering
University of Salento
via per Monteroni
73100 Lecce
Italy
e-mail: laura.delorenzis@unile.it

T. J. Stratford
The University of Edinburgh
School of Engineering and
  Electronics
The King's Buildings
Mayfield Road
Edinburgh
EH9 3JL
UK
e-mail: tim.stratford@ed.ac.uk

L. C. Hollaway
Faculty of Engineering and
  Physical Sciences
University of Surrey
Guildford
Surrey
GU2 7XH
UK
e-mail: l.hollaway@surrey.ac.uk

Chapter 2

L. C. Hollaway
Faculty of Engineering and
    Physical Sciences
University of Surrey
Guildford
Surrey
GU2 7XH
UK
e-mail: l.hollaway@surrey.ac.uk

Chapter 3

A. R. Hutchinson
School of Technology
Oxford Brookes University
Wheatley Campus
Oxford
OX33 1HX
UK
e-mail: arhutchinson@brookes.
    ac.uk

Chapter 4

J. G. Teng*
Department of Civil and Structural
    Engineering
The Hong Kong Polytechnic
    University
Hong Kong
China
e-mail: cejgteng@polyu.edu.hk

S. T. Smith
Department of Civil Engineering
The University of Hong Kong
Pokfulam Road
Hong Kong
China
e-mail: stsmith@hku.hk

J. F. Chen
The University of Edinburgh
School of Engineering and
    Electronics
The King's Buildings
Mayfield Road
Edinburgh
EH9 3JL
UK
e-mail: J.F.Chen@ed.ac.uk

Chapter 5

J. F. Chen*
The University of Edinburgh
School of Engineering and
    Electronics
The King's Buildings
Mayfield Road
Edinburgh
EH9 3JL
UK
e-mail: J.F.Chen@ed.ac.uk

J. G. Teng
Department of Civil and Structural
    Engineering
The Hong Kong Polytechnic
    University
Hong Kong
China
e-mail: cejgteng@polyu.edu.hk

Chapter 6

J. G. Teng* and T. Jiang
Department of Civil and Structural
    Engineering
The Hong Kong Polytechnic
    University
Hong Kong
China
e-mail: cejgteng@polyu.edu.hk and
    tony.jiang@polyu.edu.hk

Chapter 7

J. G. Teng*
Department of Civil and Structural
  Engineering
The Hong Kong Polytechnic
  University
Hong Kong
China
e-mail: cejgteng@polyu.edu.hk

S. T. Smith
Department of Civil Engineering
The University of Hong Kong
Pokfulam Road
Hong Kong
China
e-mail: stsmith@hku.hk

J. F. Chen
The University of Edinburgh
School of Engineering and
  Electronics
The King's Buildings
Mayfield Road
Edinburgh
EH9 3JL
UK
e-mail: J.F.Chen@ed.ac.uk

Chapter 8

T. J. Stratford
The University of Edinburgh
School of Engineering and
  Electronics
The King's Buildings
Mayfield Road
Edinburgh
EH9 3JL
UK
e-mail: tim.stratford@ed.ac.uk

Chapter 9

L. De Lorenzis
Department of Innovation
  Engineering
University of Salento
via per Monteroni
73100 Lecce
Italy
e-mail: laura.delorenzis@unile.it

Chapter 10

L. Canning* and S. Luke
Mouchel
Ground Floor
Square One
4 Travis Street
Manchester
M1 2NF
e-mail: lee.canning@mouchel.com

Chapter 11

T. E. Boothby
Department of Architectural
  Engineering
104 Engineering Unit A
The Pennsylvania State University
University Park, PA 16802
USA
e-mail: tebarc@engr.psu.edu

C. E. Bakis*
Department of Engineering
  Science and Mechanics
212 Earth-Engineering Science
  Building
The Pennsylvania State University
University Park, PA 16802
USA
e-mail: cbakis@psu.edu

Chapter 12

J. L. Clarke
Concrete Society
Riverside House
4 Meadows Business Park
Station Approach
Blackwater
Camberley
Surrey
GU17 9AB
UK
e-mail: j.clarke@concrete.org.uk

Chapter 13

L. C. Hollaway
Faculty of Engineering and
 Physical Sciences
University of Surrey
Guildford
Surrey
GU2 7XH
UK
e-mail: l.hollaway@surrey.ac.uk

Preface

Over the last two decades there has been a growing awareness amongst civil/structural engineers of the importance of the unique mechanical and in-service properties of advanced fibre reinforced polymer (FRP) composites. They have emerged as an attractive competitor to the more conventional civil engineering materials for the creation of new structures and the strengthening/rehabilitation of existing ones. For new structures the material is used for reinforcing and/or prestressing concrete structures as well as constructing all FRP or hybrid FRP structures such as FRP bridge decks and concrete-filled FRP tubular columns and piles. Currently, one of the main uses of FRP composites is the strengthening/rehabilitation of structures which were erected post Second World War. From the mid-1980s and continuing to the present time a vast number of concrete, metallic and masonry structures were/are in urgent need of repair/strengthening/rehabilitation, due to either a change in use or structural degradation. Furthermore, many concrete, metallic, timber and masonry structures were built prior to the introduction of modern design codes, and hence do not meet modern design requirements.

The extraordinary properties of FRP composites of lightweight, high strength-to-weight ratio, corrosion resistance, potentially high overall durability, tailorability and high specific attributes enable them to be used in areas where the conventional construction materials might be restricted.

Over the past three decades, the fabrication technologies for the production of FRP composites have been revolutionised by sophisticated manufacturing techniques. These technologies have enabled FRP composites to be produced to high quality with minimal voids and accurate fibre alignment. In addition, a number of design guidance documents have been produced and, with rational design methods, a safe and economic utilisation of this relatively new technology can be assured.

The aim of this book is to provide under one cover a background to the physical and mechanical properties, the use and the design of FRP composites in upgrading structures in the civil infrastructure. Strengthening techniques and design methods for rehabilitating reinforced concrete, metallic, timber and masonry structural members are presented and some case study examples are provided to illustrate the use and versatility of the material. Moreover, the book discusses the current fabrication techniques for the material. It introduces the fibres and matrices that are used in the production of FRP composites for the rehabilitation of civil structures, their mechanical and in-service properties and their long-term durability and loading characteristics; the important topic of the surface preparation of the two dissimilar adherends in the rehabilitation technique is discussed at length. A chapter is also devoted to quality assurance/quality control, maintenance and repair of concrete structures rehabilitated with FRP composites.

This book has been written mainly with practitioners (including designers, engineers and contractors) in mind. Therefore, the book attempts to present information that is of direct interest to practitioners instead of providing a comprehensive or exhaustive summary of all published research. Nevertheless, the book will also provide research engineers, academics, and research students working in the field of rehabilitation of structural members in the civil infrastructure with a useful and fundamental guide to the latest structural design techniques and the latest utilisation of advanced composite materials.

The editors would like to express their sincere thanks to the authors of the chapters in this book and to the past and present members and collaborators of their respective research groups at The University of Surrey, UK and at The Hong Kong Polytechnic University, China who have directly or indirectly helped to produce this book. The first editor acknowledges the financial support of the EPSRC, Industry and the University of Surrey which has made it possible for him to undertake research work in the area of advanced polymer composites for construction. The second editor is grateful to The Hong Kong Polytechnic University, the Research Grants Council of the Hong Kong Special Administrative Region and the Natural Science Foundation of China for supporting his FRP composites research programme. Particular thanks go to the engineers who willingly supplied details of the case studies concerned with the durability of rehabilitated structures for which they were responsible.

L. C. Hollaway
J. G. Teng

1
Structurally deficient civil engineering infrastructure: concrete, metallic, masonry and timber structures

L. DE LORENZIS, University of Salento, Italy;
T. J. STRATFORD, University of Edinburgh, UK;
L. C. HOLLAWAY, University of Surrey, UK

1.1 Introduction

The repair of deteriorated, damaged and substandard civil infrastructure has become one of the important issues for the civil engineer worldwide. The rehabilitation of existing structures is fast growing, especially in developed countries, which completed most of their infrastructure in the middle period of the last century. Furthermore, structures which were built after World War II had little attention paid to durability issues, and the USA and Japan had inadequate knowledge of seismic design. In 1995, the Hyogoken–Nanbu earthquake caused a great disaster to the city of Kobe, Japan. As a result, in September 1999, the Japan Building Disaster Prevention Association (JBDPA, 1999) published the Seismic Retrofitting 'Design and Construction Guidelines for Existing Reinforced Concrete Bridges with Fibre Reinforced Polymer Materials'. In the European Union nearly 84 000 reinforced and prestressed concrete bridges require maintenance, repair and strengthening with an annual budget of £215 M, excluding traffic management cost (Leeming and Derby, 1999). In the USA, infrastructure upgrading of structures has been estimated as $20 trillion (NSF, 1993).

Within the scope of rehabilitation of concrete structures and metallic and timber systems, it is essential to differentiate between the terms *repair*, *strengthening* and *retrofitting*; these terms are often erroneously interchanged, but they do refer to three different structural conditions. In 'repairing' a structure, the composite material is used to improve a structural or functional deficiency such as a crack or a severely degraded structural component. In contrast, the 'strengthening' of a structure is specific to those cases where the addition or application of the composite would enhance the existing designed performance level. The term 'retrofit' is specifically used to relate to the seismic upgrade of facilities, such as in the case of the use of composite jackets for the confinement of columns. Strengthening/rehabilitating/retrofitting existing structures, manufactured from the more conventional materials, by utilising advanced fibre reinforced polymer

(FRP) composites is a powerful and viable alternative to the use of steel. Since the 1980s, the realisation amongst civil/structural engineers of the importance of the specific weight and stiffness, the resistance to corrosion, durability, tailorability and ease of installation is encouraging the use of FRP composites in the rehabilitation of structures throughout the world. Externally bonded FRP composite strengthening is particularly attractive where there are severe access restraints or high cost associated with installation time. In addition, the capacity of FRP composite strengthening to extend the life of historic structures with minimum disruption to users makes for genuinely sustainable engineering solutions. Furthermore, the fabrication technologies for the production of FRP composites have been revolutionised by sophisticated manufacturing techniques. These methods have enabled polymer composite materials to produce good-quality laminates with minimal voids and accurate fibre alignment.

This book will discuss the mechanical and in-service properties and the relevant manufacturing techniques and aspects related to externally bonded FRP composites to strengthen/rehabilitate/retrofit civil engineering structural materials. The book concentrates on:

1. the mechanical properties of the FRP materials used;
2. the analysis and design of strengthening/rehabilitating/retrofitting beams and columns manufactured from *reinforced concrete (RC)*, *metallic*, *masonry* and *timber* materials;
3. the failure modes of strengthening systems;
4. the site preparation of the two adherend materials;
5. the durability issues;
6. the quality control, maintenance and repair of structural systems;
7. some case studies.

This chapter gives an overview of the forms and properties of concrete, metallic, masonry and timber structures that may need rehabilitation. The ways in which externally bonded FRP can be used to extend the lives of these structures are described. An introductory section describes general structural deficiencies and the closing section discusses the reasons for using FRP strengthening rather than conventional strengthening techniques. Subsequent chapters describe the use of FRP composite strengthening in more detail, including aspects of design, durability, and inspection of a strengthening scheme.

1.2 Structural deficiencies

The world's infrastructure comprises a wide range of structures, constructed over many years and from a variety of materials. Any of these structures might be structurally deficient and in need of strengthening to allow their continued use. The reasons for structural deficiency in the civil infrastructure can be split into two broad groups:

- *changes in the use of a structure*, so that it needs to carry different loads from those originally specified;
- *degradation of a structure*, so that it cannot carry the loads for which it was originally intended.

Both of these broad classifications of structural deficiency can be addressed using FRP composites. The term 'strengthening' is commonly used to describe rehabilitation of a structure, even though it might not be an increase in strength that is required.

Each construction material has different properties and different strengthening requirements. Hence, structural deficiencies are discussed for each of the materials below. Some structural deficiencies, however, are common to any type of structure.

1.2.1 Changes in the use of a structure

Civil infrastructure routinely has a serviceable life in excess of 100 years. It is inevitable that the structure will be required to fulfil a role not envisaged in the original specification. The structure is often unable to meet these new requirements, and consequently needs strengthening. Changes in use of a structure include:

- *Increased live load.* For example, increased traffic load on a bridge; change in use of a building resulting in greater imposed loads.
- *Increased dead load.* For example, additional load on underground structures due to new construction above ground.
- *Increased dead and live load.* For example, widening a bridge to add an extra lane of traffic.
- *Change in load path.* For example, by making an opening in a floor slab to accept a lift shaft, staircase or service duct.
- *Modern design practice.* An existing structure may not satisfy modern design requirements; for example, due to development of modern design methods, or due to changes in design codes.
- *New loading requirements.* For example, a structure may not have originally been designed to carry blast or seismic loads.

1.2.2 Degradation of a structure

The condition of a structure deteriorates with time, due to the service conditions to which the structure is subjected. In some cases this deterioration might be slowed or rectified by maintenance (for example, periodic painting); however, if the deterioration is unchecked the structure will become unable to perform the purpose for which it was originally designed.

- *Corrosion* is the most common mechanism of structural degradation, particularly where a member is exposed to an aggressive environment, such as the de-icing salts used on highways. Corrosion can lead to a loss of member cross-section, and a consequent reduction in the capacity of the member.
- *Fatigue* is a second cause of structural degradation, which can govern a structure's remaining life.
- *Hazard events*. Structural degradation can also result from hazard events, such as impact (for example, 'bridge bashing' by over-height vehicles), vandalism, fire, blast loading or inappropriate structural alterations during maintenance. A single event may not be structurally significant, but multiple events could cause significant cumulative degradation to a structure.
- *Design or construction errors* due to poor construction workmanship and management, the use of inferior materials, or inadequate design, also result in deficient structures that cannot carry the intended loads.

1.3 Structural deficiencies in concrete structures

A vast number of RC structures are in urgent need of repair and strengthening, due to either a change in use or structural degradation. Many concrete structures were built prior to the introduction of modern design codes, and hence do not meet modern design requirements. Strengthening is often needed due to modern loading requirements (as discussed above), or due to structural modifications such as the formation of openings. Structural assessment may highlight inadequate reinforcement within the concrete, when compared to today's more stringent requirements.

A particular concern is the seismic performance of structures originally designed for only gravity loads. RC frames that were not designed for seismic loads can have inadequate ductility and a lack of robustness. Seismic upgrade of these structures can have profound economic and social implications. The need for economically viable seismic retrofit systems has driven research in Japan into FRP strengthening following the disastrous Hyogoken–Nanbu earthquake which occurred in 1995.

Reinforced concrete structures built between World War II and the 1980s were often designed with little attention to durability issues, and have thus suffered severe structural degradation. Material deterioration in concrete structures typically involves corrosion of the internal steel reinforcement due to long-term chloride ingress or carbonation of the concrete. This deterioration results in poor service load performance (viz. excessive cracking, spalling of the concrete cover and increased deflections) and a reduction in the ultimate capacity of the structure. Material deterioration is particularly likely in aggressive marine or industrial environments, and in cold regions where de-icing salts are used.

Concrete structures are also susceptible to hazard events, and to design or construction errors (as discussed above). Prestressed concrete (PC) members are susceptible to steel strand fatigue and may require strengthening to prevent further loss of prestress.

1.4 Strengthening concrete structures using FRP composites

1.4.1 Flexural strengthening

Beams

The first application of FRP strengthening was to beams, using wet lay-up sheets or pre-cured plates bonded to the tension face of the beam with the fibre direction aligned to the beam axis. This non-metallic version of 'beton plaqué' gives all the advantages of the high strength-to-weight ratio and good corrosion resistance of FRP materials with respect to steel. The effectiveness of flexural strengthening of RC beams with FRP is evident from the large database of experiments, reported by Smith and Teng (2002) among others.

Under service loads, the effectiveness of a passive (i.e. non-prestressed) FRP system is usually limited. Conversely, a notable increase in the ultimate moment of the cross-section can be obtained. The analysis of strengthened members at the ultimate limit state may follow well-established procedures valid for RC members, with the exceptions that: (i) the contribution of the FRP must be properly accounted for, and (ii) the issue of bond between FRP and concrete must be given particular care in design and execution. More details on all aspects of design concerned with the upgrading of reinforced concrete will be given in Chapters 4–7.

Bond between FRP and concrete has a profound impact on the failure mode and, consequently, on the ductility and the failure load of the strengthened member. Possible failure modes have been identified as: *classical* failure modes, whereby full composite action is maintained between FRP and concrete until concrete crushing or FRP rupture, and *debonding* failure modes, consisting in loss of composite action prior to attainment of any of the classical modes; this topic is discussed generally in Chapters 4–7.

Debonding failures are undesirable as they are brittle in nature and occur at lower load levels than classical failures, penalising the exploitation of the FRP. It is then necessary either to *prevent* them with anchoring devices and detailing, or *predict* them and adopt suitable factors of safety. From the standpoint of the bond performance a key role is played by surface preparation of the substrate, discussed in Chapter 3.

Anchoring techniques such as steel bolting and the use of bonded FRP U-shaped channels or jackets at the end of the beam and/or at intermediate locations have been developed to limit debonding failures (Quantrill *et al.*, 1996). However, bolting may create damage to the internal steel bars or to the FRP laminate, and jacketing is usually impractical to implement. More recently, the use of FRP anchor spikes has been proposed by Teng *et al.* (2000) and Eshwar *et al.* (2003). On the modelling side, several approaches with different levels of complexity have been proposed and adopted from the available design guidelines.

The ductility of a flexural member generally decreases as a result of strengthening, especially if the controlling failure mode is debonding or FRP rupture. To guarantee adequate ductility of a strengthened cross-section, the strain level of the internal steel reinforcement at ultimate should considerably exceed the steel yield strain, as indicated by available design recommendations [e.g. FIB Task Group 9.3 (*fib*, 2001) and ACI 440 (ACI, 2002)]. ACI 440 also suggests that the lower ductility should be compensated with a higher reserve of strength through the use of a lower overall strength reduction factor.

The ductility index of an RC section is usually defined as the ratio of the ultimate curvature to the curvature corresponding to yielding of the steel tension reinforcement. De Lorenzis *et al.* (2004), noting that this definition is not appropriate for FRP-strengthened cross-sections, defined the ductility index as the ratio of the energy dissipated by the element at ultimate, to the total energy furnished to the element through the work of the external loads. In general, the reduction in this ductility index as a result of strengthening was estimated as acceptably low, as this index relates the dissipated energy to the total energy and both values are lower for the strengthened section. However, this evaluation did not account for the brittleness of the failure mode.

The loss in ductility of FRP-strengthened cross-sections, and particularly the possibility of brittle debonding failures, has led various design guidelines to prohibit or discourage the application of moment redistribution into or out from strengthened cross-sections, leading to onerous conditions for such strengthening, particularly when the original design was based on moment redistribution. The research carried out on this subject is still rather limited (El-Refaie *et al.*, 2003). Ibell and Silva (2004) related moment redistribution in FRP-strengthened concrete members to the level of ductility at critical sections, and proposed that moment redistribution into FRP-strengthened zones should be permitted, and that moment redistribution out of FRP-strengthened zones should be permitted if the curvature ductility ratio across the critical section is at least 2.50.

Slabs

The effectiveness of strengthening slabs using FRP has been demonstrated by numerous experimental investigations. The simple extrapolation of results obtained from beams is inappropriate for a number of reasons, among which are the biaxial load-bearing response of some slabs, their lack of shear reinforcement and the influence of the spacing of the FRP strips.

Tests performed by Seim *et al.* (2001) on *one-way slabs* strengthened with pre-cured strips and fabric demonstrated that the load capacity can be increased by up to 370%. However, the failure mode of the slab changed from the conventional ductile mode to a more sudden failure associated with debonding of the FRP composite or FRP rupture. The use of fabric covering the entire width of the slab produced more uniform shear stresses, resulting in significantly higher load capacity. The use of fabric yielded 50–70% higher levels of deformation than that of pre-cured FRP strips. Mosallam and Mosalam (2003) reported that the FRP systems upgraded the structural capacity of *two-way slabs* by up to 200%. Failure was preceded by large deformations providing adequate visual warning. Crushing of the concrete was the common failure mode, with localised debonding close to the ultimate load.

Lam and Teng (2001) investigated the strengthening of RC *cantilever slabs* with glass FRP (GFRP) strips, focusing on anchorage of the strips to the fixed end and to the slab. Inserting GFRP strips into narrow epoxy-filled rectangular holes in the supporting wall provided a strong fixed-end anchorage system. Debonding of the GFRP strips from the concrete was arrested or prevented with the use of fibre anchors, allowing a significant increase in the ultimate load and ductility.

A significant application of FRP strengthening is around new openings in slabs. Strengthening with FRP strips of one-way and two-way slabs with an opening in the positive moment region can be used to effectively recover the strength of the slab prior to making the cut-out and to increase the stiffness (Casadei *et al.*, 2003a). However, for specimens with an opening in the negative moment region and strengthened using top-surface carbon FRP (CFRP) laminates, Casadei *et al.* (2003b) found shear failure to occur at a lower load than for the unstrengthened specimen, as the CFRP laminates increased the shear demand on the concrete.

Columns

Columns may need flexural strengthening, especially for seismic upgrade of structures originally designed for gravity loads. In this case, the maximum bending moments are usually attained at the upper and lower cross-sections of the column, and hence the strengthening system requires anchorage to the adjacent beams. This anchorage may be implemented (for example) with the 'U-anchor' system (see the section on shear strengthening) (Khalifa *et al.*, 1999). Alternatively, the use of near-surface mounted (NSM) bars, discussed later in this chapter, may result in more practical anchorage by drilling holes in the adjacent members and epoxy-bonding the bar extremities into these holes.

Flexural strengthening with mechanically fastened FRP laminates

A recently proposed technology is flexural strengthening using a mechanically fastened FRP system. The first version of this system, which attaches the FRP plates to concrete using closely spaced steel power-actuated fastening 'pins' and a limited number of steel expansion anchors (mechanically fastened-FRP, or MF-FRP), was developed by Lamanna *et al.* (2004). The installation of an MF-FRP system is fast and easy, requiring only unskilled labour and common hand tools. Epoxy adhesive is not required and surface preparation can be reduced to the removal of sizeable protrusions. A similar method was developed by Rizzo (2005), using concrete wedge bolts and anchors instead of pins. Tests also showed that using a steel washer improves the performance of the connection by spreading the clamping load on a bigger surface of FRP and transferring a portion of the load by friction. In 2004 three off-system bridges in Missouri were strengthened using the MF-FRP system, proving its cost benefits.

Flexural strengthening with near-surface mounted FRP bars

An alternative technique to externally bonded FRP laminates is NSM reinforcement. The reinforcement is embedded into a groove cut on the surface of the member, and bonded into this groove with an appropriate binder (usually high-viscosity epoxy or cement paste). A state-of-the-art review on NSM reinforcement can be found in De Lorenzis and Teng (2007).

The NSM reinforcement method offers several advantages with respect to externally bonded FRP. However, it is applicable only if the cover of the internal steel reinforcement is sufficiently thick for the groove size to be accommodated. Test results indicate that NSM reinforcement can significantly increase the flexural capacity of RC elements (e.g. De Lorenzis *et al.*,

2002; El-Hacha and Rizkalla, 2004). As in the case of externally bonded laminates, bond may be the limiting factor on the efficiency of this technology. El-Hacha and Rizkalla (2004), by strengthening reinforced concrete T-beams with the same CFRP strips used both as NSM and externally bonded reinforcement, obtained a strength increase 4.8 times higher in the first case, due to early debonding failure of the external FRP as opposed to the tensile rupture of the NSM strips.

Flexural strengthening of prestressed concrete members

Limited research has been produced on strengthening PC members, parallel with the scarcity of *in situ* installations [as reported by FIB Task Group 9.3 (*fib*, 2001) less than 10% of FRP-strengthened bridges as of 2001 are prestressed]. Strengthening usually takes place when all long-term phenomena (creep, shrinkage, relaxation) have fully developed, which may complicate the preliminary assessment of the existing conditions. Apart from this, the conventional verification procedures adopted for PC can be applied, provided that the FRP contribution is appropriately considered. As in RC strengthening, the required amount of FRP will generally be governed by the ultimate limit state design in PC members. However, additional failure modes controlled by rupture of the prestressing tendons must also be considered, and consideration should be given to limitations on cracking. In this latter case, the possibility of admitting tensile stresses in the PC section after FRP strengthening is the subject of ongoing debate. At the same time, the role of non-prestressed externally bonded FRP in reducing cracking in tensile regions still needs to be quantified (*fib*, 2001).

1.4.2 Shear strengthening

To strengthen RC beams or columns in shear, FRP laminates are bonded to the sides of the member. Efficient design requires the principal fibre direction to be parallel to that of the maximum principal tensile stresses, i.e. (in the most common cases) at approximately 45° to the member axis. However, for practical reasons it is usually preferred to attach the external FRP reinforcement with the principal fibre direction perpendicular to the member axis.

Different strengthening patterns can be used, both along the axis of the beam and in the plane of the cross-section. Along the axis of the beam, the strengthening system can either be continuous or discontinuous. The use of a continuous pattern may limit the migration of moisture and hence should be considered with caution. In the cross-sectional plane, wet lay-up sheets can be completely wrapped along the cross-section, wrapped on three sides

(U-wrap), or bonded on two opposite sides. The first pattern is obviously the most efficient, and is typically adopted for the shear strengthening of columns. However, it is impractical for strengthening beams in the presence of an integral slab. Strengthening on three sides is less efficient, while strengthening only on the two side faces is the least efficient scheme. Precured plates, which typically have larger thickness, cannot be bent around corners and can then only be bonded to the sides. However, prefabricated L-angles, specifically suited for shear strengthening, are also manufactured. A wide database of experimental research has demonstrated that FRP laminates can significantly enhance the shear capacity of an RC member (Khalifa et al., 1998). As in the case of flexural strengthening, bond between FRP and concrete plays a crucial role in the failure mode and ultimate load of a shear-strengthened member. The typical shear-controlled failure modes of strengthened members are debonding of the FRP laminate from the concrete and FRP fracture, both associated with concrete diagonal tension. If the load associated with these failure modes is large enough, the critical mode may shift to concrete diagonal compression, or even to flexural failure, which is more ductile and hence more desirable.

Various researchers (Triantafillou, 1998a; El-Hacha and Rizkalla, 2004) and current design recommendations (e.g. El-Refaie et al., 2003; Ibell and Silva, 2004) show that an FRP-shear-strengthened member can be modelled in accordance with Mörsch's truss analogy. The external FRP reinforcement may be treated by analogy to the internal steel (recognising that the FRP carries only normal stresses in the principal fibre direction), assuming that at the instant of shear failure by concrete diagonal tension the FRP develops in the principal fibre direction a tensile 'effective strain'. The effective strain is, in general, less than the ultimate strain in uniaxial tension for both cases of FRP debonding and fracture, reflecting the experimental evidence. When full wrapping is not feasible, the debonding failure mechanisms may be delayed or prevented by using mechanical anchorages at the termination of U-wrapped or side-bonded laminates, or by bonding the ends of the strips in core holes through the flange of a T-beam. The 'U-anchor' system consists of embedding a bent portion at or near the end of the FRP reinforcement into a slot cut into the concrete substrate and filled with epoxy paste (Khalifa et al., 1999). The slot can be at the corner between web and flange, and may also contain an FRP bar around which the FRP laminate is wrapped. This system uses only FRP materials, eliminating possible concerns regarding galvanic corrosion in the use of metallic anchoring bolts, and it has been shown to be very effective in delaying shear debonding failure. It can also be used as an anchorage system for other applications.

NSM FRP reinforcement can also be used effectively to enhance the shear capacity of RC beams. In this case, the bars are embedded in grooves cut on the sides of the member at the desired angle to the axis. Test results with NSM round bars showed an increase in capacity as high as 106% in the absence of steel stirrups, and a significant increase also in the presence of internal shear reinforcement (De Lorenzis and Nanni, 2001). One of the observed failure modes was debonding of one or more FRP bars, associated with concrete diagonal tension. This mechanism can be prevented by providing a larger bond length by either anchoring the NSM bars in the beam flange (for T-beams) or using 45° bars at a sufficiently close spacing. Once debonding of the bars is prevented, splitting of the concrete cover of the longitudinal reinforcement may become the controlling mechanism, due to the fact that, unlike internal steel stirrups, NSM rods are not able to exert any restraining action on the longitudinal reinforcement subjected to dowel forces.

The truss analogy has been used to compute the shear capacity of a member strengthened in shear with NSM reinforcement, accounting for debonding of the NSM bars (De Lorenzis and Nanni, 2001). The basic assumption is that, at the instant of failure, bond stresses are evenly distributed along the bars crossed by the critical shear crack, and are equal to the bond strength. This approach compared favourably with test results. However, further research is needed to improve the accuracy and reliability of the model and to assess its wider validity.

1.4.3 Confinement

Numerous experiments since the 1980s have demonstrated the effectiveness of FRP composites for confining RC columns (see, for example, the review in De Lorenzis and Tepfers, 2003). For confinement, wet lay-up laminates, filaments or prefabricated FRP composites can be used. Automated filament winding, first developed in Japan in the early 1990s and then in the USA, involves continuous winding of wet fibres under a slight angle around columns using a robot, resulting in good-quality control and rapid installation. Finally, prefabricated (pre-cured) elements in the form of shells or jackets can be bonded to the concrete and to each other to provide confinement [ACI 440, (ACI 2002)]. Both wet lay-up and prefabricated systems are normally used with the principal fibre direction perpendicular to the axis of the member; they can be applied either continuously over the surface (which does, however, pose the problem of moisture migration) or as strips with certain width and spacing.

The use of FRP offers several advantages over steel for this application. If the ratio of circumferential to axial fibres is large, the FRP axial modulus is small, allowing the concrete to take essentially the entire axial load; the tensile strength in the circumferential direction is very large and essentially independent of the value of the axial stress; ease and speed of application result from the FRP low weight; the FRP minimal thickness does not alter the shape and size of the strengthened elements; and the good corrosion behaviour of FRP materials makes them suitable for use in coastal and marine structures.

The FRP confinement action is passive, i.e. it arises as a result of the lateral expansion of the concrete core under axial load. The confining reinforcement develops a tensile stress balanced by pressures reacting against the concrete lateral expansion. FRP displays an elastic behaviour up to failure and therefore exerts a continuously increasing confining action, until failure normally results by tensile rupture of the FRP. Hence, the confined concrete strength is closely related to the tensile rupture strain of the FRP on the confined element. Experimental evidence shows that this failure strain is usually lower than the ultimate strain obtained by standard tensile testing of the FRP sheet. There are several reasons for this reduction (De Lorenzis and Tepfers, 2003).

The effectiveness of FRP confinement on strength and ductility enhancement of concrete columns depends on several factors, amongst which are: (i) the confinement strength; (ii) the confinement stiffness; (iii) the cross-sectional shape of the column; (iv) the use of continuous vs discontinuous confinement along the member axis; and (v) the fibre orientation. This topic is discussed further in Chapter 6.

In summary:

1. Among the available FRP materials, CFRP is to be preferred if a strength increase is sought, GFRP or aramid FRP (AFRP) if the main objective is an increase in ductility.
2. Confinement is most effective for circular columns, as the confinement pressure is, in this case, uniform. Both strength and ductility can be significantly enhanced. In the case of rectangular columns, the confining action is less efficient. The achievable increase in strength is usually modest or negligible, but a ductility enhancement can still be obtained. The effectiveness decreases as the cross-section aspect ratio increases.
3. The use of discontinuous confining devices has reduced effectiveness compared to the equal continuous device, as portions of the column between adjacent strips remain unconfined.
4. The optimal fibre orientation is perpendicular to the member axis, and different angles result in reduced confinement strength and stiffness.

1.4.4 Seismic retrofit

The seismic performance of RC structures is governed by their strength hierarchy: by boosting the strength of those members in which failure is not desirable, it is possible to achieve a global performance characterised by the failure of more ductile and energy-dissipating components. In the behaviour of an under-designed frame, the lower bound equates to column failure. Upgrading columns by providing them with confinement and/or more flexural reinforcement can initiate failure in the joints. The upgrading of both columns and joints will transfer failure to the upper bound level of the hierarchy of strength. The formation of plastic beam hinges is a ductile mechanism, allowing the global integrity of the structure to be maintained. Seismic upgrade with FRP-based technologies can be used to move up along the hierarchy of strength from column to joint and/or from joint to beam failure, depending upon the strengthening parameters. Research conducted so far has concentrated on each of these possible upgrades in the controlling failure mechanism. A brief outline is reported below.

FRP confinement can be used to change the column failure mode from shear to flexural failure, or even to transfer failure to the joint. Researchers studying seismic retrofitting found that the ductility of concrete columns is significantly increased by FRP wrapping, due to confinement of concrete and by preventing buckling of the longitudinal rebars (Saadatmanesh, 1995). Tests on bridge columns showed that CFRP jacket retrofits can be just as effective as equivalent steel jacket retrofits (Seible *et al.*, 1997). An FRP jacket thickness of only 0.4 mm was required over the shear critical centre region of the column to prevent brittle shear failure (associated with a displacement ductility level of 2.0) and produce stable hysteresis loops up to a displacement ductility level of 10.5.

Reinforced concrete beam-column joints typically fail in diagonal-tension shear, due to inadequate transverse reinforcement, or by bond failure of rebars, due to inadequate anchorage. The simplest way to strengthen such joints is to bond FRP sheets or strips to the joint region with the fibres in the two orthogonal directions of the beams and columns. The FRP system in the joint also acts as shear reinforcement. FRP-upgraded RC joints have been studied by several investigators (Gergely *et al.*, 2000), who recorded increases in strength, stiffness and ductility. In specimens designed to fail by shear in the joint (Antonopoulos and Triantafillou, 2003), debonding dominated the behaviour of external reinforcement unless very low area fractions were employed or proper mechanical anchorages were provided. In particular, wrapping the longitudinal FRP sheets with transverse layers proved to be a highly effective anchorage system.

Recently, Prota *et al.* (2004) proposed the combined use of FRP laminates and NSM bars for upgrading beam–column connections. FRP laminates were used to confine the column and upgrade the shear capacity of the joint, and NSM bars were applied to the columns to increase their flexural capacity. The simultaneous presence of FRP confinement by the laminates prevents the NSM reinforcement from becoming ineffective as a result of load reversals. Varying the amount of FRP reinforcement, its location (column or column plus joint) and the reinforcement type (laminates, bars or their combination) resulted in a different level of the strength hierarchy and hence in a distinct failure mode and associated level of ductility.

1.4.5 Strengthening corroded reinforced concrete (RC) structures

FRP jacketing can be used to strengthen RC structures suffering corrosion-induced deterioration, and to reduce the rate of corrosion (Pantazopoulou *et al.*, 2001; Tastani and Pantazopoulou, 2004). FRP jacketing slows down iron depletion due to continued post-repair exposure. The hardened resin matrix acts as a diffusion barrier to further ingress of the agents (oxygen, chlorides and water) necessary to sustain corrosion. Furthermore, the FRP jacket limits expansion due to the corrosion products, and hence modifies the rate of rust production from a Faraday model (the depleted mass of metal varies linearly with time of exposure), to a Boltzman-type law (the rate of depletion decays exponentially with time). A 47% reduction of corrosion rate (providing no moisture is trapped behind the jacket at cover replacement) has been reported using GFRP jacketing (Pantazopoulou *et al.*, 2001), and 85% where low permeability mortar was used for patch repair under the FRP jacket (Tastani and Pantazopoulou, 2004).

The seismic performance of severely corroded column specimens under cyclic loading in double curvature showed a significant loss of stiffness due to bond degradation during tests, giving increased pinching in the hysteresis loops as loading progressed (Lee *et al.*, 2000). Failure of the corroded unrepaired specimens occurred by hoop fracture, buckling of compression reinforcement and brittle shear disintegration. Repair with FRP jacket acting as confining and shear reinforcement enhanced the ductility of the specimens up to the levels attained by identical specimens in an uncorroded condition.

When bonded transversely to corroded anchorages, FRP sheets can also recover bond development capacity, as they limit propagation of cover splitting cracks. Success of this repair is conditional upon cover replacement prior to application of the sheets, so that any existing splitting cracks are eliminated.

1.4.6 'Active' strengthening

Active flexural strengthening

If the objective of strengthening is an increase in flexural stiffness under service loads, or a substantial enlargement of the service load range, active strengthening solutions must be sought. Prestressed FRP strengthening is inherently more complicated than unstressed strengthening (Ning *et al.*, 2004), but methods have been developed to prestress the FRP composites under real-life conditions (see Chapters 10 and 13). Amongst the advantages of prestressing compared to passive strengthening are: (i) the stiffness at service increases considerably; (ii) the load at first yield of the steel increases, prolonging the service load range; (iii) crack formation is delayed and the cracks are more finely distributed and narrower; pre-existing cracks can be closed; (iv) the shear resistance of the member is improved, provided that the concrete remains uncracked (*fib*, 2001). However, the technique is more expensive and slower than normal FRP bonding due to the greater number of operations and equipment required. Furthermore, the design and construction of the end zones requires special attention, as the development of high bond shear stresses may cause failure of the beam at release of the prestressing force (Triantafillou *et al.*, 1992). A technically and economically rational prestress would require a degree of prestressing in the range of 50% of the FRP tensile strength, only achievable by the use of special anchorages applying vertical confinement. Such systems have been developed for practical applications as well as research purposes. NSM can also be post-tensioned, as shown by Nordin and Täljsten (2003). Where there is no access to the ends of the beam, a tensioning/anchoring device for NSM bars has been proposed by De Lorenzis *et al.* (2002).

In all applications of active strengthening, attention should be given during design to the issue of creep rupture and to the effects of creep and stress relaxation (which still require further investigation).

Active shear strengthening

The larger ultimate strain of FRP composites with respect to the yield strain of steel has two undesirable consequences on shear behaviour of shear-strengthened members: (i) at large strains, the contribution of concrete to the shear capacity of the beam through aggregate interlock may be reduced or lost; and (ii) the strengthening design may be controlled by shear crack widths under service loads, as the FRP can be mobilised only with significant crack widths. To overcome these problems, Lees *et al.* (2002) explored the use of prestressed FRP straps. In their method, unbonded FRP straps are wrapped around the beam and then prestressed to the desired level. Limited experimental results are available on this topic.

Active confinement

Experiments have shown that concrete circular columns confined by FRP laminates, when loaded in uniaxial compression, display a distinct bilinear stress–strain response with transition zone around the strength of unconfined concrete; this is discussed in detail in Chapter 6. The slope of the branch after the transition zone depends on the volumetric ratio and stiffness of the confining device and is always lower than that of the initial branch. The possibility of adopting service stress levels above the transition zone is questionable for confined columns, because of internal damage to the concrete, low tangent stiffness and reduced Euler buckling load. The load level of the transition zone can be raised through active confinement, i.e. by prestressing the confining device using different methods: pretensioning the fibre bundles during filament winding (Rousakis *et al.*, 2003), injecting resin under pressure in the gap between an unstressed FRP jacket and concrete (Saadatmanesh, 1995) or using expansive grout (Mortazavi *et al.*, 2003).

1.4.7 Improvement in fatigue resistance

Externally bonded FRP laminates can significantly improve the fatigue resistance of RC beams (Shahawy and Beitelman, 1999), as they cause a reduction in stress in the reinforcing steel and reduce crack propagation. Grace (2004) applied a fatigue loading for two million cycles on beams strengthened with CFRP plate and fabric. The maximum cyclic load was equal to 40% of the ultimate load and no significant effects on the load-carrying capacity of such beams were observed. Failure of RC beams strengthened with FRP laminates under higher fatigue loading was primarily initiated by failure of the steel reinforcement followed by FRP debonding as a secondary failure mode.

1.5 Metallic materials used in civil infrastructure

The world's infrastructure includes a range of metallic structures, made from a range of metallic materials. The structural forms used in bridges and buildings have evolved to take advantage of the materials available at the time. A comprehensive review of these different metallic structures can be found in Bussell (1997).

1.5.1 Cast iron

Cast iron was the earliest metallic material to be used structurally (from around 1780), and many examples of its use are still in service today. Cast iron girders were often used to support brick jack arch infill (Fig. 1.1) so as to form a bridge deck or the floor of a large building. Cast iron columns were used in both buildings and bridge piers. Cast iron arches allowed longer spans to be crossed, typified by the earliest example of a cast iron bridge at Ironbridge. Cast iron is brittle and weak in tension, and its proper-

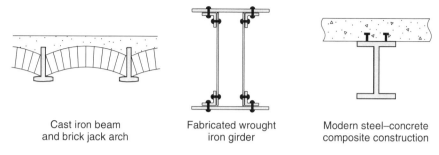

1.1 Typical uses of metallic members in construction.

ties can vary widely according to the method of production and manufacturing defects.

1.5.2 Wrought iron

Wrought iron (made by repeatedly heating and reworking cast iron) has a far higher tensile strength and is more ductile than cast iron. However, the manufacturing process gives wrought iron a laminar structure, so that its out-of-plane properties are often inferior.

Wrought iron was labour-intensive to manufacture and consequently was often used in combination with cast iron, or girders were fabricated by riveting together smaller wrought iron sections (Fig. 1.1). These sections were used in girder and truss bridges (examples of which remain common on the UK rail network). Wrought iron was also used to form roof trusses and tie rods and to support brick jack arches.

1.5.3 Steel

Carbon steel rapidly replaced wrought iron (in about 1870) and has since been prevalent in all forms of structures, such as bridges, building frames, pipes, steel–concrete composite floors, piles, etc. (Fig. 1.1).

Steel has been developed continually since its first use. Early steel (roughly pre-1950) has many similarities with wrought iron: large sections were fabricated by riveting together smaller sections, and steel plates were often laminar in nature. Engineers are most familiar with modern steel, which is ductile, available in a variety of quality-controlled grades, and can be joined by welding or high-strength friction-grip bolts.

1.6 Structural deficiencies in metallic structures

Metallic structures commonly require strengthening to carry increased loading requirements, as discussed above. Many cast iron structures were constructed before the modern appreciation of structural mechanics had developed and hence strengthening is often required to satisfy modern design requirements; for example, cast iron beam–columns often rely on mechanical interlock joints that place small brittle sections under high bending stresses.

Metallic structures are also susceptible to structural degradation. In particular, protection against *corrosion* of the structural section is often reduced locally in regions that are difficult to paint; around connection details; in members that trap water; or where a member is subject to frequent water flow (due to leaking drainage arrangements or bridge expansion joints). Corrosion is likely to be most significant in modern thin-walled members, although it can cause delamination of wrought iron and early rolled steels. Cast iron is generally more corrosion-resistant than steel (due to the silica-rich surface left by the sand mould used in casting); however, cast iron corrodes rapidly if the outer layer is damaged and the interior metal becomes exposed.

The remaining life of a metallic structure can often be governed by *fatigue*. Fatigue cracks propagate from stress concentrations, such as around openings, welded connections or rivet and bolt holes. The fatigue life of a structure may be controlled by crack propagation from the stress concentration around a particular detail; however, many structures contain a large number of stress raisers, such as the rivet holes in fabricated wrought iron or early steel members. If such a structure's fatigue life is to be extended, fatigue crack propagation must be arrested at all fatigue-critical positions. Cast iron structures are particularly vulnerable to *hazard events*. Unlike ductile steel, brittle cast iron will crack under impact or thermal shock.

1.7 Strengthening metallic structures using FRP composites

The use of FRP composites to strengthen metallic structures is a relatively recent development compared to its use with concrete structures; however, this use is developing rapidly. There have been a number of applications of FRP to metallic structures that have shown that the technique can have significant benefits over alternative methods of strengthening.

The largest number of applications of FRP to metallic structures to date has been in the UK, and the experience gained on these projects has led to two comprehensive guidance documents being published: (i) ICE Design & Practice Guide – 'FRP Composites – Life Extension and Strengthening of Metallic Structures (Moy, 2001) and (ii) CIRIA Report C595 – 'Strengthening Metallic Structures using Externally-Bonded FRP' (Cadei *et al.*, 2004). Design guidance has recently also been published by the Italian National Research Council (CNR, 2004) and by Schnerch *et al.* (2006).

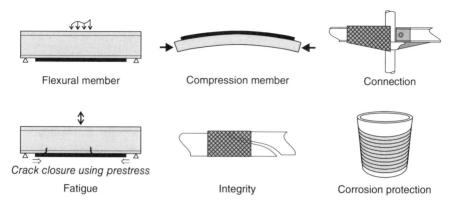

1.2 Applications of FRP to rehabilitate metallic members.

FRP strengthening can be used to address any of the structural deficiencies described above. The reasons for using FRP to rehabilitate a metallic or concrete structure may be similar; however, the way in which the FRP works with an existing metallic structure can often be very different to that in a concrete structure.

The following sections describe a variety of ways in which FRP can be applied to metallic structures (see Fig. 1.2). As the technique remains under development, care has been taken to distinguish between applications that addressed real structural deficiencies, proof of concept applications and research.

1.7.1 Flexural members

Cast iron beams and arches

The largest number of applications of FRP strengthening to metallic structures has been to increase the flexural capacity of cast iron beams. Bonded FRP strengthening is particularly suited to this application, because:

- the modular ratio of CFRP to cast iron is higher than for steel, making the CFRP more effective;
- the adhesive connection allows distributed load transfer between the FRP and the cast iron, unlike conventional mechanical connections (such as bolts or clamps) which cause stress concentrations that can cause failure of the brittle cast iron;
- FRP strengthening can be used to relieve the cast iron of permanent dead load stresses.

A number of highway bridges in the UK have been strengthened using this technique, allowing them to carry modern traffic loads. These bridges typically comprise cast iron beams, supporting brick jack arches, forming the bridge deck. For example, Hythe Bridge, on a major traffic artery into Oxford, was strengthened in 2000 using prestressed high-modulus CFRP plates. This allowed a severe weight restriction to be removed and 40 tonne trucks to be carried (Cadei *et al.*, 2004; this bridge is further discussed in Chapter 13). Maunders Road Bridge in Stoke on Trent (strengthened in 2001) was strengthened using ultra-high-modulus (UHM) CFRP (Smith, 2004). Before strengthening, the bridge was propped at mid-span and jacked to reduce the dead load carried by the cast iron beams. These dead load stresses were transferred to the CFRP after the adhesive had cured (Cadei *et al.*, 2004*)*.

Other cast iron girders strengthened using CFRP include King Street Bridge (Farmer and Smith, 2001), Bid Road Bridge, Bow Road Bridge and Redmile Canal Bridge (Cadei *et al.*, 2004), Covered Ways 12 & 58 and Bridge EL31 on the London Underground (Church and Silva, 2002), Nunnery Bridge and New Moss Road Bridge (Luke and Canning, 2004). CFRP strengthening has also been used to increase the load capacity of the arched Tickford Bridge (Milton Keynes, UK) the oldest cast iron bridge still in service (Cadei *et al.*, 2004).

Steel and wrought iron beams

FRP strengthening has also been used to increase the flexural strength of steel beams, for which different constraints govern the use of FRP strengthening:

- The modular ratio of FRP to steel is lower than for FRP to cast iron; consequently a large amount of strengthening is required to significantly affect a structure's elastic response.
- It is desirable to allow yield of the steel at ultimate, but this is only possible if premature failure of the adhesive joint can be prevented, for example, by additional mechanical restraints (Sen *et al.*, 2001).
- Thin-walled steel beams may require additional restraint against buckling after strengthening.

A comprehensive programme of research and development in Delaware, USA, concluded with a demonstration project on Interstate 95 (Miller *et al.*, 2001). CFRP plates were bonded to one of the steel girders on this bridge (which has a non-composite concrete deck slab), where it has been monitored since installation in 2000. The technique was used in 2002 to strengthen the steel cross girders of the Ashland Bridge in Delaware, USA, a through girder bridge (Chacon *et al.*, 2004). CFRP plates have also been used to strengthen a steel–concrete composite highway bridge in Iowa, USA (Phares *et al.*, 2003).

Instead of bonding preformed FRP elements to the structure, the FRP composite can be formed and bonded to the structure in a single, *in situ* operation. *In situ* techniques are particularly useful where the geometry of the existing member is complex. For example, a principal curved steel beam in the Boots Building (Nottingham, UK) was strengthened using a low-temperature moulding, pre-impregnated CFRP, which was vacuum consolidated and cured under elevated temperature (Cadei *et al.*, 2004). *In situ* wet lay-up strengthening has also been investigated for hollow section monopole mobile communication masts (Schnerch *et al.*, 2004).

Laminated and corroded wrought iron or early steel structures could have insufficient surface shear strength to allow bonded FRP strengthening to be applied. For this reason, bonded FRP strengthening has not so far been applied to wrought iron structures, although recent work indicates that wrought iron in a good condition can have sufficient shear strength for bonding (Moy, 2004).

An alternative to bonded strengthening (which avoids some of the problems listed above) is to post-tension the bottom flange of a girder using FRP rods, cables or ropes attached to mechanical anchorages near the end of the girder. CFRP rods have been used to post-tension a bridge in Iowa, USA (Phares *et al.*, 2003).

1.7.2 Compression members

FRP strengthening has been applied to struts and columns carrying compression. Cast iron cruciform-section struts in a ventilation shaft on the London Underground were strengthened using UHM CFRP (Moy and Lillistone, 2006) where the CFRP was designed to increase the capacity of the tensile zone of the strut during buckling. The use of FRP sheets to increase the local buckling strength of hollow square section columns has been investigated (Shaat and Fam, 2006).

The capacity of a metallic compression member is often governed by buckling, and strengthening is required to provide an increase in *flexural rigidity*. As mentioned above, the effectiveness of strengthening using composites is governed by the modular ratio of the FRP to the metallic member. The FRP can be used more effectively in combination with a spacer or filler material.

1.7.3 Connections

The capacity of a metallic structure may be governed by the strength of connections between members. This is particularly likely in cast iron members joined by mechanical interlock, but is also true of many riveted and bolted connections. Over-wrapping the connection detail using FRP is an effective and simple method of increasing the strength of a connection detail, if the connection is accessible on all sides. This method was used to strengthen a bolted connection in the Boots building (mentioned in Section 1.7.1 above).

1.7.4 Fatigue life extension

Research work has shown that FRP composites can be used to extend the fatigue life of structures. A demonstration project on Acton Bridge (an early steel underbridge on the London Underground) used UHM CFRP strengthening to reduce the magnitude of the cyclic live load stresses in the bridge's cross-girders by 25%, and hence extend the fatigue life of the structure (Moy and Bloodworth, 2007). Tests have shown the beneficial effect of a CFRP patch by comparing the *S–N* curves of non-retrofitted and retrofitted damaged steel beams (Tavakkolizadeh and Saadatmanesh, 2003). Laboratory investigations into fatigue life extension using CFRP have also been undertaken in China (Ning *et al.*, 2004).

Prestressed FRP strengthening is particularly advantageous for slowing fatigue crack development, as it applies compressive stress across the fatigue details, so as to close the fatigue cracks. Laboratory tests have demonstrated the use of prestressed FRP strengthening for extending the fatigue life of riveted steel railway bridge girders (Bassetti *et al.*, 2000).

1.7.5 Structural integrity

The Corona Bridge in Venice (built in 1852) is vulnerable to boat impact and contained serious structural cracks and corrosion. Aramid FRP has been bonded to the surfaces of the bridge to improve its structural integrity, to increase its resistance to impacts and to provide corrosion protection. The work was carried out after laboratory tests to assess the mechanical performance of the strengthening and its durability, and into non-destructive methods for assessing the quality of the adhesive bond (Bastianini *et al.*, 2004).

1.7.6 Corrosion protection

Over-wrapping using FRP protects the surface of a metallic member from corrosion. FRP over-wrapping can provide a thicker, more durable surface protection than paint systems, and has been used to protect marine piles on a jetty in China.

1.8 Overview of masonry structures

Unreinforced masonry (URM) buildings account for a large proportion of buildings around the world. In the past, these buildings were often constructed with unsafe features, such as unbraced parapets and inadequate connections to the roof. Effective and affordable retrofitting techniques are therefore urgently needed for masonry elements. Most applications of FRP strengthening to masonry to date have studied URM structures, as these represent the majority of masonry buildings, and FRP can provide a significant structural contribution by introducing tensile capacity to brittle URM elements. This section therefore discusses only URM structures.

The international and historical development of masonry has resulted in diverse materials and bonding patterns due to the local availability of different materials. Consequently, there is notable difficulty in generalising results obtained on one system of masonry to another. URM members in modern structures are typically walls, either load-bearing or infill, which may be designed to resist lateral and/or gravity loads. Historic masonry structures often made use of structural arches and vaults. Strengthening such structures also involves challenges such as reversibility, minimal intrusion and the expectation of durability over a century timescale.

1.9 Structural deficiencies in masonry structures

1.9.1 Change in use of the structure

URM structures can be subjected to extreme loading due to changing traffic loads or water loads, or structural modifications that overload existing elements. Historic structures generally have high self-weight and low levels of material stress and, typically, overloading is not a problem for these structures. However, substantial changes (such as the removal of buttresses or tie-rods, construction of new columns or walls on the extrados of a vault, new openings and removal or lightening of the spandrel fill) are often made to satisfy new architectural requirements, and such changes can lead to a significant reduction in safety margins.

A masonry vault with tie-rods or non-slender piers only collapses if its loading is severely asymmetrical. Safety of a common masonry vault consequently depends strongly on the live-to-dead loads ratio. If this ratio is low, modern loads are supported because of the outstanding combination of mechanical properties that masonry vaults can rely on to carry symmetrical loads. If the ratio is high, modern loads have the potential for causing masonry vaults to collapse because the live loads distribution may be asymmetrical.

Earthquakes are a major threat to URM structures. In the USA, organisations such as The Masonry Society and the Federal Emergency Management Agency have identified failure of URM walls as resulting in the most material damage and loss of human life. This was evident from post-earthquake observations in Northridge, California (1994) and Turkey (1999). The weak anchorage of load-bearing walls to adjacent concrete members, or the absence of anchorage for infill walls, can cause these walls to fail and collapse under the combined effects of out-of-plane and in-plane loads generated by seismic forces. During an earthquake, failure of URM infill panels in RC structures is a potential threat to bystanders; hence such elements could need strengthening despite their non-structural role.

1.9.2 Structural degradation

Masonry materials can *decay* through environmental attack from moisture, air pollution, corrosion, freeze/thaw cycles, chemical or biological attack. Material degradation can cause localised damage (such as localised water erosion or intrusion), as well as global damage (such as the widespread decay of the surface of a stone façade due to chemical attack). Localised material damage can lead to long-term structural damage by destabilising key structural elements.

Imposed displacements due to foundation movements, material creep or temperature variations can destabilise an URM structure. Historic structures are particularly sensitive to these actions, as the stability of arches and vaults is usually geometry-based and small changes in geometry may drastically alter the equilibrium conditions leading to collapse.

1.10 Strengthening masonry structures using FRP composites

The use of FRP composites for retrofitting of masonry structures offers various advantages compared to conventional techniques (Triantafillou, 1998b). In particular, the low size and weight of the FRP does not significantly modify the dynamic response of the structure, which occurs when masonry–RC composite walls or ferrocement overlays are used. From the architectural point of view, the application of FRP composites has minimal aesthetic impact and has been proved reversible, by raising the temperature of the FRP above the glass transition temperature of the resin. It should be noted, however, that the long-term durability and compatibility of FRP with the masonry substrate in various moisture and temperature conditions still requires thorough investigation. This is a major obstacle preventing

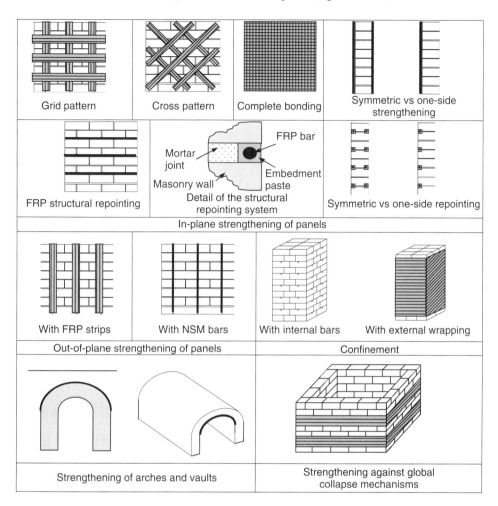

1.3 Example applications of FRP to strengthen masonry members.

the widespread use of FRP, especially for strengthening historic structures.

In the following subsections, the main possible applications of FRPs to strengthen masonry structures are introduced briefly (see also Fig. 1.3). Chapter 9 reports more details on the best-established applications and, when available, on the relevant methods of analysis and design.

1.10.1 Unreinforced masonry panels under out-of-plane loading

Several experimental programs have demonstrated the effectiveness of FRP laminates as flexural strengthening for URM walls. As for RC beams, either FRP sheets impregnated using the wet lay-up technique, or pre-cured FRP plates bonded in place with a suitable adhesive can be used. The strengthening system is applied to the tension face of the panel with the fibre direction aligned with the principal tensile stress.

Tests on FRP-strengthened URM walls loaded in four-point bending and simply supported at the ends showed extreme increases in ultimate load (typically of one order of magnitude or more), which is easily explained by the virtual absence of tension resistance in the unstrengthened walls. The possible failure modes of the system are: (i) debonding of the FRP laminate from the masonry substrate; (ii) flexural failure (rupture of the FRP laminate in tension or crushing of masonry in compression); and (iii) masonry shear failure. The available literature indicates that FRP rupture may be achieved with low area fractions of reinforcement and lower-strength materials such as GFRP (Ehsani and Saadatmanesh, 1996), whereas high FRP amounts and the use of high-strength composites such as CFRP and AFRP may lead to crushing of masonry in compression or brittle shear failure. In the intermediate (and most common) condition, the controlling mode is debonding of the FRP, giving a brittle behaviour. Proper masonry design philosophy dictates that brittle shear failure should be avoided by ensuring that the masonry flexural capacity is exceeded by its shear capacity. On the other hand, debonding of the FRP is also brittle and, as such, should be avoided. In design, Tumialan and Nanni (2001) suggested that debonding be taken into account by introducing a bond-dependent coefficient penalising the ultimate strain of the FRP laminate. For more details see Chapter 9.

FRP composites are highly effective for walls that can be treated as simply supported (i.e. walls with a high slenderness ratio). For a wall with low slenderness ratio built between rigid supports, the wall is restrained from free rotation at its extremities as the out-of-plane deflection increases. This action induces an in-plane compressive force, which (depending on the degree of support fixity) can greatly increase the ultimate capacity of the URM wall. This arching mechanism means that the increase in flexural capacity for walls strengthened with FRP laminates may be considerably less than expected, as the ultimate behaviour is governed by crushing of the masonry units in the boundary regions (Galati, 2002).

Out-of-plane strengthening has also been investigated under cyclic loading, so as to improve the ductility of URM walls during earthquakes. Substantial increases in strength and deformation capacity of URM walls retrofitted with GFRP strips on both faces have been reported (Velazquez-Dimas et al., 2000). These walls resisted pressures up to 24 times the weight of the wall and deflected as much as 5% of the wall height. To avoid very stiff behaviour and improve the hysteretic response, it was recommended that the reinforcement ratio be limited to two times the value at balanced section condition. Although the brittle URM walls were retrofitted with a linear elastic material, the FRP-masonry system was capable of dissipating some energy and exhibited system non-linearity. Framing with FRP laminates of openings in out-of-plane loaded masonry walls has been found to reduce the negative effects of openings on the out-of-plane load-carrying capacity of the walls, and results in significant improvements in their ductility (Cadei *et al.*, 2004).

The NSM method has recently been extended to URM walls. The bars are embedded into epoxy-filled grooves cut on the surface of the wall in the appropriate direction. To avoid the creation of local damage, special care needs to be taken during grooving of masonry walls, especially if these are made of hollow concrete masonry units (CMU). CMU masonry panels strengthened with NSM bars were tested by De Lorenzis *et al.* (2000) under four-point bending, giving strength increases of one order of magnitude. Load cycles on the masonry assemblage revealed an elastic behaviour until slip of the reinforcement in the grooves started taking place. After this, energy dissipation was developed by friction and the flexural deformation increased. Collapse occurred by debonding of the bars. By using epoxy strengthened with short fibres, Bajpai and Duthinh (2003) were able to prevent debonding of NSM GFRP bars and consistently rupture the bars in flexural tests of masonry walls. This method resulted in higher wall strength and a more brittle behaviour.

1.10.2 Unreinforced masonry panels under in-plane loading

A number of experimental results show that FRP laminates can be used to effectively strengthen URM walls carrying in-plane loads. They increase the load-carrying capacity and ductility of the system, and generate a uniform distribution of cracks. The strengthened walls are typically stable after failure (unlike URM walls), which could avoid injuries or loss of human life in a real building.

FRPs can be used in the form of wet lay-up sheets, pre-cured plates, pre-cured grids and NSM bars. Each of these systems is bonded in place with a suitable adhesive, which can be epoxy- or cement-based (however, if dry carbon or glass fibers are embedded into a cementitious matrix the system is no longer 'FRP' as the matrix is no longer a polymer). Sheets, plates and bars are bonded to one or both sides of the panel, with the fibre orientation following different possible patterns: horizontal, vertical, horizontal and vertical (grid pattern) or diagonal (cross-pattern).

Researchers comparing different strengthening patterns found the cross-pattern layout to be the most effective configuration (Schwegler, 1995). Also, by comparing walls strengthened in one side and two sides, it has been observed that single-side strengthening is much less effective than double-side (Schwegler, 1995) and can even be detrimental (Valluzzi et al., 2002).

The possible failure modes of the strengthened wall are: (i) debonding of the FRP laminate from the masonry substrate; (ii) splitting of masonry; and (iii) sliding shear along a mortar joint. As mentioned earlier, debonding typically occurs cohesively within the masonry substrate and depends upon the tensile strength of masonry and the substrate surface characteristics. In some cases, anchorage measures can be used (see Chapter 9).

Structural repointing can be considered as a variant of the NSM reinforcement technique, applied to masonry (Tumialan and Nanni, 2001). It consists in placing FRP rods into grooves cut into the joints, and bonding them to the wall using a suitable paste. In addition to improved structural performance, this technique offers several advantages over alternative methods, such as reduced surface preparation and minimal aesthetic impact and invasiveness. For more details see Chapter 9.

FRP structural repointing can significantly increase the shear capacity of URM walls, as shown by tests on CMU walls (Tumialan et al., 2001). Strengthened walls exhibit post-failure stability (i.e. no loose material was observed), which reduces the risk of partial or total collapse. Walls with reinforcement on both wall faces exhibited the largest pseudo-ductility. In tests by Tinazzi et al. (2000) on unreinforced clay brick masonry panels, failure was by sliding of a mortar joint. Strengthening each joint with FRP rods gave increases in capacity of about 45%, and the failure mode changed to sliding of the masonry–paste interface. Tumialan et al. (2003) tested full-scale concrete masonry infill walls subjected to in-plane cyclic load. Specimens strengthened with FRP structural repointing could reach lateral drifts of 0.7% without losing lateral carrying capacity and, for this drift level, the degradation of lateral stiffness in the strengthened walls did not imply degradation of lateral carrying capacity. The word 'lateral' in this paragraph and the next refers to the in-plane shear behaviour.

The effectiveness of in-plane strengthening of masonry walls with FRP systems has been evaluated with a number of full-size tests. *In situ* tests were performed by Corradi et al. (2002), on masonry walls damaged by recent earthquakes and retrofitted with CFRP and GFRP unidirectional laminates. The shear capacity of the panels was significantly increased by the FRP materials. Moon et al. (2003) tested a full-size two-storey URM brick building strengthened using several FRP techniques under lateral loads. GFRP sheets were epoxy bonded vertically on the inside face of one wall, while NSM GFRP rods were epoxy bonded into horizontal bed joints on the exterior face. This two-way retrofit increased the lateral strength, caused the cracks to be well distributed and produced a ductile failure mode with broad hysteresis loops and considerable energy dissipation. Tumialan and Nanni (2001) performed *in situ* tests on walls strengthened by FRP structural repointing and registered approximately 100% increments in the shear wall capacity, although it was recognised that this increase could be less when the wall panel interacts with a surrounding structural frame.

1.10.3 Confinement

Masonry columns can be confined by wrapping the columns with wet lay-up FRP sheets and/or by using internal FRP bars, acting as 'anti-expulsion' ties. The research to date on masonry columns is still very limited, compared with the vast amount of data on confinement of RC columns. However, it indicates that FRP confinement can substantially increase the stiffness, load-carrying capacity and ductility of masonry columns (Masia and Shrive, 2003; Micelli *et al.*, 2004; Aiello *et al.*, 2007).

1.10.4 Strengthening arches and vaults

Curved masonry elements typically fail by formation of a sufficient number of non-dissipative hinges activating collapse mechanisms. The formation of these hinges is due to the limited tensile strength of the masonry material. The application of FRP sheets at the intrados or at the extrados of an arch or vault modifies the static behaviour of the element by inhibiting the formation of the hinges and, consequently, may greatly increase the collapse multiplier of the structure. If the hinge formation is completely prevented, structural collapse is controlled by alternative failure modes depending on material strengths: crushing of masonry, shear sliding at mortar joints, debonding or tensile rupture of the FRP (Valluzzi *et al.*, 2001; Foraboschi, 2004). Several practical applications of FRP laminates for repair and strengthening of masonry arches and vaults have been carried out recently, especially in Italy. The most famous intervention of this type is probably repair with AFRP laminates of the vaults of St Francis Basilica in Assisi, damaged by the earthquake (Croci and Viskovic, 2000). Borri and Giannantoni (2004) repaired masonry vaults and arches of an earthquake-damaged building using CFRP sheets applied at the extrados, intended to prevent the most probable failure mechanisms. During post-strengthening measurements under dynamic excitation, the resulting peak displacements of the vault were significantly lower than in the original condition as a result of application of the FRP (Bastianini *et al.*, 2005). Valluzzi *et al.* (2004) repaired the cross vault of a church, suffering excessive deformations due to thrust lines close to critical conditions, using GFRP laminates applied at the extrados.

When curved members are strengthened with FRP sheets, a combination of shear and normal stresses is transferred across the bonded interface. The normal stresses are compressive or tensile depending on the substrate surface being convex or concave, respectively (e.g. the extrados or intrados of an arch). The latter is obviously the worst condition, as a combination of tensile and shear stresses at the interface produce, in many cases, premature bond failure (Aiello *et al.*, 2004). In these cases, construction details such as the inclusion of FRP anchor spikes are important measures against early debonding and have been proved effective by recent tests on masonry vaults (De Lorenzis *et al.*, 2007). FRP bars can also be used as tie-rods in masonry arches and vaults in place of steel bars (La Tegola *et al.*, 2000).

1.10.5 Other applications

Use of FRP for preventing 'global' collapse mechanisms

Triantafillou and Fardis (1997) proposed using FRP ties in the form of external unbonded tendons to tie different elements of a masonry building together into a three-dimensional structural system. The ties would be anchored to the masonry only at their ends, applied to the external face of the structure and post-tensioned to provide horizontal confinement and prevent collapse mechanisms by overturning under seismic actions. The effectiveness of this strengthening technique was established both analytically, for structures with simple geometries, and numerically, for a real three-dimensional structure with openings. The effects of temperature change on the tendons and the masonry were shown to be negligible. An attractive feature of this method in the case of historic structures is its reversibility, as the bonded portion is limited to the anchorage zones.

Blast strengthening

FRP-strengthening against blast loads has recently been investigated. In tests by Muszynski (1998), CFRP-strengthened walls subjected to blast loading were damaged by debonding of the FRP or composite tensile failure at wall mid-height, but had less residual displacement after the loads were released than the bare control walls. Carney and Myers (2003) studied the connection between FRP-strengthened infill walls and concrete boundary, examining two connection details. Increases in strength and strain energy in the order of two to three times were observed for the walls with the boundary connection compared to the walls with no connection. All these studies highlighted that FRP retrofits prevented loose fragments and scatter of debris, an important safety factor for building occupants.

Settlement repair

Hartley *et al.* (1996) tested two full-size concrete block walls, to investigate the feasibility of using unidirectional CFRP sheets to repair settlement damage. Settlement loads were first applied to induce characteristic step cracking. CFRP was then applied to one surface, and the wall was retested, recording strength gains of over 50%.

1.11 Overview of timber structures

Timber was the only material able to carry both tensile and compressive stresses for thousands of years, and for this reason has been extensively used to build horizontal members (beams, slabs, roofs, etc.), ties (often inserted in masonry walls or located at the base of arches and domes), foundation piles or arches and vaults (as an alternative to masonry). In modern structures, timber is used as an economic alternative to concrete and steel for residential, commercial, industrial and infrastructure applications. Also, the strength and size limitations of sawn timber have been substantially improved through the use of glue-laminated (glulam) members.

The first studies into using FRP materials to reinforce timber beams date back to the 1960s. At that time GFRP was mostly investigated as a strengthening system for laminated and solid wood members. The research interest increased in the 1990s, when CFRP and AFRP composites were used in some research projects. Since then, the falling cost associated with the increasing popularity of FRPs has led to an increasingly vital interest in their use in combination with timber. However, a major concern that remains to be solved before FRP-reinforced wood beams are widely used for *in situ* applications is the long-term durability of the FRP–wood interface under in-service hygro-thermal mechanical stresses. The interactions between moisture, temperature and fatigue and their effect on bond strength and creep behaviour of the system are not entirely understood.

1.12 Structural deficiencies in timber structures

1.12.1 Change in use of the structure

Timber structures may be required to carry increased loads, or subjected to revised design methods (as discussed in the first section). One example is

1.12.2 Structural degradation

The material degradation of wood results from biological attack, such as fungi, woodworm and termites. The durability of wood is heavily influenced by the micro-climate parameters, as it is hygroscopic and hence dependent upon the presence of internal water, the amount of which varies with the season and the thermo-hygrometric conditions of the air. Moisture dramatically influences the mechanical behaviour and durability of wood, and periodic and seasonal variations of the moisture level accelerate degradation phenomena. For instance, foundation piles may undergo deterioration as a result of periodic variation of the water level (such as in Venice); but also the supports of a wooden beam on masonry walls may degrade with time as a result of the moisture level in masonry. Material degradation can result in both a reduced size of the cross-section of the member and decay in its mechanical performance (strength and stiffness).

Shrinkage in timber members can produce cracking and excessive deformations. The pronounced anisotropy of wood results in high tensile shrinkage stresses and longitudinal cracking. In new structures, the effect of shrinkage can be lessened by prolonged seasoning, until the material reaches a moisture content matching the service moisture level (although this may not be economical). In old structures, timber members have often been seasoned under applied load. This is visible from the presence of cracks not related to the external loads (especially the typical longitudinal cracks), and from excessive deformations due to the viscoelastic behaviour of wood during loss of moisture under load. Numerous cracks may result in discontinuities between fibres, which in turn may compromise the shear strength of the member and its overall capacity.

1.13 Strengthening timber structures using FRP composites

1.13.1 Flexural strengthening

FRP materials can be used for strengthening solid and glulam timber beams in flexure. Wet lay-up sheets, pre-cured plates bonded in place using adhesive or NSM bars can be used. Sheets or plates are bonded to the tension face (sometimes also the compression face) of the beam, with the fibre direction aligned with the axis of the member. Bars are adhesive-bonded into grooves parallel to the axis of the member cut either on the soffit (sometimes also on the extrados) or on the lateral surfaces of the beam.

The advantages of using FRP over equivalent steel plates or [...] high strength-to-weight ratio and good corrosion resistance of [...] als, as described in Chapter 2. There may be additional advan[...] to aesthetic aspects and possibly to a better thermal compati[...] can be achieved by properly tailoring the composite material. The use of NSM bars rather than laminates may offer the further advantages of invisibility and better fire-proofing, and possibly also of faster installation. Practical applications of FRP laminates and bars for the flexural upgrade of wood structures include the strengthening of an historic wood bridge near Sins in Switzerland (this structure is discussed in Chapter 13), the prestressing of timber beams at the Horrem Community Centre in Germany with GFRP tendons and the prestressing of wood beams in a Japanese tea house with CFRP tendons.

Laboratory tests on flexural strengthening have mostly been conducted in Europe and North America and have encompassed a wide range of timber species (from low-grade to high-grade varieties, both in solid and glulam forms), types of FRP (mostly GFRP or CFRP, in the form of sheets, plates and bars) and reinforcement ratios. Despite this variety of parameters, the test results appear to be very consistent.

Strengthened beams show substantial gains in capacity. If the reinforcement area ratio in the tension zone is large enough, the failure mode usually changes from brittle tensile–flexural failure preceding to pseudo-yielding of the compression fibres in the plain timber specimens, to a more ductile compression–flexural failure in the strengthened specimens (i.e., using a terminology familiar for RC, the cross-section turns over-reinforced). The region of wood yielding in compression propagates from the top of the beam down until the beam ultimately fails by tensile rupture of the bottom wooden fibres. Hence, more efficient use is made of the compressive strength of the wood. Also, the average ultimate tensile strain of the timber typically increases, indicating that the FRP reinforcement arrests crack opening, confines local rupture and bridges local defects in the adjacent timber. This transition in failure mode leads not only to increased strength but also to enhanced ductility of the strengthened beam. In general, the largest increases in strength are obtained with the lower grades of wood, which therefore seem to get the highest value-added benefits from this technology. This is due to these grades displaying a larger difference in relative tension/compression strength values, which can be remedied by adding FRP tension reinforcement.

In general, adequate short-term bond is observed between timber and resin (be it epoxy, polyester, vinylester or phenolic) used for impregnation and/or bonding of the FRP. This has been indicated by bond characterisation tests, but also confirmed by the rare occurrence of failures of strengthened beams due to debonding or delamination of the reinforcement.

Another positive consequence of applying FRP strengthening to timber is that the variability in strength for the reinforced beams typically decreases compared to the unreinforced beams, as the presence of the FRP lessens the effect of local defects such as deviations, knots, etc. As a result, not only the average strength of the beam increases, but also its standard deviation decreases, implying a further improvement of the characteristic value to be adopted for design.

A stiffness enhancement can also be achieved, the extent of which depends on the area fraction and elastic modulus of the FRP reinforcement as well as on the quality of wood. A larger increase in stiffness can be achieved by using prestressed FRP reinforcement, such as investigated by Triantafillou and Deskovic (1992) and Borri and Corradi (2000). The latter authors realised prestress by applying an upward load to the beams (equal to approximately one fourth of the ultimate load of plain timber beams) prior to bonding the FRP.

The capacity and stiffness of strengthened beams is estimated with sufficient accuracy by applying classical bending theory (based on the assumption that plane cross-sections remain plane), using the stress–strain model of Buchanan (1990) for the timber (linear elastic and brittle in tension; bilinear elasto-plastic in compression), and accounting for the presence of the FRP reinforcement by assuming perfect bond to the wooden substrate. The same simplifying analysis has been applied to glulam beams (neglecting slip between laminations), although more sophisticated models have also been attempted.

1.13.2 Shear strengthening

One major disadvantage of wood is its poor strength perpendicular to the grain, sometimes resulting in the shear resistance parallel to the grain being critical. This may also be the case when the flexural capacity of a timber beam has been increased by external strengthening, so that the failure becomes shear-controlled. FRPs have been successfully employed in place of steel or aluminium plates or bolts to increase the shear strength of timber.

Triantafillou (1997) studied shear strengthening of glulam members with FRP laminates bonded to the sides of the beam in the shear-critical regions, and showed that the FRP reinforcement can be very effective, depending upon the FRP configuration and area fraction. In particular, the use of the FRP material is optimised when the fibres are oriented longitudinally and the height of the externally bonded sheet or plate is a little higher than a limiting value, so as to give tensile failure of the FRP before shear failure of the wood outside the strengthened region. The capacity of the strengthened member was predicted using simple mechanics, which showed good agreement with test results. Furthermore, no debonding failures were reported. Svecova and Eden (2004) shear-strengthened timber stringers by inserting GFRP dowels in the centre of the cross-section along the length of the stringers. The GFRP reinforcement increased the strength of the stringers and reduced its variability, by bridging local defects and discontinuities of the timber.

Curved beams, supports, drilled holes and cut-outs often have critical shear strength. Larsen *et al.* (1992) increased the shear capacity of curved and cambered glulam beams using GFRP. A few researchers have used FRP for shear-strengthening glulam timber members in the region of circular and rectangular cut-outs (Kessler *et al.*, 2004). FRP reinforcement has been used to enhance load-carrying capacity, stiffness and, in some cases, ductility of bolted wood connections since 1973, when Poplis and Mintzer (1973) tested bolted connections of plywood strengthened with woven roving fibreglass overlays. More recently, Windorski *et al.* (1997) (for example) examined reinforcement of wood at bolted connections loaded both parallel and perpendicular to grain, using bidirectional fibreglass cloth and epoxy resin. The connection strength increased with the layers of fibreglass reinforcement and, for loading parallel to grain, the reinforcement changed the mode of failure from an abrupt, catastrophic type associated with tensile stresses perpendicular to grain to a ductile type associated with bearing stress.

1.13.3 Stiffening of slabs for seismic upgrade

GFRP laminates have recently been investigated as a means of increasing the in-plane stiffness and strength of wooden slabs for seismic upgrade (Borri *et al.*, 2004). Two configurations of wooden slabs typically found in Italian historical buildings (one-way slabs featuring primary beams, secondary beams and a floor made either of wooden elements nailed to the secondary beams or of simply supported clay units) have been tested under in-plane loading. A mesh of GFRP laminate strips was bonded to the floor, overlaid with either a second wooden floor (for slabs with wooden floor) or a thin layer of lime mortar (for slabs with clay unit floor). Test results showed a significant stiffness and strength enhancement of the slabs upgraded with FRP mesh of sufficiently small spacing.

1.14 Drivers for using FRP composites rather than conventional strengthening options

As illustrated by the numerous applications described in the preceding sections, externally-bonded FRP composites can be used to address a variety of structural deficiencies. Like any of the array of strengthening options available to engineers, FRP strengthening has very significant advantages in specific applications, but will not be suitable for every application.

A variety of techniques is already available for addressing structural deficiencies. The most extreme option is to replace a structure, or to duplicate it. This approach is rarely sustainable (in economic, environmental or social terms), whereas strengthening allows the use of an existing structure to be maximised and is thus inherently more sustainable. Alternative strengthening options available to an engineer include bolting or welding new structural members to the existing structure; replacing rivets with bolts; and removing dead load (for example, by replacing a concrete or brick jack arch deck with a FRP deck, Kessler *et al.*, 2004).

The dominant reason for selecting FRP strengthening over conventional techniques is the cost saving associated with reduced possession of the structure (Burgoyne, 2004). The costs associated with either complete closure of a structure or disruption to the use of the structure (for example, due to traffic management measures) very often outweigh the costs of the work being undertaken. FRP strengthening can be applied relatively rapidly compared to other strengthening options. FRP is lightweight, and can usually be manoeuvred into place without heavy lifting plant, which is particularly advantageous where the work needs to be carried out in confined spaces, or access to the site is restricted. In addition to lower possession charges, the reduction in installation time and ease of manoeuvring, FRP strengthening lead to savings in labour and plant hire charges. The additional material cost of FRP strengthening is small compared to these savings.

Other benefits of using FRP strengthening include:

1. there are health and safety benefits associated with reduced lifting requirements;
2. it does not add significant additional dead load to the structure;
3. FRP is a low-maintenance material and can be used to halt corrosion in the existing member;
4. bonded strengthening does not create large stress concentrations within the existing structure, which could lead to failure in brittle cast iron structures.

It should be noted, however, that the application of FRP strengthening is (at the time of writing) a relatively young technique. Research is still ongoing to address some of its applications. There are practical limitations that prevent FRP strengthening being applied in particular cases, some of which might well be addressed by future research:

1. The strengthening critically relies on a high-quality adhesive joint, and thus requires a high quality of workmanship and the correct environmental conditions during cure.

2. FRP composites have good fire resistance, but the glass transition temperature of the bonding adhesive can be as low as 65 °C. Careful consideration must be given to fire protection if the strengthening must remain functional during a fire (in particular, note that the metallic member will conduct heat to the back of the adhesive joint). However, the capacity of the unstrengthened member may be sufficient for the loads expected to be present during a fire without strengthening, so that it does not matter if the strengthening cannot withstand the fire.
3. Obstructions (such as rivet or bolt heads) can prevent a preformed strengthening plate being bonded to the surface of a member.
4. The sides of beams are often obstructed, for example due to bridge decks or floors.

1.15 References

ACI (2002) Design and Construction of Externally Bonded FRP Systems for Strengthening Concrete Structures, ACI 440.2R-02, Farmington Hills, MI, USA, American Concrete Institute.

AIELLO M A, DE LORENZIS L, GALATI N and LA TEGOLA A (2004) Bond between FRP laminates and curved concrete substrates with anchoring composite spikes, in La Tegola A and Nanni A (eds), *Proceedings Innovative Materials and Technologies for Construction and Restoration – IMTCR'04*, Naples, Italy, Liguori Editore, Vol. 2, 45–56.

AIELLO M A, MICELLI F and VALENTE L (2007) Structural upgrading of masonry columns by using composite reinforcements, *Journal of Composites for Construction*, ASCE, **11**(6), 650–659.

ANTONOPOULOS C P and TRIANTAFILLOU T C (2003) Experimental investigation of FRP-strengthened RC beam-column joints, *Journal of Composites for Construction*, ASCE, **7**(1), 39–49.

BAJPAI K and DUTHINH D (2003) Flexural strengthening of masonry walls with external composite bars, *The Masonry Society Journal*, August, 9–20.

BASSETTI A, NUSSBAUMER A and HIRT M Crack repair and fatigue life extension of riveted bridge members using composite materials, in Hosny A-N (ed.), *Proceedings Bridge Engineering Conference 2000*, ESE-IABSE-FIB, Sharm El-Sheikh, Cairo, Egyptian Society of Engineers, **1**, 227–238.

BASTIANINI F, CERIOLO L, DI TOMMASO A and ZAFFARONI G (2004) Mechanical and nondestructive testing to verify the effectiveness of composite strengthening on historical cast iron bridge in Venice Italy, *Journal of Materials in Civil Engineering*, **16**(4), 401–413.

BASTIANINI F, CORRADI M, BORRI A and DI TOMMASO A (2005) Retrofit and monitoring of an historical building using 'Smart' CFRP with embedded fibre optic Brillouin sensors, *Construction and Building Materials*, **19**, 525–535.

BORRI A and CORRADI M (2000) Consolidamento di strutture lignee: risultati di una sperimentazione, *L'Edilizia – Building and Construction for Engineers*, **5/6**, 62–67 (in Italian).

BORRI A and GIANNANTONI A (2004) Esempi di utilizzo di materiali compositi per il miglioramento sismico degli edifici in muratura, *Proceedings ANIDIS 2004*, Genova, Italy, Jan 25–29 (CD ROM) (in Italian).

BORRI A, CORRADI M, SPERANZINI E and VIGNOLI A (2004) Irrigidimento nel piano di solai esistenti con FRP, *Proceedings Conference on Mechanics of FRP-strengthened Masonry Structures*, Padua, Italy, Edizioni Cortina, 343–355 (in Italian).

BUCHANAN A H (1990) Bending strength of lumber, *Journal of Structural Engineering*, ASCE, **116**(5), 391–397.

BURGOYNE C J (2004) Does FRP have an Economic Future? *Proceedings Advanced Composite Materials in Building and Structures – ACMBS IV*, Calgary, Alberta, Canada, July 20–23 (CD ROM).

BUSSELL M N (1997) *Appraisal of Existing Iron and Steel Structures*, P138, Ascot, UK, Steel Construction Institute.

CADEI J M C, STRATFORD T J, HOLLAWAY L C and DUCKETT W G (2004) *Strengthening Metallic Structures using Externally Bonded Fibre-Reinforced Polymers*, Report C595, London, CIRIA.

CARNEY P and MYERS J J (2003) *Static and Blast Resistance of FRP Strengthened Connections for Unreinforced Masonry Walls*, CIES Report No. 03-46, Rolla, MO, USA, University of Missouri–Rolla.

CASADEI P, NANNI A and IBELL T J (2003a) Experimentation on two-way RC slabs with openings strengthened with CFRP laminates, in Crivelli-Visconti I (ed.), *Proceedings Advancing with Composites*, Milan, Italy, May 7–9, 239–246.

CASADEI P, IBELL T and NANNI A (2003b) Experimental results of one-way slabs with openings strengthened with CFRP laminates, in Tan K H (ed.), *Proceedings Fibre Reinforced Polymer Reinforcement for Concrete Structures – FRPRCS-6*, Singapore, World Scientific, 1097–1106.

CHACON A, CHAJES M, SWINEHART M, RICHARDSON D and WENCZEL G (2004) Applications of advanced composites to steel bridges: A case study on the Ashland bridge (Delaware-USA), *Proceedings Advanced Composite Materials in Building and Structures – ACMBS IV*, Calgary, Alberta, Canada, July 20–23 (CD ROM).

CHURCH D G and SILVA T D D (2002) Application of carbon fibre composites at covered ways 12 and 58 and bridge EL, in Shenoi R A, Moy S S J and Hollaway L C (eds), *Advanced Polymer Composites for Structural Applications in Construction*, London, UK, Thomas Telford, 15–17.

CNR (2004) *CNR-DT200/2004 Guide for the Design and Construction of Externally Bonded FRP Systems for Strengthening Existing Structures – Materials, RC and PC structures, masonry structures*, Rome, Italy, Consiglio Nazionale delle Ricerche (English version).

CORRADI M, BORRI A and VIGNOLI A (2002) Strengthening techniques tested on masonry structures struck by the Umbria–Marche earthquake of 1997–1998, *Construction and Building Materials*, **16**, 229–239.

CROCI G and VISKOVIC A (2000) L'uso degli FRP di fibra aramidica per il rinforzo della Basilica di San Francesco di Assisi, *Proceedings Conference on Mechanics of FRP-strengthened Masonry Structures*, Padua, Italy, Edizioni Cortina (in Italian).

DE LORENZIS L and NANNI A (2001) Shear strengthening of RC beams with near surface mounted FRP rods, *ACI Structural Journal*, **98**(1), 60–68.

DE LORENZIS L and TENG J G (2007) Near-surface mounted reinforcement: an emerging technique for structural strengthening, *Composites Part B: Engineering*, **38**(2), 119–143.

DE LORENZIS L and TEPFERS R (2003) A comparative study of models on confinement of concrete cylinders with FRP composites, *Journal of Composites for Construction*, ASCE, **7**(3), 219–237.

DE LORENZIS L, TINAZZI D and NANNI A (2000) NSM FRP rods for masonry strengthening: bond and flexural testing, *Proceedings Conference on Mechanics of FRP-strengthened Masonry Structures*, Padua, Italy, Edizioni Cortina (in Italian).

DE LORENZIS L, MICELLI F and LA TEGOLA A (2002) Passive and active near surface mounted FRP rods for flexural strengthening of RC beams, *Proceedings International Conference on Fibre Composites in Infrastructure – ICCI-02*, San Francisco, CA, USA, June 10–12 (CD ROM).

DE LORENZIS L, GALATI D and LA TEGOLA A (2004) Evaluation of stiffness and ductility of RC beams strengthened with FRP composites, *Structures and Buildings*, **157**(SB1), 31–51.

DE LORENZIS L, DIMITRI R and LA TEGOLA A (2007) Reduction of the lateral thrust of masonry arches and vaults with FRP composites, *Construction and Building Materials*, **21**(7), 1415–1430.

EHSANI M R and SAADATMANESH H (1996) Seismic retrofit of URM walls with fibre composites, *The Masonry Society Journal*, **14**(2), 63–72.

EL-HACHA R and RIZKALLA S H (2004) Near-surface-mounted fibre-reinforced polymer reinforcements for flexural strengthening of concrete structures, *ACI Structural Journal*, **101**(5), 717–726.

EL-REFAIE S A, ASHOUR A F and GARRITY S W (2003) Sagging and hogging strengthening of continuous reinforced concrete beams using CFRP sheets, *ACI Structural Journal*, **100**(4), 446–453.

ESHWAR N IBELL T and NANNI A (2003) CFRP strengthening of concrete bridges with curved soffits, in Forde M C (ed.), *Proceedings Tenth International Conference on Structural Faults and Repair*, Edinburgh, UK, Technics Press (CD ROM).

FARMER N and SMITH I (2001) King Street Railway Bridge – strengthening of cast iron girders with FRP composites, in Forde M C (ed.), *Proceedings Ninth International Conference on Structural Faults and Repair*, Edinburgh, UK, Technics Press (CD ROM).

FIB (2001) *Externally Bonded FRP Reinforcement for RC Structures*, Task Group 9.3, Bulletin No. 14, Lausanne, Switzerland, Federation Internationale du Beton.

FORABOSCHI P (2004) Strengthening of masonry arches with fibre-reinforced polymer strips, *Journal of Composites for Construction*, ASCE, **8**(3), 191–202.

GALATI N (2002) *Arching Effect on Masonry Walls Strengthened with FRP Materials*, MSc Thesis, University of Missouri-Rolla, Rolla, MO, USA.

GERGELY J, PANTELIDES C P and REAVELEY L D (2000) Shear strengthening of RC T-Joints using CFRP composites, *Journal of Composites in Construction*, ASCE, **4**(2), 56–64.

GRACE N F (2004) Concrete repair with CFRP, *Concrete International*, **26**(5), 45–52.

HARTLEY A, MULLINS G and SEN R (1996) Repair of concrete masonry block walls using carbon fibre, in El-Badry M (ed.), *Proceedings Advanced Composite Materi-*

als in Building and Structures – ACMBS II, Montreal, Quebec, Canada, Canadian Society for Civil Engineering, 795–802.

IBELL T J and SILVA P F (2004) A theoretical strategy for moment redistribution in continuous FRP-strengthened concrete structures, in Hollaway L C, Chryssanthopoulos M K and Moy S S J (eds), *Advanced Polymer Composites for Structural Applications in Construction: ACIC 2004, Guildford, UK*, Cambridge, UK, Woodhead, 144–151.

JBDPA (1999) *Seismic Retrofitting Design and Construction Guidelines for Existing Reinforced Concrete (RC) Buildings with Fibre Reinforced Polymer (FRP) Materials*, Tokyo, Japan Building Disaster Prevention Association (in Japanese).

KESSLER K, LESKO J and COUSINS T (2004) Rehabilitation design and evaluation of the Hawthorne Street Bridge FRP deck installation, in Hollaway L C, Chryssanthopoulos M K and Moy S S J (eds), *Advanced Polymer Composites for Structural Applications in Construction: ACIC 2004, Guildford, UK*, Cambridge, UK, Woodhead, 738–746.

KHALIFA A, GOLD W J, NANNI A and ABDEL AZIZ M I (1998) Contribution of externally bonded FRP to shear capacity of RC flexural members, *Journal of Composites in Construction*, ASCE, **2**(4), 195–202.

KHALIFA A, ALKHRDAJI T, NANNI A and LANSBURG S (1999) Anchorage of surface mounted FRP reinforcement, *Concrete International*, **21**(10), 49–54.

LAM L and TENG J G (2001) Strength of RC cantilever slabs bonded with GFRP strips, *Journal of Composites in Construction*, ASCE, **5**(4), 221–227.

LAMANNA A J, BANK L C and SCOTT D W (2004) Flexural strengthening of RC beams by mechanically attaching FRP strips, *Journal of Composites in Construction*, ASCE, **8**(3), 203–210.

LARSEN H J, GUSTAFSSON P J and ENQUIST B (1992) *Tests with Glass-fibre Reinforcement of Wood Perpendicular to the Grain*, Report TVSM7067, Division of Structural Mechanics, Lund Institute of Technology, Sweden.

LA TEGOLA A, LA TEGOLA AL, DE LORENZIS L and MICELLI F (2000) Applications of FRP materials for repair of masonry structures, *Workshop on Advanced FRP Materials For Civil Structures*, Bologna, Italy, Oct 19.

LEE C, BONACCI J, THOMAS M, MAALEJ M, KHAJENPOUR S, HEARN N, PANTAZOPOULOU S and SHEIKH S (2000) Accelerated corrosion and repair of RC columns using CFRP sheets. *Canadian Journal of Civil Engineering*, **27**(5), 949–959.

LEEMING M B and DERBY J J (1999) Design and specifications for FRP plate bonding, in Hollaway L C and Leeming M B (eds), *Strengthening of Reinforced Concrete Structures Using Externally-bonded FRP Composites in Structural and Civil Engineering*, Cambridge, UK, Woodhead, 242–269.

LEES J M, WINISTOERFER A U and MEIER U (2002) External prestressed carbon fibre-reinforced polymer straps for shear enhancement of concrete, *Journal of Composites in Construction*, ASCE, **6**(4), 249–256.

LUKE S and CANNING L (2004) Strengthening highway and railway bridge structures with FRP composites – case studies, in Hollaway L C, Chryssanthopoulos M K and Moy S S J (eds), *Advanced Polymer Composites for Structural Applications in Construction: ACIC 2004, Guildford, UK*, Cambridge, UK, Woodhead, 745–754.

MASIA M J and SHRIVE N G (2003) Carbon fibre reinforced polymer wrapping for the rehabilitation of masonry columns, *Canadian Journal of Civil Engineering*, **30**(4), 734–744.

MICELLI F, DE LORENZIS L and LA TEGOLA A (2004) FRP-confined masonry columns under axial loads: analytical model and experimental results, *Masonry International*, **17**(3), 95–108.

MILLER T C, CHAJES M J, MERTZ D R and HASTINGS J N (2001) Strengthening of a steel bridge girder using CFRP plates, *Journal of Bridge Engineering*, ASCE, **6**(6), 514–522.

MOON F L, YI T, LEON R T and KAHN L F (2003) Large-scale tests of an unreinforced masonry low-rise building, *Proceedings Ninth North American Masonry Conference*, Clemson, SC, USA, June 1–3 (CD ROM).

MORTAZAVI A A, PILAKOUTAS K and SANG SON K (2003) RC column strengthening by lateral pre-tensioning of FRP, *Construction and Building Materials*, **17**, 491–497.

MOSALLAM A S and MOSALAM K M (2003) Strengthening of two-way concrete slabs with FRP composite laminates, *Construction and Building Materials*, **17**, 43–54.

MOY S S J (ed.) (2001) *FRP Composites – Life Extension and Strengthening of Metallic Structures*, London, UK, Thomas Telford.

MOY S S J (2004) The Strengthening of wrought iron using carbon fibre reinforced polymer composites, in Hollaway L C, Chryssanthopoulos M K and Moy S S J (eds), *Advanced Polymer Composites for Structural Applications in Construction: ACIC 2004, Guildford, UK*, Cambridge, UK, Woodhead, 258–265.

MOY S S J and BLOODWORTH A G (2007) Strengthening a steel bridge with CFRP composites, *Proceedings Institution of Civil Engineers, Structures and Buildings*, **160**(SB2), 81–93.

MOY S S J and LILLISTONE D (2006) Strengthening cast iron using FRP composites, *Proceedings Institution of Civil Engineers, Structures and Buildings*, **159**(SB6), 309–318.

MUSZYNSKI L C (1998) Explosive field tests to evaluate composite reinforcement of concrete and masonry walls, in Saadatmanesh H and Ehsani M R (eds), *Proceedings International Conference on Fibre Composites in Infrastructure – ICCI-98*, Tucson, AZ, USA, Jan 5–7, 276–284.

NING Z, QINGRUI Y, YONGXIN L, LIQIANG H, FUMING P, PENG C, YAN Z, GUANGZHONG W and YUNSHEN Z (2004) Research on the Fatigue Tests of Steel Structure Member Reinforced with CFRP, *Industrial Construction*, **35**(4), 19–30 (in Chinese).

NORDIN H and TÄLJSTEN B (2003) Concrete beams strengthened with prestressed near surface mounted reinforcement, in Tan K H (ed.), *Proceedings Fibre Reinforced Polymer Reinforcement for Concrete Structures – FRPRCS-6*, Singapore, World Scientific, 1077–1086.

NSF (1993) *NSF 93-4 Engineering Brocure on Infrastructure*, Arlington, VA, USA, US National Science Foundation.

PANTAZOPOULOU S J, BONACCI J F, SHEIKH S, THOMAS M D A and HEARN N (2001) Repair of corrosion-damaged columns with FRP wraps, *Journal of Composites in Construction*, ASCE, **5**(1), 3–11.

PHARES B M, WIPF T J, KLAIBER F W, ABU-HAWASH A and LEE Y-S (2003) Strengthening of steel girder bridges using FRP, *Proceedings Mid-Continent Transportation Research Symposium*, Ames, IA, USA, Aug 21–22.

POPLIS J and MITZNER R (1973) *Plywood Overlaid with Fiberglass-reinforced Plastic: Fastener tests*, APA Laboratory Report 119, Part 2, Tacoma, WA, USA, American Plywood Association.

PROTA A, NANNI A, MANFREDI G and COSENZA E (2004) Selective upgrade of underdesigned RC beam-column joints using CFRP, *ACI Structural Journal*, **101**(5), 699–707.

QUANTRILL R J, HOLLAWAY L C and THORNE A M (1996) Experimental and analytical investigation of FRP strengthened beam response, *Magazine of Concrete Research*, **48**(177), 331–342.

RIZZO A (2005) *Application in Off-System Bridges of Mechanically Fastened FRP (MF-FRP) Pre-Cured Laminates*, MSc Thesis, University of Missouri-Rolla, Rolla, MO, USA.

ROUSAKIS T, YOU C S, DE LORENZIS L, TAMUZS V and TEPFERS R (2003) Concrete cylinders confined by prestressed CFRP filament winding, subjected to monotonic and cyclic axial compressive load, in Tan K H (ed.), *Proceedings Fibre Reinforced Polymer Reinforcement for Concrete Structures – FRPRCS-6*, Singapore, World Scientific, **1**, 581–590.

SAADATMANESH H (1995) Wrapping with composite materials, in Taerwe L (ed.), *Proceedings Second International RILEM Symposium (FRPRCS-2): Non-metallic (FRP) Reinforcement for Concrete Structures*, London, UK, E and F N Spon, 593–600.

SCHNERCH D, STANFORD K, SUMNER E A and RIZKALLA S (2004) Strengthening steel structures and bridges with high modulus carbon fibre reinforced polymers: resin selection and scaled monopole behaviour, *TRB 2004 Annual Meeting*, Washington, DC, USA, Jan 11–15 (CD-ROM).

SCHNERCH D, DAWOOD M and RIZKALLA S (2006) *Design Guidelines for the use of HM Strips: Strengthening of Steel Concrete Composite Bridges with High Modulus Carbon Fibre Reinforced Polymer (CFRP) Strips*, Technical Report No. IS-06-02, Constructed Facilities Laboratory, Raleigh, NC, USA, North Carolina State University.

SCHWEGLER G (1995) Masonry construction strengthened with fibre composites in seismically endangered zones, in Duma G (ed.), *Proceedings Tenth European Conference on Earthquake Engineering*, Rotterdam, The Netherlands, Balkema, 2299–2303.

SEIBLE F, PRIESTLEY M J N, HEGEMIER G A and INNAMORATO D (1997) Seismic retrofit of RC columns with continuous carbon fibre jackets, *Journal of Composites in Construction*, ASCE, **1**(2), 52–62.

SEIM W, HÖRMAN M, KARBHARI V and SEIBLE F (2001) External FRP poststrengthening of scaled concrete slabs, *Journal of Composites in Construction*, ASCE, **5**(2), 67–75.

SEN R, LIBY L and MULLINS G (2001) Strengthening steel bridge sections using CFRP laminates, *Composites: Part B*, **32**, 309–322.

SHAAT A and FAM A (2006) Axial loading tests on short and long hollow structural steel columns retrofitted using carbon fibre reinforced polymer, *Canadian Journal of Civil Engineering*, **33**(4), 458–470.

SHAHAWY M and BEITELMAN T E (1999) Static and fatigue performance of RC beams strengthened with CFRP laminates, *Journal of Structural Engineering*, **125**(6), 613–621.

SMITH I (2004) Maunders Road Overbridge – the behaviour and in-service performance of cast iron bridge girders strengthened with CFRP reinforcement, in Hollaway L C, Chryssanthopoulos M K and Moy S S J (eds), *Advanced Polymer*

Composites for Structural Applications in Construction: ACIC 2004, Guildford, UK, Cambridge, UK, Woodhead, 711–718.

SMITH S T and TENG J G (2002) FRP-strengthened RC beams. II: assessment of debonding strength models, *Engineering Structures*, **24**, 397–417.

SVECOVA D and EDEN R J (2004) Flexural and shear strengthening of timber beams using glass fibre reinforced polymer bars — an experimental investigation, *Canadian Journal of Civil Engineering*, **31**(1), 45–55.

TASTANI S P and PANTAZOPOULOU S J (2004) Experimental evaluation of FRP jackets in upgrading RC corroded columns with substandard detailing, *Engineering Structures*, **26**(6), 817–829.

TAVAKKOLIZADEH M and SAADATMANESH H (2003) Fatigue strength of steel girders strengthened with carbon fibre reinforced polymer patch, *ASCE Journal of Structural Engineering*, **129**(2), 186–196.

TENG J G, LAM L, CHAN W and WANG J S (2000) Retrofitting of deficient RC cantilever slabs using GFRP strips, *Journal of Composites in Construction*, ASCE, **4**(2), 75–84.

TINAZZI D, ARDUINI M, MODENA C and NANNI A (2000) FRP structural repointing of masonry assemblages, in Humar J and Razaqpur A G (eds), *Proceedings Advanced Composite Materials in Building and Structures – ACMBS III,* Montreal, Quebec, Canada, Canadian Society for Civil Engineering, 585–592.

TRIANTAFILLOU T C (1997) Shear reinforcement of wood using FRP materials, *Journal of Materials in Civil Engineering*, ASCE, **9**(2), 65–69.

TRIANTAFILLOU T C (1998a) Shear strengthening of reinforced concrete beams using epoxy-bonded FRP composites, *ACI Structural Journal*, **95**(2), 107–115.

TRIANTAFILLOU T C (1998b) Strengthening of masonry structures using epoxy-bonded FRP laminates, *Journal of Composites in Construction*, ASCE, **2**(2), 96–103.

TRIANTAFILLOU T and DESKOVIC N (1992) Prestressed FRP sheets as external reinforcement of wood members, *Journal of Composites in Construction*, ASCE, **118**(5), 1270–1284.

TRIANTAFILLOU T C and FARDIS M N (1997) Strengthening of historic masonry structures with composite materials, *Materials and Structures*, **30**, 486–496.

TRIANTAFILLOU T C, DESKOVIC N and DEURING M (1992) Strengthening of concrete structures with prestressed fibre reinforced plastic sheets, *ACI Structural Journal*, **89**(3), 235–244.

TUMIALAN J G and NANNI A (2001) *In-plane and out-of-Plane Behaviour of Masonry Walls Strengthened with FRP Systems*, CIES Report No. 01-24, Rolla, MO, USA, University of Missouri–Rolla.

TUMIALAN J G, MORBIN A, NANNI A and MODENA C (2001) Shear strengthening of masonry walls with FRP composites, *Proceedings Composites 2001*, Composites Fabricators Association, Tampa, FL, USA, Oct 3–6 (CD ROM).

TUMIALAN J G, SAN BARTOLOME A and NANNI A (2003) Strengthening of URM infill walls by FRP structural repointing, *Proceedings Ninth North American Masonry Conference*, Clemson, SC, USA, June 1–3 (CD ROM).

VALLUZZI M R, VALDEMARCA M and MODENA C (2001) Behaviour of brick masonry vaults strengthened by FRP laminates, *Journal of Composites in Construction*, ASCE, **5**(3), 163–169.

VALLUZZI M R, TINAZZI D and MODENA C (2002) Shear behaviour of masonry panels strengthened by FRP laminates, *Construction and Building Materials*, **16**(7), 409–416.

VALLUZZI M R, DA PORTO F and MODENA C (2004) Strengthening of a masonry vault with GFRP laminates: S. Fermo Church in Verona, Proceedings IMTCR'04, Lecce, Italy, 430–439.

VELAZQUEZ-DIMAS J, EHSANI M and SAADATMANESH H (2000) Out-of-plane behaviour of brick masonry walls strengthened with fibre composites, *ACI Structural Journal*, **97**(3), 377–387.

WINDORSKI D F, SOLTIS L A and ROSS R J (1997) Feasibility of fibreglass-reinforced bolted wood connections, FPL Research Paper FPL–RP–562, Madison, WI, USA, Forest Products Laboratory.

2
Fibre-reinforced polymer (FRP) composites used in rehabilitation

L. C. HOLLAWAY, University of Surrey, UK

2.1 Introduction

This chapter provides an introduction to the advanced polymer composite (APC) materials used in the rehabilitation of reinforced concrete (RC), prestressed concrete (PC) and metallic and masonry structures. The type of upgrading considered is that required for flexural and shear resistance of beams and wrapping of RC columns. The chapter will discuss separately the mechanical and in-service properties of the two major components of the composite (viz. the polymer matrix and the fibre) and will continue by considering these properties in terms of fibre-reinforced polymer (FRP) composites used in construction and exposed to various civil engineering environments. As the composite plate or the FRP composite wraps used in the rehabilitation of structures are fabricated from the same materials and manufactured by the same techniques as those for the FRP civil engineering structural materials, their mechanical and in-service properties will be of similar values.

APC materials consist of strong stiff fibres in a polymer matrix and require scientific understanding from which design procedures may be developed. The mechanical and physical properties of the composite are controlled by the constituent properties and by the microstructural configurations. The matrix must bond well with the fibre surface to enable transfer of stresses efficiently between the fibres. Fibre alignment, fibre content and the strength of the fibre–matrix interface all influence the performance of the composite. Furthermore, highly specialised processing techniques are used which take account of the handling characteristics particularly of carbon fibres, a material now commonly used to upgrade or strengthen structural systems.

The materials used for composite plate bonding of beams or the wrapping of columns could be fabricated by a number of techniques, but two methods are mainly used; these are the pultrusion and the pre-impregnated fibre in a resin (prepreg). This latter method can be a factory-made preformed plate or an uncured prepreg used in conjunction with a film adhesive. Both components are wrapped around the structural member on site and then fully cured under pressure and elevated temperature in one operation. The advantage of utilising the latter method over that of the two former methods is that the structural member can be of any geometrical shape. In certain circumstances, other systems have been used; these include: (i) the wet lay-up system for plate bonding or column wrapping, (ii) the resin transfer moulding (RTM) systems or variations of these, mainly for plate bonding, such as resin infusion under flexible tooling (RIFT), (iii) the XXsys Technologies (which is used only for specialised work) or variations of this method, mainly for column wrapping. In the above three processing methods, the polymer of the composite also acts as the adhesive between the composite and the structural members; they have been fully described in Hollaway and Head (2001).

The successful strengthening of structural members with FRP materials is dependent upon the quality and integrity of the composite adherent, the effectiveness of the adhesive used and the surface preparation of the two adherents to be joined. The basic requirements for the creation of a satisfactory bonded joint are:

- selection of a suitable adhesive;
- adequate preparation of the adhesive surface;
- appropriate design of the joint;
- controlled fabrication of the joint;
- a post-bonding quality assurance.

This chapter will discuss some aspects of bonding composites to the more conventional civil engineering materials.

The long-term durability of FRP composite materials for rehabilitation of structural systems is often quoted as the main reason for utilising the material. However, the durability of the composite material is highly dependent upon the choice of the constituent materials, the methods of their fabrication and the environment into which they are placed throughout their lives. Although FRP composite materials do not corrode they do undergo chemical and physical changes over a period of time. In addition, the durability of adhesives may be affected by the environment into which they are placed. The chemical degradation of the matrix component of the composite, of the adhesive, of the substrate and of chemical bonds across the interface as a result of interaction with water and other chemicals are all possible; however, the matrix component and the structural adhesives are selected with essentially hydrolysis-resistant chemistry and so chemical attack is not generally an important degradation mechanism. Any significant weakening of adhesive joints as a result of swelling would be associated with the absorption of large amounts of water, and such materials are considered unsuitable for structural applications. However, the rehabilitation of structures is not generally associated with immersion in water or chemicals, but there may be occasions on which a high humidity associated with the construction exists. If these conditions do exist the resin manufacturer should be consulted.

This chapter will detail the materials that are used for the upgrading of structures and will address the issues raised in this introduction. It will show that advanced polymer composites utilised in the rehabilitation of structural members can provide long lifetimes with very little maintenance.

2.2 Component parts of composite materials

The advanced polymer composite essentially consists of two component materials: (i) the matrix material or polymer, which is generally the low-strength and low-modulus component and (ii) the fibre, which is the relatively high-strength and high-modulus component. Under stress, the fibre utilises the plastic flow of the matrix to transfer the load to the fibre; this results in a high-strength and high-modulus composite. The primary phase, the fibres of high aspect ratio, must be well dispersed and bonded into the secondary phase, the matrix. The principal constituents of the composite are, therefore, the fibre, the matrix and the interface. This last component is an anisotropic transition region with a graduation of properties. The interface is required to provide adequate chemical and physical bonding stability between the fibre and the matrix in order to maximise the coupling between the two phases and thus allow stresses to be dispersed through the matrix, and thus transferred to the reinforcement. By wetting the reinforcement with the matrix in the liquid or low-viscosity state, coupling between the two components is provided. In the plate bonding technique, the fibre array is invariably aligned along the longitudinal direction of the beam or aligned transversely around the column. Consequently, the main function of the matrix is to combine and to protect the fibre against the external environment into which the composite will be placed. The two component parts of the polymer composite will now be discussed.

2.3 Properties of matrices

The polymer is an organic material composed of molecules made from many repeats of the same simpler unit called the monomer. There are many different polymer matrices used in advanced polymer composites but, within the composite family, there are two major types, the *thermosetting* and the *thermoplastic* binders.

The thermoplastic polymer is not used as the matrix material for the rehabilitation of structural members, but the thermoplastic polyacrylic fibre, which is formed from the thermoplastic polymer, polyacrylonitrile, is used as the PAN precursor for the manufacture of high-strength and high-modulus carbon fibre (see Section 2.4). Further information on the thermoplastic polymer may be obtained from Hollaway and Head (2001).

The thermosetting polymers are those which form the matrix material of the composite used in the rehabilitation of structures. They are manufactured from liquid or semi-solid precursors, which harden irreversibly. On completion of the chemical reaction, the liquid resin is converted to a hard solid by chemical cross-linking, which produces a tightly bound three-dimensional network of polymer chains. The chemical reaction is known as polycondensation, polymerisation or curing. The molecular units forming the network and the length and density of the cross-links of the structure will influence the mechanical properties of the resulting material; *the cross-linking is a function of the degree of cure*. The polymer matrix plays a number of vital roles in the formation of the characteristics of a composite. As has been mentioned the fibre generally has a high stiffness and strength but tends to be brittle. The matrix protects the reinforcement against abrasion or environmental corrosion, both of which can initiate fibre fracture. The load, which is carried by the fibre, is distributed through the matrix to adjacent fibres. The matrix should be both chemically and thermally compatible with the fibres. The most common thermosetting resin systems used in civil engineering generally and specifically in the rehabilitation of structures are the *epoxies* and the *vinylesters*.

The most important epoxy resins used in the construction industry are oligomers (low molecular weight polymers), produced from the reaction of bisphenol A and epichlorohydrin; this reaction provides an excellent balance between physical, chemical and electrical properties. The epoxies range from medium-viscosity liquids through to high melting solids. The mix rates of the base resin and the curing agents must carefully follow the manufacturer's instructions. Varying the proportions of the two components will form different polymers. The higher performance epoxies require *elevated temperature post-cure*. Generally, epoxies have high specific strengths, dimensional stability, temperature resistance and good resistance to solvents and alkalis, but most have weak resistance to acids. The toughness of the epoxies is superior to that of the vinlyester resins and therefore the epoxies will operate at higher temperatures. They also have good adhesion to many subtrates and are used as the adhesives in plate bonding; in addition, they have low shrinkage during polymerisation. The elevated temperature cure epoxies, which would include those used in the pultrusion technique, one of the methods used in plate bonding (see Section 2.5.4), have a high temperature resistance and can be used at temperatures up to 177 °C, with some epoxies having a maximum temperature range up to 316 °C.

Vinlyesters are unsaturated esters of epoxy resins; consequently, they posses mechanical and in-service properties similar to those of epoxy resins but, because their processing and curing techniques are similar to those of the polyesters, they are often identified as a class of unsaturated polyester. They can be processed at both room and elevated temperatures *but the room temperature process method does require post-cure.* Vinylesters have good wetting characteristics and bond well to glass fibres. The cured polymer has considerable resistance to strong acids, alkalis and chemicals and offers some resistance to water absorption and shrinkage. Irrespective of the cure mechanism used, vinylesters will cure to a level of about 95% during the initial cure stage; beyond this period, the cure will continue very slowly. Because of incomplete cure, the mechanical properties, moisture absorption and susceptibility to moisture-induced degradation of the resin and the fibre matrix interface can adversely affect a FRP composite made from vinylester.

2.3.1 Properties of the relevant thermosetting polymer matrix materials

Discussions on the properties of thermosetting polymers have been given in Hollaway and Head (2001) and the following text has been based upon these. The properties are relevant to the thermosetting polymers concerned with the upgrading of conventional civil engineering materials.

Mechanical properties

- *Stiffness.* The stiffness of a thermosetting polymer is a function of its degree of cure which, in turn, is a function of the degree of cross-linking of the three-dimensional network of polymer chains, and its long-term stiffness will be dependent upon its durability in the site environment.
- *Strength.* The tensile, compressive and flexural strength characteristics of a polymer are, in general, different. In addition, the strength properties of a polymer vary depending upon the loading to which it is subjected; if, for example, it is under long-term loading, dynamic or impact loading, the ultimate strength values under these situations will all be lower than those under short-term loading. All loading is dependent upon the bonding, the length, the density and the degree of cross-linking of the molecular structure of the polymer and, generally, the higher the degree of cross-linking the higher will be the strength. If, however, substantial heating is developed during the fatiguing of the material, all factors that influence thermal stability will also influence fatigue strengths.

1 Typical mechanical properties for the two polymers

Material	Specific weight	Ultimate tensile strength (MPa)	Modulus of elasticity in tension (GPa)	Coefficient of linear expansion ($10^{-6}/°K$)
Thermosetting				
Vinylester (Palatel A430-01)[1]	1.07	90	4.00	80
Epoxy	1.03	90–110	3.50–7.00	48–85

[1] Palatel® BASF AG, Germany

- *Toughness.* The ability of a material to absorb energy is known as toughness and is defined as the work required to rupture a unit volume of the material. It is proportional to the area under the load–deflection curve from the origin to the point of rupture. A method of increasing toughness of a polymer is to blend fill or co-polymerise a brittle but higher stiffness polymer with a tough one. As an increase in the stiffness of the material will result in a decrease in its toughness, it is necessary to compromise between stiffness and toughness.

  Typical short-term mechanical properties of the two main polymers used in plate bonding on to the more conventional civil engineering materials are given in Table 2.1.

- *Creep characteristics of polymers.* Polymer materials have mechanical characteristics of both elastic solids and viscous fluids; consequently, they are classified as viscoelastic materials (Hollaway and Head, 2001). The polymer composites used in the rehabilitation of structural members operate at ambient temperatures which are close to their viscoelastic phase; therefore, the creep properties of the polymer are an important characteristic in assessing its long-term load-carrying capacity. Experimentally, creep data are obtained by plotting the applied strain to the sample or member against log time over which the test extends, thus forming an isostress creep curve. By cross-plotting from the isostress creep curve, a family of isochronous stress–strain data curves, each associated with a specific loading duration, can be obtained. Figure 2.1 represents the family of creep curves that are obtained by varying the stress. The creep modulus, which is the slope of this curve is not constant; consequently, it will be necessary to specify at which point in the curve the slope has been determined. Thus the total strain at the end of a particular time can be plotted against the corresponding stress level. The British Standards Specification, BS 4618 (BSI, 1975), requires that the constant load tests are carried out under controlled conditions for the following durations 60 s, 100 s, 1 h, 2 h, 100 h, 1 year, 10 years and 80

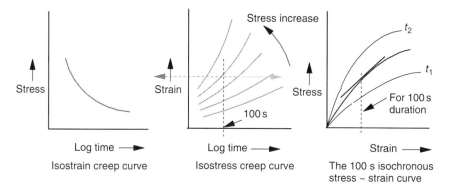

2.1 Creep curves, isochronus stress, strain curves for polymers. As fibre/polymer composites contain a polymer the material will creep and will exhibit similar forms of curves to the above but the polymer will be stabilised by the fibres which do not creep (adapted from Hollaway and Head, 2001).

years. The topic of the creep of FRP composite materials is more complicated than that of the polymer component and is dealt with in Section 2.6.3.

A great deal of time is required to obtain sufficient data to provide an estimate for the creep of composite materials. It is possible to undertake accelerated tests where the data are obtained over a few weeks or months and an extrapolation is made beyond the test duration by mathematical formulation. The extrapolation period should not be extended beyond three times the period of experimental testing, because the same law that exists over the test period will invariably not be relevant at the end of the extended period. For a greater creep-time period, a more sophisticated approach must be used, such as the time–temperature superposition principle (Aklonis and MacKnight, 1983). By undertaking a series of tests at different temperatures, activation energy may be determined and, through a kinetic approach of temperature–time superposition plots, a master curve can be developed and predictions made. The time, applied stress, superposition principle (Cessna, 1971) can also be used for polymers and polymer composite materials.

In-service properties

Thermal properties

- *Glass transition temperature of polymers.* The temperature at which an *amorphous* polymer changes from a brittle or vitreous state to a plastic state on a rising temperature is known as the glass transition

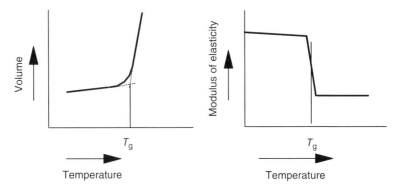

2.2 Relationship of volume and modulus of elasticity of polymer against temperature rise. (i) The thermosetting polymers below the glass transition temperature are rigid and frequently brittle in nature but are soft elastomers or viscous liquids above that temperature. (ii) As the hot cured composites are manufactured (and cured at ≃125 °C–140 °C) under factory conditions they will have a higher glass transition temperature compared with those which are cured at ambient temperature (cold cure polymers). The value of the glass transition will depend upon the temperature of cure, the higher the cure temperature the higher will be the glass transition temperature. (Adapted from Hollaway, 2008a, b).

temperature (T_g) (Fig. 2.2). It will be seen that the T_g takes place over a temperature range, and the mid-point of this range is the T_g. A reverse state of the polymer will take place on a falling temperature.

The actual value will depend upon the detailed chemical structure of the polymer. *Crystalline* polymers do not have a glass transition temperature but have a heat distortion temperature point, and possess some degree of an amorphous structure; this portion usually makes up 40–70% of the polymer sample. Thus, a polymer can have both a glass transition temperature (the amorphous portion) and a melting temperature (the crystalline portion). The effect of *crystallinity* on properties is observed when the polymer is at a temperature between the T_g and the crystalline melting point. All polymers below the T_g, whether crystalline or not, are rigid and frequently brittle; therefore, they have both stiffness and strength. Above the T_g, the amorphous polymers are soft elastomers or viscous liquids; consequently, they have no stiffness or strength, but the crystalline polymers will range in properties from a soft to a rigid solid, depending upon the degree of crystallinity. The epoxies used in construction would generally be in the amorphous state, with a small amount of crystalline structure. The numerical value of the T_g may be quoted in the literature with slightly different values, depending upon the testing technique used; these methods are the dynamic mechanical thermal analysis (DMTA) and the differential scanning calorimetry (DSC). The drastic changes in mechanical/physical properties which any polymer will experience at the T_g are shown as typical properties in Fig. 2.2.

- *Thermal conductivity.* The thermal conductivity of all polymers is low; consequently they are good heat insulators. The thermal conductivity may be increased by adding metallic fillers to the chemical formulation. It can be decreased further still by adding a blowing agent to chemical formulation and causing the material to expand and to increase its original volume many times by the formation of small cells; these systems are known as foamed polymers. This latter system, however, is not relevant to the strengthening of plate bonding but might be used as a fire-resistant external component to the upgrade.
- *Coefficient of thermal expansion of polymers* (α). The values of the coefficient of thermal expansion of epoxies vary between 48 and $85 \times 10^{-6}/°K$; vinylesters (BASF) are about $80 \times 10^{-6}/°K$, and these values are much higher than those of conventional materials (steel $10.8 \times 10^{-6}/°K$ and concrete $10.6 \times 10^{-6}/°K$). When bonded to concrete or steel for rehabilitation of a structural system, the differential in the value of α for different materials must be considered in design. The value of the coefficient of thermal expansion of polymers varies with temperature; it may be determined by the secant gradient of the thermal expansion curve between a reference temperature (generally room temperature) and the working temperature. The values of the coefficient of linear expansion are shown in Table 2.1. The degree of cross-linking of the polymer will influence the rate of the thermal expansion.

Physical property

Dimensional stability is the most important thermal property for polymers as they are unable to function above the glass transition temperature (T_g); at this temperature they will lose dimensional stability. All polymers at their T_g will either soften, decompose or both, and the temperature at which this occurs will depend upon the detailed chemical structure of the polymer. Consequently, the upper temperature limit for the use of many thermosetting polymers lies between 100 and 200°C; therefore, the strength of the polymer at this temperature will be dependent upon its method of manufacture and the degree of its cross linking. Hot cured thermosetting polymers would normally have a higher T_g value than the room temperature cured ones; their actual values would generally be about 15–20°C above the curing temperature. The hot cured polyesters, vinyl esters and epoxies all begin to weaken and break down at about 200°C. The T_g of *some* low temperature cure moulded composites can be increased in value by further post-curing the polymer at a higher temperature, but there is a maximum value of the T_g, irrespective of the post-cure temperature value. It is critical that a FRP composite has a T_g value of 30 °C above the maximum service temperature to allow for any potential depression in value due to some external environmental influence. There are a number of specialised polymers (mainly thermoplastic polymers) which can be used at much higher temperatures (up to 400 °C). These have been developed mainly for use in aerospace applications and are very expensive; therefore, although they would be acceptable for upgrading structures, economics rules them out.

Durability property

All engineering materials are sensitive to environmental changes in different ways, and the ageing and degradation of FRP composites is a slow and irreversible variation of a polymer's material structure, morphology and/or composition, leading to a deleterious change of use. However, composite materials do offer some significant durability advantages over the more conventional construction materials, such as their good resistance to corrosion and their magnetic neutrality.

The instability of a material in conditions under which it is employed and/or its interaction with the environment may be the cause of its durability, which is a difficult problem to study because it proceeds very slowly (typical lifetime in years). Furthermore, it is particularly difficult with respect to polymer composites used in construction as there are many different polymers on the market and the majority will have additives incorporated into their chemical systems which may have been added to enhance curing or to improve some specific mechanical or physical property. These additives are impurities in the polymer which will tend to reduce its mechanical and physical properties. There are two methods which are used to obtain data on the durability of engineering composites; these are: (i) accelerated tests on coupon specimens or parts of the structural component and (ii) field tests of the structure in real time. Accelerated testing can be undertaken to build kinetic models that describe the time changes of the material's behaviour and then to use these models to predict the durability from a conventional lifetime criterion. It is then necessary to show the pertinence of the choice of accelerated aging conditions; the mathematical form of the kinetic model and the lifetime criterion have to be proved. Empirical models are highly questionable because they have to be used in extrapolations for which they are not appropriate. Hollaway (2007) has discussed field surveys, where monitoring of some structures has been undertaken for the last ten years; the field work is still continuing. These surveys have been performed on FRP composite materials throughout the world. These field examples have shown that FRP composites do not degrade so rapidly and severely as the results of laboratory composite samples which have undergone accelerated testing would indicate, particularly if these latter coupons have been heated to accelerate a reaction more quickly. It is necessary to treat with caution/care the results of accelerated experimental testing, particularly if the FRP composites are heated during the testing.

Chemical properties

The ability of a polymer to resist chemical attack, whether this is from a natural or chemical environment, is known as chemical resistance and depends upon the chemical composition and bonding in the monomer. The environmental effect on this material is an important issue to be addressed and will be discussed in Section 2.7.

The two properties which are discussed here are the *solubility* and the *permeability*; both relate to the breakdown of the polymer and are very relevant to FRP plate bonding.

- *Solubility* is the ability of a solvent to diffuse into the polymer. In plate bonding, possibly the most severe chemical solution to which the polymer will be exposed is moisture. The composite systems used in construction have shown that polymers can be engineered to provide resistance against moisture and aqueous solutions diffusing into them; however, moisture will eventually diffuse into all organic polymers leading to changes in thermophysical, mechanical and chemical characteristics. The primary effect of diffusion is through hydrolysis and plasticisation, and this process may cause reversible and irreversible changes to a polymer. The characteristic of the polymer which may occur is a lowering of the T_g value. A possible solution for the reduction of diffusion through the polymer is to apply a protective coating/gel coat on to the system. Therefore, when selecting a polymer for rehabilitating a structural member, attention should be given to all environmental influences with which the polymer composite may come into contact. It is worth mentioning here that a polymer will not dissolve in a solvent unless the chemical structure of its monomer is similar to that of the solvent. In order to alter the property of the polymer or to improve its processing workability, a plasticiser may be added to it. It should be realised, however, that such an addition is likely to lower the hardness, stiffness, temperature resistance and tensile strength of the polymer, although its toughness may be increased.
- *Permeability* is the ability of bodies to allow gasses and fluids to permeate through them. It is important that a polymer composite (i.e. the protective polymer) should have a low permeability value to prevent or to reduce the rate of any deleterious material passing through it. Poly-

mers (i) with a high degree of cross-linking, (ii) cured at an elevated temperature, (iii) with factory fabrication under controlled conditions, will reduce permeability by ensuring low void ratios, full cure and high levels of integrity. Currently research work is being undertaken to investigate the possibility of decreasing the permeability of polymers by incorporating nanoparticles into the polymer. Chemically treated layered silicates (clays) can be mixed with polymer matrix materials to form a nanocomposite in which clay layers are evenly distributed through the material. Research has shown that these high-aspect ratio clays alter the properties of the composite by a number of mechanisms, reducing permeability, increasing the strength and improving fire resistance properties (Haque and Shamsuzzoha, 2003; Liu et al., 2005; Hackman and Hollaway, 2006). This technique has not yet been incorporated into the FRP material used in plate bonding; further research is required in this area and, at present, the cost of incorporating nanocomposites into the polymer is too high.

Additives and fillers

Various chemicals and fillers can be included in resin formulations to tailor their performance; these fillers include:

- UV stabilisers;
- smoke and flame spread inhibitors;
- viscosity modifiers (for example, thixotropic and thickening additives may be needed if laminating on a vertical surface on site);
- nanoparticles which in the future may improve: (i) the mechanical properties of adhesive polymers, (ii) the barrier properties of the polymer by reducing its permeability, (iii) the fire properties of the polymer to a limited degree. This is a new technique and it is currently expensive to incorporate the nanoparticles into the polymer. Until a more appropriate and less expensive method is developed it is unlikely that nanoparticles will be incorporated into civil engineering composites in the near future.

2.4 Filaments and fibres

The three fibres, which are used for the rehabilitation of structural components, are (i) the carbon fibre, (ii) the aramid fibre and (iii) the glass fibre. Probably, the most used of these for flexural, shear and wrapping of beams and columns is the carbon fibre, but the fibre finally chosen would depend upon the material of the structural unit to be upgraded and the required strength, stiffness, in-service properties and cost. Glass is the least expensive fibre, currently costing between $1.5/kg and $10/kg, depending upon the specific type of glass fibre. Carbon fibre is currently significantly more expensive than glass; prices range from $40–$100/kg (48–80 K heavy tow) to $20/kg (12 K tow). Aramid fibres currently cost $24–$60/kg. It should be realised, however, that these costs should not be compared on a weight basis with the more conventional civil engineering materials as the structural design techniques, transportation and erection of units on site for the two types of materials are not comparable; the higher specific strength and stiffness of the composite will not require as much material to form the structural system, will require only lightweight lifting equipment on site and will need less falsework than that for the traditional civil engineering materials.

A wide range of amorphous and crystalline materials can be used as the fibre component of the advanced composite. The process of making a fibre is one that involves axial alignment of molecules of the material; the high tensile strength is associated with improved intermolecular attraction resulting from the alignment.

Fibres generally consist of a number of long filaments, which have exceptionally high specific stiffness and strength. Their diameters are of the order of 10 µm with an aspect ratio, of length to diameter, between 1000 and infinity for continuous fibres. The filaments are extremely fragile and should be handled with extreme care. The terminology for identifying these bundles depends upon the type of fibre. For instance, a strand is a bundle of continuous glass or aramid filaments and 200 filaments will form one strand of continuous glass. A glass 'roving' refers to a collection of untwisted strands. Untwisted carbon filament bundles are usually called 'tows'. For all fibres 'yarns' are collections of filaments or strands that are twisted together.

2.4.1 Carbon fibres

The manufacturing process for carbon fibres consists of a sequence of procedures, namely, stabilisation, carbonisation, graphitisation and surface treatment. These procedures depend upon the pyrolysis and crystallisation of certain organic precursors. Most elements, other than carbon, are removed during the manufacturing process and carbon crystallites are preferably orientated along the fibre length. At temperatures above 2000 °C the size of the carbon crystallites increases and their orientation improves. Carbon fibre filaments are typically between 5 and 8 µm in diameter and the tows contain 5000–12 000 filaments, which can be twisted into yarns and woven into fabrics. Figure 2.3 shows a diagrammatic representation of the manufacture of carbon fibre.

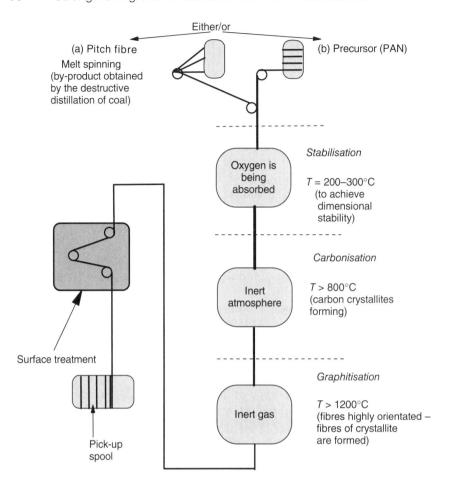

Precursors to the carbon fibre used in civil engineering.
(a) Pitch fibre produced from the distilation of coal. (Used for the manufacture of ultra-high modulus carbon fibres)
(b) Polyacryonitrile (PAN) fibre (Used in the manufacture of high modulus carbon fibres)

2.3 Schematic representation of the manufacture of carbon fibre. (Adapted from Hollaway and Head, 2001).

The crystallinity of the fibre increases, with a corresponding decrease in its amorphous characteristic, as the heat treatment temperature to the carbon fibre increases. This causes the modulus of elasticity to rise exponentially throughout the temperature range as shown in Fig. 2.4. This figure also illustrates that the tensile strength increases in value as the temperature rises to a maximum of about 1600 °C and then falls to a constant value as the temperature continues to rise to the required value of the modulus.

Fibre-reinforced polymer composites used in rehabilitation 59

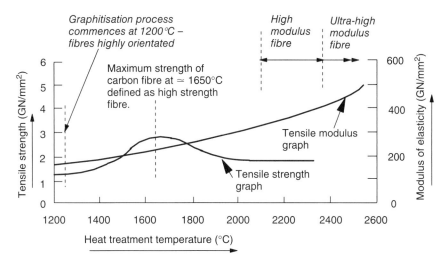

2.4 Carbon fibre – stiffness/strength vs temperature rise. (Adapted from Hollaway and Head, 2001).

The carbon fibre is formed from a precursor, and it is necessary to ensure that the precursor fibre carbonises and does not melt at the high temperatures required for manufacture. Two types of precursors are used to produce carbon fibres for the construction industry:

- *PAN precursors.* The majority of carbon fibres currently commercially available for the construction industry are manufactured from PAN precursors. After carbonisation, about 50% of the original fibre mass remains. The final manufactured fibres have higher tensile strengths than other precursor-based fibres. The commercial production of the PAN precursor carbon fibres is either by a spinning technique or by a melt-assisted extrusion as part of the spinning operation, and the cross-sections produced are either round, rectangular, I-type or X-type sections. An advantage of the latter geometries is that they allow a closer fibre packing in the composite.
- *Pitch precursors* are the by-product of the destructive distillation of coal. The fibres are relatively low in cost and high in carbon yield, but from batch to batch the fibres tend to be non-uniform in their final cross-section. This does not generally cause a problem in civil engineering but they would not be used in the aerospace industry.

Carbon fibres are available as 'tows', and a 12 K tow has 12 000 filaments. The fibres are commonly sold in a variety of modulus categories:

1. standard or low modulus $\simeq$ 200 GPa;
2. high strength (the modulus is of the order of 220 GPa) and the strength is 3 GPa;

Table 2.2 Typical properties of carbon fibres

Material	Fibre	Elastic modulus (GPA)	Tensile strength (MPa)	Ultimate strain (%)
Pan-based carbon fibres				
Hysol Grafil Apollo[1]	IM[4]	>300	5200	1.73
	HM[5]	<400	3500	0.88
	HS[6]	260	5020	2.00
BASF AG Celion[2]	T300	300	4960	1.66
	T400	572	1860	1.66
Toray Industries Torayca[3]	T300	234	3530	1.51
Pitch-based carbon fibres				
Hysol Union Carbide	T300	227.50	2758	1.76
	T400	724	2200	0.31
	T500	241.30	3447.50	1.79
	T600	241.30	4137.00	1.80
	T700	248.20	4550.70	1.81

[1] Apollo® Hysol Grafil Inc., USA
[2] Celion® BASF Structural Materials Inc., USA
[3] Torayca® Toray Industries Inc., Japan
[4] Intermediate modulus (European terminology – high modulus)
[5] High modulus (European terminology – Ultra-high modulus)
[6] High strain

3. high modulus (this category of fibre is called intermediate modulus in the USA, Canada and some Asian Countries) 220–300 GPa;
4. ultra-high modulus (this category of fibre is called high modulus in the USA, Canada and some Asian Countries) > 450 GPa.

For plate bonding of RC and PC structural members item (3) would generally be used. These items would be manufactured from the PAN precursor and item (4) would be manufactured from the pitch precursor (for upgrading mainly metallic members). An ultra-high modulus fibre would invariably be produced by the PAN precursor for the aerospace industry. The carbon fibre produced from the pitch precursor tends to be non-uniform in cross-section from batch to batch, which is not generally a problem for the construction industry but is for the aerospace industry. The pitch fibres are relatively low in cost and high in carbon yield; this fibre is used for the upgrade of metallic systems for the construction industry. The important typical mechanical properties of carbon fibres are given in Table 2.2. It is important for engineers to define precisely which carbon fibre is

Table 2.3 Typical properties of aramid fibres

Material	Fibre	Elastic modulus (GPA)	Tensile strength (MPa)	Ultimate strain (%)
Aramid	49	125	2760	2.40
Aramid	29	83	2750	4.00

used in the polymer composite when upgrading structural members; it is clear from the above that carbon fibres are not unique materials.

2.4.2 Aramid fibres (aromatic polyamide)

An extrusion and spinning process is typically used to produce an aromatic fibre. A solution of the polymer in a suitable solvent at a temperature of between −50 and −80 °C is extruded into a hot cylinder which is at a temperature of 200 °C; this causes the solvent to evaporate and the resulting fibre is wound onto a bobbin. To increase its strength and stiffness properties, the fibre undergoes a stretching and drawing process, thus aligning the molecular chains which are then made rigid by means of aromatic rings linked by hydrogen bridges. There are two grades of stiffness available; one has a modulus of elasticity of about 130 GPa and is the one used in polymer composites for upgrading structural civil engineering systems and the other has a modulus of elasticity of 60 GPa and is used in bullet-proof systems.

Aramid fibres are resistant to fatigue, both static and dynamic, and have high tensile elastic characteristics. They have a ductile compressive characteristic, but their ultimate compressive strength is low. Furthermore, they have low ultimate shear strengths. Unlike carbon fibres they are non-conductive and therefore can be used adjacent to overhead electrical rail cables. Typical important mechanical properties of aramid fibres are given in Table 2.3. Aramid fibres/epoxy polymer composites have been used for the wrapping of concrete columns.

2.4.3 Glass fibres

Glass fibres have a very high specific strength and are one of the strongest and most commonly used structural materials. Due to their low cost compared with the other two civil engineering fibres they are beginning to be used more than previously in the upgrading of concrete structures, particularly in Canada. Carbon fibres are almost invariably used in the UK for the rehabilitation of structural members. The commercial grades of glass have strength values up to 4800 MPa.

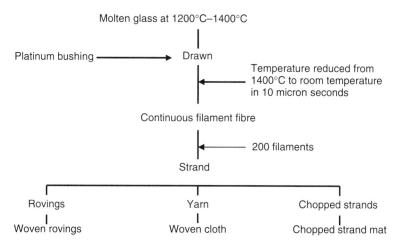

2.5 Manufacturing process of glass fibre.

The commercial manufacturing process for the production of glass fibre is by the direct melt process in which fine filaments of diameters 3–24 μm are produced by continuous and rapid drawing from the melt. Two hundred individual filaments are formed into one strand and, during the production stage, many strands are formed. Figure 2.5 shows a diagrammatic representation of the production process. To prevent abrasion-related degradation of the glass fibre against neighbouring fibres, a surface treatment (or sizing) is applied before the fibres are gathered into strands and wound onto a creel. The size also binds the filaments together and provides a chemical coupling link between the glass fibre and the polymer during impregnation.

The most important grades of glass for the rehabilitation of structural members are as follows:

- *E-glass* which has a low alkali content of the order of 2%; it is used for general-purpose structural applications and is the major glass fibre used in the retrofitting of structural components. It also has good heat and electrical resistance.
- *S-glass* which a stronger and stiffer fibre compared to the E-glass; having typically 40% greater strength at room temperature and greater corrosion resistance. It also has good heat resistance. The S-2-glass has the same glass composition as S-glass but differs in its coating. The S-2-glass has good resistance to acids such as hydrochloric, nitric and sulphuric acids.

Other less important glass fibres for rehabilitation are:

Table 2.4 Typical properties of civil engineering glass fibres

Material	Fibre	Elastic modulus (GPa)	Tensile strength (MPa)	Ultimate strain (%)
Glass	E	72.40	2400	3.50
	A	72.40	3030	3.50
	S-2	88.00	4600	5.70

- *E-CR-glass* which has a high resistance to acids and bases and has chemical stability in chemically corrosive environments.
- *R-glass* which has a higher tensile strength and tensile modulus compared to the E-glass fibre and also has greater resistance to fatigue, aging and temperature corrosion.
- *A-glass* which has high resistance to alkali attack and is used as the reinforcement for glass fibre reinforced cement (GFRC). When glass fibres are used to reinforce cement, degradation in strength and toughness occurs when the material is exposed to normal weathering environments, but especially in humid conditions. This process can also take place with AR-glass, albeit at a much slower rate.
- *T-glass* which has an improved performance compared to that of E-glass, with a 36% increase in tensile strength and 16% increase in tensile modulus.
- *N-Varg* which has a high resistance to alkali attack and is used mainly in composites with a cement-based matrix.

Table 2.4 gives the physical and mechanical properties of the most important glass fibres used in civil engineering.

The fibres most used to manufacture plates for rehabilitating structural members are:

- High stiffness carbon fibres for reinforced and prestressed concrete beams and columns.
- Ultra-high stiffness carbon fibres mainly for steel systems but have also been used for upgrading RC systems (these fibres have very low strain to failure values of the order of 0.4%).
- E-glass or AR-glass. These fibres have a very low modulus of elasticity and therefore a number of laminates have to be incorporated into the lay-up to approach the equivalent area stiffness value of a carbon fibre lamination. The Canadian authorities are using these fibres for upgrading concrete structural members by incorporating three times the amount of longitudinal fibres compared with that which would be used for a carbon fibre composite.
- The aramid fibre for reinforced concrete beams and columns.

2.5 The advanced polymer composite

To form a structural polymer composite, long fibres are introduced into the resin matrix. This combination of two dissimilar materials leads to a component that has enhanced strength, stiffness and toughness over the properties of the individual parts.

The mechanical and in-service properties of the final fibre–matrix composite will be dependent upon the individual properties of the two component parts with the former property (viz. mechanical) being largely dependent upon the fibre and the latter property (viz. in-service) being largely dependent upon the matrix. Therefore, the following will affect the overall mechanical properties and to a lesser extent the in-service properties of the composite:

- the type of the fibre used (viz. carbon, glass or aramid fibres);
- the relative proportions of the polymer and fibre (fibre volume fraction);
- the orientation of the fibre (viz. unidirectional, bidirectional aligned or randomly orientated);
- the method of manufacture;
- the temperature and duration of the cure cycle;
- the age of the polymer composite.

2.5.1 The type of fibre used

As discussed earlier, carbon, aramid and glass fibres are the ones generally used for plate bonding. Their strengths and stiffnesses are different, although the former property for the three fibres is of the same order. Consequently, the higher the stiffness of the fibre the greater will be the stiffness of the composite. Fibres can be hybridised by a mix of different fibre types in the same fabric. The aim would be to achieve a balance of properties, for example blending glass and aramid will combine strength and toughness and will reduce cost. However, the stiffness will differ from either component and a consideration must be given to the loading of the component. In producing hybridised components a greater success will be achieved by segregating fibres into discrete layers than with fibre mixing in the same layers.

The fibre used to rehabilitate RC structural members is generally the high stiffness carbon fibre (modulus of elasticity values in the order of 220–300 GPa). Glass fibre is being used in Canada, due to economical and environmental considerations (Demers *et al.*, 2005). The ultra-high stiffness carbon fibre composite is not normally used to upgrade RC structures due to its low value of ultimate strain which is of the order of 0.4%.

2.5.2 The relative proportions of the fibre and matrix

As the fibre is the load-carrying component of the composite material, the greater the volume of fibre in the composite (i.e. the higher the fibre volume fraction) the stronger will be the composite. For the rehabilitation of structural members, composites with fibre weight fractions of between 55 and 60% would generally be used.

2.5.3 The orientation of the fibre

The properties of long fibre–polymer composites will generally be anisotropic unless an orientated angle ply system of fibre arrays of 0°, 30°, 60° or 90° is used, or randomly orientated short fibre arrays are used in which case the composite will exhibit quasi-isotropic properties. The direction and volume fraction of the fibres in composites will determine the strength and stiffness of the composite. In flexural plate bonding solutions, the fibres are unidirectionally aligned along the longitudinal length of the member. In order to maintain the alignment, particularly during the manufacturing procedure, it may be necessary to add a small percentage of fibres in the transverse direction. For shear strengthening, again the unidirectionally aligned fibres are used. For the rehabilitation or the retrofitting of RC columns, continuously-wound fibres in the hoop direction are applied to the column at a slight positive angle. These fibres confine the concrete in the columns and, as the tensile fibres can resist a much greater tensile strength compared with that of the concrete, the column will contain a much higher compressive load (see Chapter 6 for FRP confined columns).

2.5.4 The method of manufacture of the composite

There are a number of techniques used to manufacture advanced polymer composites, each of which will have an influence on the mechanical properties of the final component. The main reasons for this variation are: (a) the resin used, whether it is a high temperature cure or an ambient cure; (b) the degree of compaction; an automated hot process will produce a lower void ratio component compared to a product produced at manual ambient temperature. The five methods which are invariably considered for use in plate bonding and column wrapping of structures are: (i) the wet lay-up (site fabrication); (ii) the pultrusion (preformed plates manufactured under factory conditions and site bonded); (iii) the hot melt, factory-made, pre-impregnated fibre composite (the prepreg) preformed rigid plates manufactured under controlled temperature and pressure conditions and site bonded with a two-part cold cure adhesive; (iv) the hot melt, factory-made, pre-impregnated fibre (prepreg) site fabricated and site cured (in conjunction with film adhesive) under a vacuum assisted pressure of 1 bar and cured at 65 °C for 16 hours or 80 °C for four hours; (v) the vacuum-assisted resin transfer procedures (RTM) (site fabricated and cured). These methods have been discussed and illustrated in Hollaway and Head (2001).

The first and fifth techniques are a manual site operation whereas the second and third techniques are factory automated and site bonded and the fourth is a semi-automated factory/site method and site bonded. In all cases, except case (iv), it is advisable to post-cure the *adhesive* as the polymer used in this operation is site cured and therefore its glass transition temperature, assuming the ambient temperature at cure of the adhesive, say 20 °C, will be no greater than 50 °C. If, however, an adhesive film were utilised, such as might be the case when bonding a site cured, hot melt prepreg composite to a structural member (case (iv) above), a site elevated temperature and pressure would be used to cure and compact the prepreg and the adhesive film in one operation; a vacuum of 1 bar for compaction in this case would be applied. Method (iv) has been described in Hollaway *et al.* (2006) and is illustrated as a site procedure in the example given in Chapter 13, Section 13.7.1. All the above methods have been used for the rehabilitation of structural members by externally bonded composites, whether these are by the utilisation of precast plates or composite wraps. Another system which has been used is the near surface mounted (NSM) bars which are manufactured by the pultrusion technique and bonded, by an adhesive paste, into grooves cut into the soffit or vertical sides of reinforced concrete beams. The advantage of this system is that the composite is not exposed to environmental influences. De Lorenzis *et al.* (2000, 2002) and El-Hacha and Rizkalla (2004) give further information on this system.

2.5.5 The temperature and duration of cure cycle

The factory automated production methods, mentioned above, are manufactured at various elevated temperatures depending upon the thickness, the volume of the part and the manufacturing techniques used. For instance, the temperature of the die for the manufacture of the pultrusion plate technique will be of the order of 120–135 °C and the T_g of the polymer will be of the order of 125–140 °C. If necessary, the pultruded component would then be post-cured for a certain time depending on the temperature used at post-cure; however, this operation is not usually undertaken. A hot curing polymer is used in the pultrusion manufacture. A site fabricated technique such as the wet lay-up method would utilise a cold cure polymer which would also act as the adhesive.

2.5.6 The age of the polymer composite

All polymers (the component of the composite protecting the fibre), will degrade over time, depending upon the environment into which the composite is placed. As degradation continues over the life of the composite so will the mechanical properties decrease in value.

FRP composites in plate bonding can be either unstressed or prestressed at the time of bonding of the composite to the system to be rehabilitated (Garden *et al.*, 1998). The advantage, from a structural point of view, of using a prestressed system is that a more efficient use is made of the composite material. The prestressing force applied at the time of bonding could be up to 40% of the ultimate strength of the composite which implies that at the failure of the upgraded structural system the FRP composite will be close to its failure value. The two plate bonding systems have been illustrated in Chapter 13, in the examples given in Sections 13.7.1 and 13.8.1 respectively.

2.6 The mechanical properties of advanced polymer composites

2.6.1 Tensile properties

It has been stated above that the fibre strength and stiffness, volume fraction and orientation will determine the mechanical properties of the final composite; this is illustrated in Table 2.5 which gives typical properties of composites manufactured by the *pultrusion* technique using epoxy resin and long directionally aligned fibre reinforcement of glass, aramid and carbon with a fibre–matrix ratio by weight of 60%.

Table 2.5 Typical mechanical properties of long directionally aligned fibre–matrix composites (fibre weight fraction 58%) manufactured by the pultrusion technique, (the matrix material is epoxy)

Composite material	Specific weight	Tensile strength (MPa)	Tensile modulus (GPa)	Flexural strength (MPa)	Flexural modulus (Gpa)
E-glass	1.90	750–1050	40.00	1450	40.00
S-2 glass	1.80	1650	55.00	–	–
Aramid 49	1.45	1150–1400	70–110	–	–
Carbon (PAN)	1.60	2670–1950	150–220	1600	–
Carbon (pitch)	1.80	1400–1500	280–350	*Failure strain* $\simeq 0.40 > 330$	

2.6.2 Compressive properties

The integrity of both the fibre and matrix in a composite under a compressive load is far more critical than in tensile loading. The fibres are the principal load-bearing elements in a composite and are supported by the matrix to provide local stability and prevent a micro-buckling failure of the composite. Consequently, local resin and interface damage would lead to fibre instability that is more severe than the fibre isolation mode, which occurs in tensile loading. The possible failure modes for FRP under longitudinal compression are (i) transverse tensile failure, (ii) fibre micro-buckling and (iii) shear failure; the mode of failure will depend upon the items mentioned in Section 2.6.5. The compressive strengths of FRP composites normally increase as the tensile strengths increase; however, of the three fibre composites used in construction, aramid fibre reinforced polymer (AFRP) exhibits non-linear behaviour in compression at a relatively low level of stress. Typical values of the compressive moduli of elasticity and compressive strengths of glass fibre reinforced polymer (GFRP), AFRP and carbon fibre reinforced polymer (CFRP) composite materials compared with their tensile strengths are generally lower by approximately 80%, 100% and 85% and 55%, 20% and 78%, respectively; the test samples contain 55–60% weight fraction of continuous fibres in a vinylester resin matrix. Plate bonding FRP composite materials to upgrade, or strengthen RC, PC or metallic structural members in the compression region must be analysed and designed with extreme care.

2.6.3 Creep characteristics of FRP composites under a tensile load

The two components of the composite, namely, the polymer and the fibre, have vastly different creep characteristics; the polymer is a visco-elastic material and, therefore, will creep under load (see Section 2.3.1), but the carbon, glass and aramid fibres have virtually no creep component. Consequently, the fibres have a stabilising effect on the creep characteristics of the polymer; therefore, the overall creep rate characteristics of the FRP composite materials and the final creep value will be much lower than that of the polymer. The actual creep value of the FRP composite will depend upon the fibre volume fraction, the fibre orientation and the type of fibre used.

Creep curves for composite materials cannot be produced in the way that polymer creep curves are developed. Consequently, the mechanical behaviour of composite materials is established by applying constant loads over long periods of time. Curves of elongation against time at different stress levels are developed and, although they are not able to produce data that may be converted directly into stress–strain curves, constant time sections

Table 2.6 Typical coefficient of thermal expansion values for certain fibre arrangements in composites

Material ($\times 10^{-6}/°K$)	Arrangement	Coefficient of thermal expansion	
		0° axis	90° axis
Glass FRP	Unidirectional (0°)	8.40	20.91
	Quadraxial (0°, +45°, −45°, 90°)	11.00	11.00
HS carbon FRP	Unidirectional (0°)	0.52	25.50
	Quadraxial (0°, +45°, −45°, 90°)	9.76	9.76
UHM carbon FRP	Unidirectional (0°)	−0.07	25.68
	Quadraxial (0°, +45°, −45°, 90°)	9.60	9.60
Resin (typical value)		65	

HS = high strength
UHM = ultra-high modulus

through families of such creep curves can produce isochronous stress–strain curves.

FRP composites for strengthening concrete structures usually contain a very high proportion of unidirectionally aligned fibres; in addition, the permanent stress levels in externally bonded FRP strengthening systems are generally low enough for creep to be negligible. This is particularly important for GFRP composites where stress corrosion could take place with high permanent stress values. These stress limits for GFRP composites are of the order of 20–25% of its ultimate value. On the other hand, the prestressed composite in plate bonding will be stressed to a much higher value than that of the unstressed composite plate; in this case GFRP composites would not be used. The initial prestress value of the CFRP composite will generally be in the range of 25–50% of the ultimate strength of the composite material. As the composite will contain a fibre (unidirectional) weight fraction of the order of 55–60%, the creep characteristics will be small, but they should be checked.

2.6.4 Coefficient of thermal expansion

The coefficient of thermal expansion of an FRP composite depends upon both the fibre used and its lay-up in the matrix. The manufacturer should supply the values of the coefficient of thermal expansion for their specific composite products. Table 2.6 gives typical coefficient of thermal expansion values for certain fibre arrangements in composites.

Where the manufacturer does not provide a value of the coefficient of thermal expansion, it may be computed from the following equations. The transverse coefficient of thermal expansion is only approximate and it is recommended that experimental tests are performed to obtain a more accurate value.

$$\alpha_{longitudinal} = \frac{\alpha_{matrix}(1-V_{fibre})E_{matrix} + \alpha_{fibre}V_{fibre}E_{fibre}}{(1-V_{fibre})E_{matrix} + V_{fibre}E_{fibre}}$$

$$\alpha_{transverse} = (1+\nu_{matrix})\alpha_{matrix}(1-V_{fibre}) + (1+\nu_{fibre})\alpha_{fibre}V_{fibre}$$
$$- \alpha_{longitudinal}[\nu_{fibre}V_{fibre} + \nu_{matrix}(1-V_{fibre})]$$

where α = coefficient of thermal expansion, E = modulus of elasticity, ν = Poisson's ratio and, V = volume fraction.

2.6.5 Fatigue resistance of advanced polymer composites

The fatigue behaviour of fibrous composite materials is more complex than that of metals due to the anisotropic nature of the material. Complex failure mechanisms and excessive, damage modes can be caused by fatigue cycles and their possible interaction. Generally, polymeric composites experience progressive fatigue degradation due to failure of the fibres, fibre stacking sequence and type of fatigue loading. However, the magnitude of peak stress in a load cycle is usually a small proportion of the ultimate stress in practical applications and the in-plane fatigue endurance of composites is generally good. The most fatigue-sensitive modes of failure are matrix-dominated modes but, if the failure strain of the matrix exceeds that of the fibre, fibre fracture will dominate fatigue failure; an example of this is the ultra-high modulus carbon fibre composite where the fibre has a very low strain to failure. The interface between the fibre and matrix plays an important role in the fatigue behaviour of the FRP composite. A strong interface delays the occurrence of fibre ridging and longitudinal matrix cracking. At low cycles interfacial shear and matrix cracking will predominate.

Under fatigue loading there are four basic failure mechanisms of polymeric composites:

- *Fibre breakage interface debonding;*
- *Fibre-matrix interface failure;*
- *Delamination.* The mainly directional fibre composites used in the rehabilitation of structural systems contain a small percentage of transverse fibres for stability of the laminates. In the longitudinal direction, the composites will have strength and stiffness, but they tend to be weak in the through-thickness direction. This weakness can increase the likeli-

hood of delamination between the layers of the laminates particularly in areas of high interlaminar shear.
- *Matrix cracking.* This mechanism does not usually apply to FRP composites used for rehabilitating structural systems due to the mainly unidirectionally aligned fibre composites. Cracking will normally occur in the off-axis plies and is usually the first damage mechanism due to lack of fibres orientated in the direction of the applied load, thus causing more load to be distributed to the matrix.

Either a fatigue or a static loading may cause similar damage to develop in a FRP composite, except that under fatigue loading, to any given stress level, additional damage is caused and the degree of damage is a function of the number of cycles.

2.7 The in-service properties of the advanced polymer composite

The polymer component of the composite material protects the fibre from external and environmental damage and, therefore, the in-service properties of advanced polymer composites will be wholly dependent upon those of the polymer to protect the fibre and the interface between the polymer and fibre.

2.7.1 Durability

The long-term durability of a FRP composite depends intrinsically upon the choice of constituent materials, processing methods and the environmental conditions to which they are exposed throughout their service life. One of the major concerns with the use of this material in plate bonding environments is associated with the durability and the long-term characteristics of the material. Exposure to a variety of adverse and sometimes harsh environmental conditions in construction could degrade the FRP composite material and thus this degradation will alter their mechanical performance. Exposure to high and low temperature variations, moisture and salt solution ingress, ultra-violet rays from the sun and fire will all lead to reduced mechanical performance. A further concern is the durability of ambient cured systems if they have not been post-cured before use, as these composites may not reach their full polymerisation; depending upon the ambient cure temperature they may have a relatively low glass transition temperature, thereby making them even more susceptible to degradation. These adverse factors, however, are not unique to the FRP materials; all engineering materials are sensitive to environmental changes in different ways.

72   Strengthening and rehabilitation of civil infrastructures

Composite materials do offer some significant durability advantages over the more conventional construction materials.

Durability from a civil engineering point of view may be divided into six groups:

- moisture and aqueous solutions;
- alkaline environment;
- age of the polymer composite;
- fire;
- thermal effects;
- ultra-violet radiation.

Each item in this durability group is relevant to the rehabilitation of structural members.

Moisture, aqueous solutions and alkaline environment

The composite plate which is bonded, say, on to bridge sections is out of the direct influence of the sun and rain, although moisture in the atmosphere and splashing from vehicles, particularly if the spray droplets contain salt solutions from de-icing salts, can affect the soffit and sides of bridge members. Changes in the mechanical characteristics of the polymer and/or polymer composite which may occur due to diffusion are (i) lowering of the T_g value of the polymer and (ii) causing deleterious effects to the fibre–matrix interface bond of the composite resulting in a loss of integrity. Possible solutions for the reduction of diffusion through the polymer are:

- to apply a protective coating/gel coat on to the system;
- to cure the product at an elevated temperature;
- factory fabrication under controlled conditions. (This ensures low void content, full cure, high levels of overall integrity, a high glass transition temperature and greater resistance to moisture degradation compared with the cold curing polymers. Another effective method of enhancing durability of composites is through the use of appropriate sizing/finishes on the fibres).

A further potential technique for enhancing the resistance of the ingress of moisture and aqueous solutions is by the addition of nanoparticles into the polymer. This technique has been discussed in Section 2.3.1 under Additives.

The age of the polymer composite.

This topic is associated with the degradation of the polymers over time and has been discussed in Section 2.3.1. As the composite ages and degradation progresses, due to moisture uptake, atmospheric conditions or any adverse environmental influences, the process will affect the mechanical properties of the components (in particular the polymer) and their interaction, which in turn will reduce the mechanical properties of the composite material. The ageing of polymers is a difficult subject to discuss in a chapter dealing with all aspects of the material because of the wide environmental influences to which it might be exposed. It is, therefore, suggested that reference should be made to Karbhari *et al.* (2003).

Fire

The matrix component of the composite used for rehabilitation is composed of carbon, hydrogen, oxygen and nitrogen atoms and will burn, but not all polymers are equally prone to ignition and fire growth on their surfaces; the degree of ignition will depend strongly upon the thermal stability of the polymer. In addition, when the material is relatively thin, it burns away quite quickly; conversely, when the material is relatively thick, it quickly forms a protective char and can then survive for relatively long periods before it loses a significant proportion of its strength. Its survival in such conditions is further helped by the fact that the pyrolysis is endothermic. The polymer composites used in plate bonding are relatively thin and therefore it follows that, if deemed necessary, polymer composite members are likely to require some form of fire protection.

There are some fire-resistant polymers, but they are generally impracticable for infrastructure use by virtue of their high cost. An exception is phenolic resin, which is both relatively inexpensive and fire-resistant but is difficult to pultrude to a sufficient quality to use in plate bonding and the in-service properties required of it in this capacity are not sufficient. It is known that vinylester and epoxies are not very resistant to fire growth. However, it is possible to incorporate additives into the resin formulations or to alter their structure, thereby modifying the burning behaviour and producing a composite with enhanced fire resistance (Hollaway, 1993).

The fire resistance of vinylester resins may be enhanced by:

- incorporating halogens (viz. the fluorine, chlorine, bromine and iodine family of chemicals);
- combining synergists in the resin (HET acid resin – saturated dicarboxylic acid anhydride, containing chlorine);
- adding chlorinated paraffin and antimony oxide;
- utilising halogenated phosphates (trichloroethylphosphate).

For instance, brominated vinylester resins are a distinct improvement but still exhibit significant fire growth (Ohlemiller *et al.*, 1996).

The fire resistance principles of epoxy resins are the same in epoxies as in polyester, but the actual formulations used are usually very different. It is important to distinguish two types of epoxy resins.

- the standard epoxy resin – as normally used in civil engineering;
- epoxy-vinylester resins (EVER).

The EVER resins are a type of hybrid product containing substantial amounts of 'polyester' material. These can be treated with the conventional 'polyester'-type flame-retardant systems such as HET-acid.

The standard epoxy resins require slightly different treatment. The flame-retardant systems include:

- halogens such as bromine and chlorine incorporated into the polymer chains, e.g. by using tetra-bromo Bisphenol A or deca-bromo diphenyl oxide;
- antimony oxide Sb_2O_3, either alone or in combination with zinc borate;
- phosphorus compounds, such as phosphorus-containing esters or ammonium polyphosphate–bis(aminophenyl) methyl phosphine oxide is used as a curing agent in order to incorporate the phosphorus into polymer chain;
- aluminium trihydrate alone or in combination with phosphorus ester.

Intumescent coatings incorporate an organic material, which will char, and evolve gases at a designed temperature so as to foam the developing char of the polymer in the composite. The properly foamed char serves as an effective thermal insulation layer, protecting the underlying material from the fire. Thin-film intumescents are sometimes favoured in *steel* plate bonding systems because of their appearance and durability. However, when used with thin polymer composites, they do not currently react sufficiently quickly and unacceptable damage takes place before the protective char can form. Thick-film intumescents have been used in practice to protect the water deluge composite fire suppression systems on offshore rigs, but this solution is unlikely to find favour in more conventional structural applications and in composites for plate bonded beams. However, research into intumescent coating technology continues to progress, and improvements in the means necessary to make a practical and effective product continue. It must be realised that the durability of a coating (in terms of long-term adhesion and weather resistance) is a key factor in its viability for infrastructure applications.

The inert fibres in a composite, particularly the high fibre volume fraction material, help resist the worst consequences of the fire. The fibres displace the polymer resin, thus making less fuel available to the fire. When the outermost layers of the composite lose the resin component, the fibres act as an insulating layer, slowing heat penetration into the depth of the composite and at the same time reducing the evolution of gases (Gibson and Hume, 1995).

To summarise, the methods of imparting flame retardancy to FRP composite used in plate bonding systems, are:

- the choice of polymer system used and the quality of the cure process;
- the application of a coating of intumescent resin;
- the addition of mineral fillers such as calcium carbonate and alumina trihydrate which impart varying degrees of flame retardancy to FRP composites – the addition of alumina trihydrate has an added advantage of providing smoke reduction due to the significant quantity of heat absorbed by the endothermic reaction during the decomposition of this material into alumina and water;
- the ability of the non-combustible fibres in a fibre/polymer composite to displace the polymer matrix and in so doing reduce the fuel available to the fire;

Moreover, it has been shown experimentally that the addition of exfoliated/intercalated nanoparticles to the resin formulation can improve fire resistance to the polymer. A considerable amount of research has been conducted into the fire properties of nanocomposites, the majority of which concentrates on the properties of heat distortion temperature (HDT), peak and mean heat release rate (HRR) and mass loss rate (MLR). Comprehensive research into the fire properties of a variety of different polymer nanocomposites has been conducted and published in various papers (Gilman, 1999; Gilman *et al.*, 2000; Wang *et al.*, 2000). It was found that the additions of nanoparticles to all polymers tested resulted in large reductions in peak and mean HRR and mean MLR. However, the specific heat of combustion for some of the nanocomposites was decreased by a small amount. These increases in fire performance have been tested under laboratory conditions and using low temperature sources. In a real fire situation the benefits of the polymer-layered silicate nanocomposite would be to form a protective network of char around the non-burnt polymer. However, this advantage would be seen only in thick sections in which an outer layer will be burnt and form a protective layer for the remaining polymer. In thin laminates, too much of the component cross-section will be lost in the creation of this char protective network to sustain any of the structural properties of the original component.

Whatever fire protection system is adopted for polymers, it is implicit that the protection is only partial. It is not feasible to prevent any temperature rise during the required survival time of the component and design may be based on the attainment of a limiting temperature distribution in the member concerned. One difficulty with the design of polymers associated with fire is that the deterioration in the material properties of polymer composites commences at much lower temperatures than other construction materials. In addition, in a fire situation invariably the polymer will be heated to a temperature above its T_g. This will reduce the modulus of elasticity and strength values of the composite. Provided that the temperature of chemical degradation is not exceeded, the loss is reversible; if, however, the temperature is exceeded, an irreversible loss of the load-bearing characteristics of the material will result due to thermal damage.

Thermal effects

The thermal conductivity and the coefficient of thermal expansion of polymers have been dealt with in section 2.3.1; the addition of the fibre will stabilise the effects of the polymer. The composite material will be affected by heating and cooling but to a lesser extent than its component polymer. The effects on the composite will be dependent upon the type of fibre used, the fibre volume fraction and the method of manufacture of the composite.

Ultraviolet radiation

The ultraviolet component of sunlight degrades the composite, and the short wavelength band at 330 nm has the most effect on polyesters. It is manifested by a discoloration of the polymer and a breakdown of the surface of the composite. To obviate this problem, an ultraviolet stabiliser can be incorporated into the polymer at the time of manufacture. Furthermore, a polyurethane lacquer can be applied to the surface of the composite to protect it from the ultraviolet light. The epoxy polymer does not seem to be affected by ultraviolet light, and the inclusion of stabilisers into epoxy resin formulations seems to have little effect regarding discoloration; furthermore, there is no evidence that continuous exposure to the sunlight affects the mechanical properties of the epoxy polymers.

2.7.2 Chemical properties

As the matrix polymer protects the fibre, the same discussion as given in Section 2.3.1 for the polymer applies to the advanced polymer composite.

The two properties of particular importance are the solubility and the permeability of the polymer; both relate to the breakdown of the polymer and are very relevant to FRP plate bonding. These topics have been dealt with in Section 2.3.1.

2.7.3 Test procedures for monitoring durability

One of the problems in acquiring data relating to the durability properties of any material is the length of time involved in gathering the relevant information. There are many different polymers on the market, and this poses a particular difficulty with respect to civil engineering polymer composites as the majority used will have additives which may have been incorporated to enhance curing or to improve some specific mechanical or physical property. Furthermore, polymers have been chemically upgraded over the years to further improve their properties and, as a consequence, their durability performance.

Many laboratory accelerated tests and field surveys which are currently being or have been undertaken throughout the world have been discussed by Hollaway (2007). For instance, to ascertain the degradation of GFRP composites in contact with RC members Bank and Gentry (1995), Bank *et al.* (1998) and Sen *et al.* (2002) experimented using the accelerated testing technique by exposing GFRP composites to a simulated concrete pore water solution of high pH values and elevated temperatures up to 80°C. These tests indicated that there was a decrease in the tensile, shear and bond strengths of the GFRP composites and, therefore, suggested that there is a case for not using GFRP composites in contact with concrete (Uomoto, 2000). However, Tomosawa and Nakatsuji (1997) have shown that after 12 months exposure to alkaline solutions at temperature between 20 and 30°C there had been no material or physical deterioration to the GFRP composite. Furthermore, Clarke and Sheard (1998) reported on the testing of samples to different specifications; these specimens were exposed for two years to a tropical climate on a test platform off the Japanese coast. The test results revealed no deterioration to the GFRP composite. Moreover, Sheard *et al.* (1997) reported that the overall conclusions of the work of the EUROCRETE project were that GFRP is suitable in a concrete environment. Furthermore, Mukherjee and Arwikar (2007a, b) have discussed the performance of externally bonded GRP composites to concrete in a tropical environment. The results from all these investigations are revealing and have provided interesting discoveries regarding the long-term resistance of FRP composites to the natural environments to which civil engineering composites are exposed.

2.8 Conclusion

This chapter has summarised the FRP composite materials used in the construction industry for upgrading RC structures, and has given information on the mechanical and in-service properties of the most relevant matrices and fibres used and the properties of the combination of these two materials to form the FRP composite. The topics covered have included most areas concerned with properties and manufacturing techniques of FRP composites in the field of rehabilitation of structural systems. With polymer composites it is essential to ensure that the service temperatures do not approach the glass transition temperature of the polymer. Evidence has shown that composite material, if appropriately designed and fabricated, can provide longer lifetimes and lower maintenance than conventional materials. However, actual data on durability are sparse, not well documented and, where available, not easily accessible to the civil engineer; this situation is currently being addressed.

2.9 Sources of further information and advice

ACI (1996) *State-of-the-Art Report on Fiber Reinforced Plastic (FRP) Reinforcement for Concrete Structures*, ACI 440R-96, Farmington Hills, MI, USA, American Concrete Institute.

BANK L C (2006) *Composites for Construction Structural Design with FRP Materials*, Hoboken, NJ, USA, John Wiley and Sons.

BANK L C, GENTRY R T, BARKATT A, PRIAN L, WANG F and MANGLA S R (1998) Accelerated aging of pultruded glass/vinylester rods, in Saadatmanesh H and Ehsani M R (eds), *Proceedings International Conference on Fibre Composites in Infrastructure – ICCI-98*, Tucson, AZ, USA, Jan 5–7, Vol. 2, 423–437.

BISBY L A, KODUR V K R and GREEN M F (2004). Performance in fire of FRP-confined reinforced concrete columns, *Proceedings Advanced Composite Materials in Building and Structures – ACMBS IV*, Calgary, Alberta, Canada, July 20–23 (CD ROM).

BLONTROCK H, TAERWE L and VANDEVELDE P (2001) Fire testing of concrete slabs strengthened with fibre composite laminates, in Burgoyne C J (ed.), *Proceedings Fibre Reinforced Polymer Reinforcement for Concrete Structures – FRPRCS-5*, London, UK, Thomas Telford, 547–556.

CADEI J M, STRATFORD T K, HOLLAWAY L C and DUCKETT W G (2004) *Strengthening Metallic Structures Using Externally Bonded Fibre-Reinforced*, Polymers Report C595, London, CIRIA.

Concrete Society (2000) *Design Guidance for Strengthening Concrete Structures Using Fibre Composite Materials*, TR55, 2nd edn, Camberley, UK, The Concrete Society.

Concrete Society (2003) *Strengthening Concrete Structures using Fibre Composite Materials: Acceptance, Inspection and Monitoring*, TR57, Camberley, UK, The Concrete Society.

GDOUTOS E E, PILAKOUTAS K and RODOPOULOS C A (2000) *Failure Analysis of Industrial Composite Materials*, New York, USA, McGraw-Hill.

HILL P S SMITH S and BARNES F J (1999) Use of high modulus carbon fibres for reinforcement of cast iron compression struts within London Underground: Project details, *Proceedings Conference on Composites and Plastics in Construction, Nov, BRE, Watford, UK. RAPRA Technology*, Paper 16, 1–6.

HOLMES T M, LEATHERMAN G L and EL-KORCHI T (1991) Alkali-resistant oxynitride glasses, *Journal of Materials Research*, **6**(1), 152–158.

MANDER J B, PRIESTLEY M J N and PARK R (1988) Theoretical stress–strain model for confined concrete, *Journal of Structural Engineering, ASCE*, **114**, 1804–1826.

MATTHEWS F L and RAWLINGS R D (2000) *Composite Materials: Engineering and Science*, Cambridge, UK, Woodhead.

MAYS G C and HUTCHINSON A R (1992) *Adhesives in Civil Engineering*, Cambridge, UK, Cambridge University Press.

MCKENZIE M (1991) *Corrosion Protection: The Environment Created by Bridge Enclosure*, Research Report 293, Wokingham, UK, Transport and Road Research Laboratory.

MCKENZIE M (1993) *The Corrosivity of the Environment Inside the Tees Bridge Enclosure*: Final Year Results, Project Report PR/BR/10/93, Wokingham, UK, Transport and Road Research Laboratory.

Modern Plastics Encyclopedia (1988) New York, USA, McGraw-Hill. October, Vol. 65, No. 11, 576–619.

MUFTI A, BENMOKRANE B, BOULFIZA M, BAKHT B and BREYY P (2005) Field study on durability of GFRP reinforcement, *International Bridge Deck Workshop*, Winnipeg, Manitoba, Canada, Apr 14–15.

MUFTI A, ONOFREI M, BENMOKRANE B, BANTHIA N, BOULFIZA M, NEWHOOK J, BAKHT B, TADROS G and BRETT P (2005) Durability of GFRP reinforced concrete in field structures, in Shield C K, Busel J P, Walkup S L and Gremel D D (eds), *Proceedings Fibre Reinforced Polymer Reinforcement for Concrete Structures – FRPRCS-7*, Farmington Hills, MI, USA, American Concrete Institute.

MUFTI A, ONOFREI M, BENMOKRANE B, BANTHIA N, BOULFIZA M, NEWHOOK J, BAKHT B, TADROS G and BRETT P (2005) Report on the studies of GFRP durability in concrete from field demonstration structures, in Hamelin P, Bigaud D, Ferrier E and Jacquelin E (eds), *Proceedings Composites in Construction – 3rd International Conference*, Lyon, France, July 11–13.

PILAKOUTAS K (2000) Composites in concrete construction, in Gdoutos E E, Pilakoutas K and Rodopoulos C A (eds), *Failure Analysis of Industrial Composite Materials*, New York, USA, McGraw-Hill, 449–497.

SAADATMANESH H and TANNOUS F E (1999) Relaxation, creep and fatigue behaviour of carbon fiber reinforced plastic tendons, *ACI Materials Journal*, **96**(2), 143–153.

STARR T F (ed.) (2000) *Pultrusion for Engineers*, Cambridge, UK, Woodhead.

2.10 References

AKLONIS J J and MACKNIGHT W J (1983) Introduction to polymer viscoelasticity, 2nd edn, New York, USA, Wiley, 36–56.

BANK L C and GENTRY R T (1995) Accelerated test methods to determine the long-term behaviour of FRP composite structures: environmental effects, *Journal of Reinforced Plastics and Composites*, **14**, 559–587.

BANK L C, GENTRY R T, BARKATT A, PRIAN L, WANG F and MANGLA S R (1998) Accelerated aging of pultruded glass/vinylester rods, in Saadatmanesh H and Ehsani M R (eds), *Proceedings International Conference on Fibre Composites in Infrastructure – ICCI-98*, Tucson, AZ, USA, Jan 5–7, Vol. 2, 423–437.

BSI (1975) *BS 4618: 1975 Recommendations for the presentation of plastics design data. Mechanical properties* (7 sections), London, UK, British Standards Institution.

ASCE (2001) Gap Analysis for Durability of Fiber Reinforced Polymer Composites in Civil Infrastructure, based upon work supported by the Federal Highway Administration under Cooperation Agreement No DTFH61-93-X-00011, Reston, VA, USA, Civil Engineering Research Foundation, American Society of Civil Engineers.

CESSNA L C (1971) Stress-time superposition for creep data for polypropylene and coupled glass reinforced polypropylene, *Polymer Engineering Science*, **13**, May 211–219.

CLARKE J L and SHEARD P (1998) Designing durable FRP reinforced concrete structures, in Benmokrane B and Rahman M (eds), *Proceedings 1st International Conference on Durability of Fibre Reinforced polymer (FRP) Composites for Construction – CDCC 1998*, Sherbrooke, Quebec, Canada, Aug 5–7, 13–24.

DE LORENZIS L, NANNI A and LA TEGOLA A (2000) Flexural and shear strengthening of reinforced concrete structures with near surface mounted FRP rods, in Humar J and Razaqpur A G (eds), *Proceedings Advanced Composite Materials in Building and Structures – ACMBS III*, Montreal, Québec, Canada, Canadian Society for Civil Engineering, 521–528.

DE LORENZIS L, MICELLI F and LA TEGOLA A (2002) Passive and active near surface mounted FRP rods for flexural strengthening of RC beams, *Proceedings International Conference on Fibre Composites in Infrastructure – ICCI-02*, San Francisco, CA, USA, June 10–12 (CD ROM).

DEMERS M, LABOSSIÈRE P and NEALE K (2005) Ten years of structural rehabilitation with FRPs – a review of quebec applications, in Hamelin P, Bigaud D, Ferrier E and Jacquelin E (eds), *Proceedings Composites in Construction – 3rd International Conference*, Lyon, France, July 11–13.

EL-HACHA R and RIZKALLA S H (2004) Near-surface-mounted fibre-reinforced polymer reinforcements for flexural strengthening of concrete structures, *ACI Structural Journal*, **101**(5), 717–726.

GARDEN H, HOLLAWAY L C and THORNE A T (1998) The strengthening and deformation behaviour of reinforced concrete beams upgraded using prestressed concrete plates, *Journal of Materials and Structures*, **31**, 247–258.

GIBSON A and HUME J (1995) Fire performance of composite panels for large marine structures, *Plastic, rubbers and composites Processing and Applications*, **23**, 175–183.

GILMAN J W (1999) Flammability and thermal stability studies of polymer layered (clay) nanocomposites, *Applied Clay Science*, **15**, 31–49.

GILMAN JW, KASHIWAGI T, MORGAN A, HARRIS R, BRASSELL L, VAN LANDINGHAM M and JACKSON C (2000) *Flammability of Polymer Clay Nanocomposites Consortium: Year One Annual Report*, Gaithersburg, MD, USA, National Institute of Standards and Technology, US Department of Commerce.

HACKMAN I and HOLLAWAY L C (2006) Epoxy-layered silicate nanocomposites in civil engineering, *Composites: Part A*, **37**, 1161–1170.

HAQUE A and SHAMSUZZOHA M (2003) S2-glass/epoxy polymer nanocomposites: manufacturing, structures, thermal and mechanical properties, *Journal of Composite Materials*, **37**(20), 1821–1837.

HOLLAWAY L C (1993) *Polymer Composites for Civil and Structural Engineering*, London, Glasgow, New York Tokyo, Melbourne, Madras, Blackie Academic & Professional.

HOLLAWAY L C (2007) Fibre-reinforced polymer composite structures and structural components: current applications and durability issues, in Karbhari V M (ed.), *Durability of Composites for Civil Structural Applications*, Woodhead Publishing Cambridge, UK, 189–224.

HOLLAWAY L C (2008a) Advanced fiber reinforced polymer composites, Chapter 5 of *Advanced Materials*, to be published by Published by World Science Publishing.

HOLLAWAY L C (2008b) Chapter 13 of 'Manual of Bridge engineering,' in Edited by G. Parke and N. Hewson to be published by Thomas Telford, London.

HOLLAWAY L C and HEAD P R (2001) *Advanced Polymer Composites and Polymers in the Civil Infrastructure*, Oxford, UK, Elsevier.

HOLLAWAY L C, ZHANG L, PHOTIOU N K, TENG J G and ZHANG S S (2006) Advances in adhesive joining of carbon fibre/polymer composites to steel members for repair and rehabilitation of bridge structures, *Advances in Structural Engineering*, **9**, 101–113.

KARBHARI V M, CHIN J W, HOUSTON D, BENMOKRANE B, JUSKA T, MORGAN R, LESKO J J, SORATHIA U and REYNAUD D (2003), 'Durability gap analysis for fiber-reinforced polymer composites in civil infrastructure' *Journal of Composites for Construction*, **7**(3), 238–247.

LIU W, HOA S and PUGH H (2005) Epoxy-clay nanocomposites: dispersion, morphology and performance, *Composites Science and Technology*, **65**, 307–316.

MUKHERJEE A and ARWIKAR S J (2007a) Performance of externally bonded GFRP sheets on concrete in tropical environments. Part 1: structural scale tests, *Composite Structures*, **81**(1), 21–32.

MUKHERJEE A and ARWIKAR S J (2007b) Performance of externally bonded GFRP sheets on concrete in tropical environments: Part II: microstructural tests, *Composite Structures*, **81**(1), 33–40.

OHLEMILLER T, CLEARY T and SHIELDS J (1996) Effect of ignition conditions on upward flame spread on a composite material in a corner configuration, *Proceedings 41st International SAMPE Symposium*, Covina, CA, USA, Society for the Advancement of Material and Process Engineering, 734–747.

SEN R, MULLINS G and SALEM T (2002) Durability of E-glass/vinylester reinforcement in alkaline solution, *ASI Structural Journal*, **99**, 369–375.

SHEARD P, CLARKE J L, DILL M, HAMMERSLEY G and RICHARDSON D (1997). EUROCRETE – taking account of durability for design of FRP reinforced concrete structures – non-metallic (FRP) reinforcement, for concrete structures, *Proceedings 3rd International Symposium on Non-metallic (FRP) Reinforcement for Concrete Structures*, Tokyo, Japan, Japan Concrete Institute, Vol. 2, 75–82.

TOMOSAWA F and NAKATSUJI T (1997) Evaluation of ACM reinforcement durability by exposure tests, Non-metallic (FRP) reinforcement for concrete structures,

Proceedings 3rd International Symposium on Non-metallic (FRP) Reinforcement for Concrete Structures, Tokyo, Japan, Japan Concrete Institute, Vol. 2, 139–146.

UOMOTO T (2000) Durability of FRP as reinforcement for concrete structures, in Humar J and Razaqpur A G (eds), *Proceedings Advanced Composite Materials in Building and Structures – ACMBS III*, Montreal, Québec, Canada, Canadian Society for Civil Engineering, 3–14.

WANG Z, MASSAM J and PINNAVAIA T J (2000) Epoxy–clay nanocomposites, in Pinnavaia T J and Beall G W (eds), *Polymer-clay Nanocomposites*, Chichester, UK, 127–150 Wiley.

3
Surface preparation of component materials

A. R. HUTCHINSON, Oxford Brookes University, UK

3.1 Introduction

One of the most important aspects in adhesive bonding and laminating is surface preparation, which is a critical topic that is often not given the attention that it requires. When surface preparation is undertaken as part of the fabrication of bonded joints, whether in a factory or on a construction site, it is vital that designers, operatives and their supervisors understand the principles behind obtaining satisfactory short- and long-term adhesion.

Adhesives and resins are frequently blamed for 'not sticking', but the source of the trouble lies generally with the surface preparation. A major barrier to the more confident use of adhesives and resins is a lack of understanding about adhesion, appropriate surface preparation techniques and their effects on initial bond strength and, to a greater degree, long-term durability. Hot/wet environments are particularly deleterious for adhesive bonds to metallic substrates and adequate surface preparation is a vital prerequisite for maintaining joint integrity.

A vital property of adhesives (and laminating resins) is that they must adhere to the relevant substrate surfaces: some generic types, and particular formulations, do this better than others. Experience has shown that the best chance of success in construction industry applications is likely to be achieved by using two-part cold-curing paste epoxy adhesives that have been specially developed for use on site. Mays and Hutchinson (1992) identified the principal requirements for bonding steel to concrete, and these are very similar for the case of bonding fibre reinforced polymers (FRP) to concrete, metals and timber:

- The adhesive should exhibit adequate adhesion to the materials involved.
- A two-part epoxy resin with a polyamine-based hardener should be used, which exhibits good moisture resistance and resistance to creep.

- The adhesive's glass transition temperature (T_g) should generally be greater than 60 °C. In some circumstances, such as bonding to metals or when bonding FRP materials to the top surface of a bridge deck that is to receive hot bituminous surfacing, an adhesive with a much higher T_g may be required (Concrete Society, 2004). It should be noted that the T_g of many adhesive and resin systems can be increased by warm curing.
- The flexural modulus of the material should fall within the range 2–10 GPa at 20 °C. Generally-held views are that the lower bound to this range might be reduced to 1 GPa.
- The bulk shear and tensile strength at 20 °C should be ≥12 MPa.
- The minimum shear strength at 20 °C, measured by the thick adherend shear test (TAST), should be 18 MPa.
- The mode I fracture toughness (K1 c) should be ≥0.5 MNm-3/2.
- The equilibrium water content (M_∞) should not exceed 3% by weight after immersion in distilled water at 20 °C. The coefficient of permeability should not exceed 5×10^{-14} m^2/s.
- It should possess gap-filling properties, be thixotropic and be suitable for application to vertical and overhead surfaces.
- It should not be sensitive to the alkaline nature of concrete (if present) and its potential effect on the durability of joints.
- It should not be unduly sensitive to variations in the quality or moisture contents of prepared surfaces.
- The working characteristics of the material should enable an adequate joint quality to be achieved with respect to mixing, application and curing.

A further item to add to this list is that the adhesive should exhibit sufficient tack or 'grab' to enable thin FRP materials to be attached to overhead or vertical surfaces without the need for temporary fixings whilst the adhesive cures.

It is essential that the adhesives used should possess high surface activity and good wetting properties for a variety of substrates. The adhesion mechanism is primarily chemical, but usually involves an element of mechanical keying. Put simplistically, the liquid adhesive will bond to the first material with which it comes into contact: if this is dust, loosely adhering paint, cement laitance or mould releases, the joint will clearly be inadequate. The adhesive should also displace air or volatiles when spreading over the surface so that there are no voids to act as flaws.

The purpose of surface preparation is to remove contamination and weak surface layers, to change the substrate surface roughness at a micro level and/or introduce new chemical groups onto the surface to link with the adhesive (or primer). The key stages involved in achieving this are:

- cleaning;
- material removal and surface modification;
- further cleaning (to remove contamination introduced by treatments, such as oil-mist, dust or chemical residues).

With metallic or porous surfaces, the application of an adhesive-compatible primer layer may be desirable.

A fundamental principle of the selection and specification of composite materials systems for strengthening existing structures is that the whole system should be *qualified* for use (www.compclass.org). This means taking the materials through a sequence of tests that ensure that the system meets the design performance requirements. The qualification procedure should include a number of tests that include a check on the adhesion between all relevant combinations of materials. This approach employs control substrate materials and test protocols that link together initial qualification, design assumptions and quality control (QC) acceptance criteria. Central to this are the assessments of adhesion and adequacy of surface preparation procedures.

3.2 Adhesion and interfacial contact

A detailed discussion of interfacial forces and adhesion mechanisms is provided by many authors (e.g. Mays and Hutchinson, 1992; Adams *et al.*, 1997; Packham, 2005). Adhesives join materials primarily by attaching to their surfaces within a layer of molecular dimensions, i.e. of the order of 0.1–0.5 nm. The term 'adhesion' is associated with intermolecular forces acting across an interface and involves a consideration of surface energies and interfacial tensions. Being liquid, adhesives flow over and into the irregularities of a solid surface, coming into contact with it and, as a result, interact with its molecular forces. The adhesive then solidifies to form the joint. The basic requirements for good adhesion are therefore intimate contact between the adhesive and substrates and an absence of weak layers or contamination at the interface. Adhesive bonding involves a liquid 'wetting' a solid surface, which implies the formation of a thin film of liquid spreading uniformly without breaking into droplets (Fig. 3.1).

Fundamentally, the surface tension of the adhesive should be lower than the surface energy of the solids involved; in this case, the treated surface of FRP and the surfaces of the parent materials. Because of the similarity of adhesive and composite matrix composition, values of surface tension and surface energy are very similar. Both compositions contain polar molecular

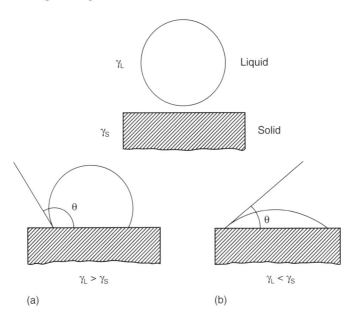

3.1 Simple representation of wetting: (a) non-wetting; (b) wetting.

γ_L = surface free energy (surface tension) of liquid
γ_S = surface free energy of solid
θ = contact angle between liquid and solid

groups that are mutually attractive and chemically compatible. Thus good adhesion is assured, provided that contamination is removed by surface preparation. Some typical values of the surface energies of various materials and the surface tension of epoxy adhesive are given in Table 3.1.

Finally, the rheological characteristics of the adhesive (or a primer) are also important and low viscosity materials are beneficial for penetrating into, and binding together, surfaces such as concrete which are porous and which may be fractured at a micro-scale. The interaction of surface energetics, tensions and rheology is a complex one.

3.3 Primers and coupling agents

Primers, being low viscosity materials, assist adhesion either by partially penetrating the pores of a porous surface or by forming a chemical link between the surface and a relatively high viscosity adhesive. Primers may also bind and reinforce weak surface layers of certain substrates such as concrete or stone. Improvement in adhesion relates directly to an improvement in bond durability, and it is evident that adhesion failure is often due to a lack of priming.

Table 3.1 Typical values of surface free energies

Material class	Surface	Surface free energy (mJ/m^2)	Remarks
Metals	Ferric oxide	1350	Under ideal conditions in a vacuum. However, realistic values in air would be 50–100 mJ/m^2
	Aluminium oxide	630	
Ceramic	Silica	290	
Water	Water	72*	Surface tension
Timber	Oak	50	Measured along the grain
	Pine	53	Measured along the grain
Thermosets	Epoxy adhesive	45*	Surface tension
	CFRP/GFRP	45–60	Vinylester, epoxy and polyester resin matrices. Surfaces pre-treated, with higher values associated with peel-ply, heavy abrasion or corona discharge treatment
	GFRP	38	As moulded
Thermoplastics	Nylon	42	
	Acrylic adhesive	35	Surface tension value
	Polyurethane adhesive	30	Surface tension value

CFRP = carbon fibre reinforced polymer
GFRP = glass fibre reinforced polymer
*Surface tension, rather than the surface free energy, of liquids is more usually quoted. Tension and energy are numerically identical but dimensionally different

The use of adhesive primers may be more critical in some instances than others, for instance in association with joints involving metallic substrates or with porous surfaces such as concrete and stone. In steel plate bonding applications it is virtually essential to prime the steel in order to generate a reliable and reproducible surface that confers good bond stability to an adhesive in the presence of moisture. The advantages to be gained by priming generally far outweigh disadvantages such as the need for an extra process, or the primer or a primer interface becoming the weakest link in the joint. The experience of paint and adhesion technologists is that primers greatly reduce the variability of subsequent interfacial bond performance, and that certain products can create a water-stable interface. Their use may also obviate the need for complex surface pretreatment procedures normally associated with metal alloys.

Any primers used must be appropriate to structural adhesive bonding rather than merely corrosion inhibition. The danger of film-forming primers

88   Strengthening and rehabilitation of civil infrastructures

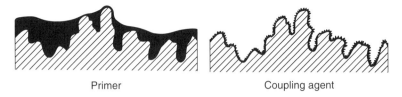

3.2 Primers and coupling agents.

is that they may become the weak link in a joint because they are themselves mechanically weak. The alternative approach utilises a very thin layer of a chemically-reactive system, as used in the defence and, to a lesser extent, marine industries. The concept is depicted schematically in Fig. 3.2. Most of these reactive products are based on silane coupling agents, applied as a dilute solution (0.1–5%). Ideally, this results in the deposition of a single layer of molecules, the respective ends of the silane molecule having a strong affinity for the surface and the adhesive. Plueddemann (1991) and Mittal (1992) review such materials in detail. The key motivation for encouraging the use of silane coupling agents is to provide high integrity bonds without the need for complex and environmentally-unfriendly chemical treatment processes. Such adhesion promoters are routinely applied to mechanically abraded surfaces and have been shown to work well on mild steel and on stainless steel (Mays and Hutchinson, 1992; DTI, 1993).

3.4   Surface preparation

The purpose of surface preparation is to remove contamination and weak surface layers, to change the substrate surface topography and/or introduce new surface chemical groups to promote bond formation. An appreciation of the effects of surface preparation may be gained from surface analytical or mechanical test techniques. Surface preparation generally has a much greater influence on long-term bond durability than it does on initial bond strength, so that a high standard of surface preparation is essential for promoting long-term bond integrity and durability (Kinloch, 1983; Mays and Hutchinson, 1992; Adams *et al.*, 1997).

In strengthening applications the parent material must be treated *in situ*, generally under less than ideal conditions. The plane of the surface(s) to be treated (horizontal, vertical, overhead, etc.) has a large bearing on the selection of an appropriate method. The choice of method, or combinations of methods, depends upon the costs, the scale and location of the operation, access to equipment and materials and health and safety conditions.

The composite reinforcement may be provided in a variety of forms, but prefabricated elements and pultruded profiles can be treated off site. This

Surface preparation of component materials

Table 3.2 Pretreatment requirements

Material	Suitability for bonding	Pretreatment required
Cast iron	*****	Cursory
Steel	****	Straightforward
Stainless steel	***	⎫
Zinc	***	⎬ Quite demanding
Aluminium	***	⎭
Concrete	****	⎫
GFRP	****	⎬ Straightforward
CFRP	***	⎭
PVC	**	Rigorous
Polyolefin	*	Complex

CFRP = carbon fibre reinforced polymer
GFRP = glass fibre reinforced polymer
PVC = polyvinylchloride

has great advantages because anything treated in a factory environment can be dealt with in a more reliable way than on site.

The methods of surface preparation can be considered under four categories (Brewis, 1982):

- solvent degreasing;
- mechanical techniques;
- chemical techniques;
- physical techniques.

The most appropriate method, or combination of methods, depends upon the nature of the substrates, but an indication of the general requirements is given in Table 3.2.

Solvent degreasing removes grease and most potential contaminants. The choice of solvent should be based on the principle that 'like dissolves like', although toxicity, flammability and cost should be taken into consideration. A volatile solvent such as acetone should always be chosen or else any residues may form a weak surface layer. For metallic substrates, alkaline cleaners and/or detergent solutions are often advised after solvent treatments, to remove dirt and inorganic solids. They may also be used instead of solvents for health and safety reasons, but should be followed by thorough rinsing and drying in hot air prior to bonding.

Mechanical treatments often cause much obvious roughening of a surface but the effect on adhesion is complex. It should be remembered that a rough surface *per se* is not a fundamental requirement for adhesion. The most important requirement of mechanical treatment is to remove weak surface layers and to expose a clean, new, surface. The various mechanical methods depend on the abrasive action of wire brushes, abrasive pads and wheels, blasting media and tools such as needle guns. Two major aspects are control of the method and assessment of the surface following treatment.

Chemical and electrochemical methods typically cause more complex changes to surfaces than do mechanical methods. In addition to the cleaning action and removal of weak layers, chemical treatments often roughen a surface microscopically. Anodising, for example, results in a very porous surface, and other techniques for metals result in a micro-fibrous topography. Treatments are designed to result in the formation of stable and coherent oxide structures. However, a significant disadvantage of chemical methods is the toxicity of the materials used and the subsequent waste disposal problem.

Physical methods include techniques that promote a strong oxidising reaction with the surfaces of materials. These include factory-based techniques such as flame treatment and corona discharge. They are very effective on inert plastics like polypropylene but also work well on thermoset–matrix composites. Flame treatment has also been applied to timber surfaces, albeit in the context of factory-based processes for painting and coating (Winfield *et al.*, 2001).

3.5 Surface preparation of concrete

The chemical and physical nature of the surface of concrete is complex and variable. The surface of this multiphase material may contain exposed aggregate, sand, unhydrated cement particles and cement gel, together with cracks and voids; the surface moisture content may also be variable. Surface treatments should remove significant contamination, cement-rich layers and traces of mould release agents. Mutual atomic or molecular attractions between an adhesive and the exposed (and probably heavily hydrated) constituents may exist, but mechanical keying or interlocking into the irregularities and pores of the surface probably plays a significant role too. It is impossible to measure the surface energy of concrete, although it should be possible to measure the energies of its constituents separately. For example, Table 3.1 indicates a value for silica.

In essence, the purpose of surface preparation is to remove the outer, weak and potentially contaminated skin together with poorly bound material, in order to expose small- to medium-sized pieces of aggregate (Fig. 3.3). This must be achieved without causing micro-cracks or other damage in the layer behind; this would lead to a plane of weakness and hence a reduction in strength of the adhesive connection. Any large depressions, blow-holes and cracks must be filled with suitable mortars whilst sharp edges and shutter marks should be removed to achieve a flat surface prior to the

Surface preparation of component materials

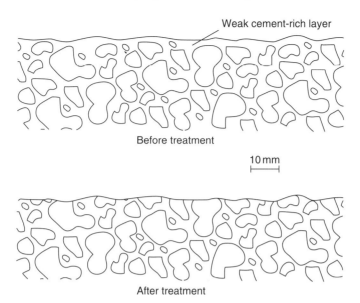

3.3 Schematic idealisation of concrete following surface preparation.

application of a structural adhesive. This will ensure a relatively uniform bondline thickness in order to maximise the efficiency of shear stress transfer. The basic sequence of steps in the process of surface preparation (Sasse and Fiebrich, 1983; Gaul, 1984; Sasse, 1986; Mays and Hutchinson, 1992; Edwards, 1993; Hutchinson, 1993; ASTM D4258, 2005a) should be:

- remove any damaged or sub-standard concrete and reinstatement with good-quality material.
- remove laitance and grease or oils, preferably by grit-blasting. Other techniques such as wire-brushing, grinding, bush-hammering, acid-etching, water-jetting or flame-blasting are not recommended for plate bonding applications. In particular, the use of pneumatic tools can cause significant damage to the underlying concrete (Sasse, 1986; Hutchinson, 1993).
- high-pressure washing may be useful in the removal of grease and oils, but more aggressive treatment includes the use of solvent-based and sodium hydroxide-based products in the form of a gel or poultice to draw out contamination. Such products must be completely removed or else bond performance may be compromised.
- remove dust and debris by brushing, air-blast or vacuum.

Additional steps could include:

- further cleaning, with a suitable solvent, to remove any remaining contaminant;

- drying the surface to be bonded, if necessary;
- application of an adhesive-compatible (epoxy) primer, if necessary.

The recommended method of laitance and oil or grease removal, by blasting, is fast, plant-intensive and operator dependent. There exist a multiplicity of types of blast media, media sizes, blast pressures and types of equipment. With dry systems, oil and water traps should be used to prevent contamination from air compressors; dry systems may be open, or closed with vacuum recovery and recycling of the blast media. Generally a fairly micro-rough but macro-smooth surface is generated, together with a lot of dust; particles of blast media may also be left embedded in the surface. This dust and debris must be removed prior to bonding. Wet blasting is another option that overcomes some of the problems associated with the dust. However, this may create a water disposal problem, and the concrete surface must be allowed to dry out to a degree that is suited to the intended adhesive system prior to bonding. The flatness of the resultant concrete surface should be such that the gap under a 1 m long straight edge does not exceed 5 mm (Concrete Society, 2004).

Slots for near-surface mounted (NSM) reinforcement are formed by making parallel cuts in the surface of the concrete to the required depth and removing the material in between. It may also be desirable to blast the cut sides of the slots if they have taken on a polished appearance. These slots must be cleaned of dust and debris using a vacuum or clean high-pressure air prior to bonding.

After preparation, the suitability of the surface should be checked using the partially-cored pull-off test procedure described in EN 1542 (EN, 1999) (Fig. 3.4). Typical pull-off strengths of 50 mm diameter steel dollies bonded to concrete provide tensile strengths in the range 1–3 MPa, where failure inevitably takes place in the concrete. Fig. 3.5 shows the possible loci of failure for the case of FRP bonded to a concrete substrate. The time lapse between preparation and bonding should be minimised in order to prevent any further contamination of the surface. Adhesion of epoxy adhesives to damp and wet concrete surfaces was investigated in the ROBUST Project, as described by Hutchinson and Hollaway (1999) and by Rahimi and Hutchinson (2001), but no significant measurable effects on bond strengths were detected.

3.6 Surface preparation of metallic materials

In joints involving metallic substrates, the adhesive 'sticks' to the metal surface oxide layer. Such joints can cause problems in service because oxide structures, and bonds to them, are susceptible to interfacial degradation. It can be quite difficult to obtain durable joints that involve high alloy metals

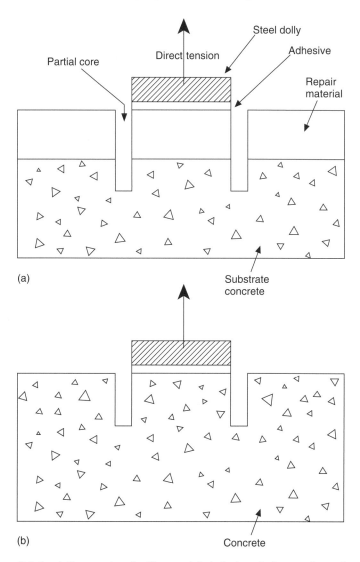

3.4 Partially cored pull-off tests: (a) dolly bonded to surface of repair materials; (b) dolly bonded to concrete surface.

that oxidise rapidly, such as stainless steels and aluminium alloys, unless complex etching and anodising procedures are adopted.

Steel surface treatments vary depending upon type – mild, stainless, and zinc-coated. Some treatments will only be appropriate for use in a factory environment whilst others are suitable for site use. More detailed guidance is given by Adams *et al.* (1997), ASTM D2651 (ASTM, 2001a), BS EN 12768 (BSI, 1997), ISO 8504 (ISO, 2000), Mays and Hutchinson (1992), NPL

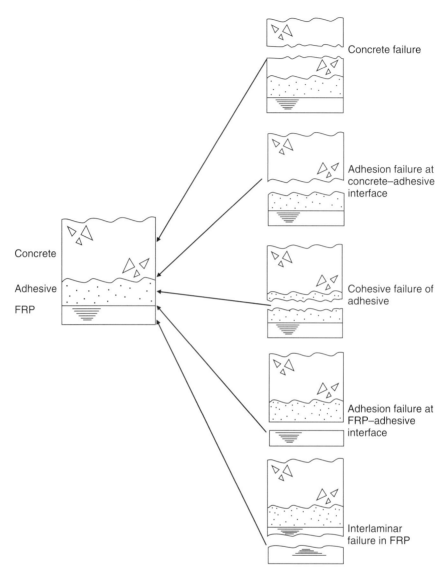

3.5 Possible modes of failure associated with joints comprising FRP adhesively bonded to concrete.

(2000), Packham (2005), and Sykes (1982), but several general points are summarised below:

- the first step is degreasing with a suitable solvent or alkaline cleaner.
- dry grit-blasting has been applied successfully to mild and stainless steels, and is particularly effective in combination with primers. Hard, sharp, media should be used and dust must be removed after the blasting

Surface preparation of component materials 95

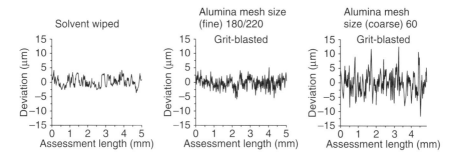

3.6 Effect of three different surface treatments on the roughness of a steel surface.

operation by vacuuming or by brushing with clean brushes. With dry blasting systems, oil and water traps should be used to prevent contamination of the surface.
- wet blasting can be used in some circumstances and has the advantage that dust is carried off the surface automatically.
- mechanical abrasion such as grinding does not provide a good surface for adhesive bonding.
- chemical treatment is not recommended because of the disposal problems associated with residues, but it is suitable for metals such as zinc and copper.
- commercially available etch primers are available to promote adhesion to some forms of zinc coated steel, although some types (e.g. electro-zinc and iron–zinc) are more amenable to bonding than others. It is commonly found that reasonable bonds to hot-dip galvanised surfaces are obtained by removing most of the zinc with a light grit blast, preferably followed by the application of a primer.

It has been suggested for many years that a surface equivalent to SA $2^1/_2$ (SIS, 0559 00) be achieved on metal (SIS, 1967). This standard was later formalised as ISO 8501-1 (ISO, 2007). However, it should be noted that SA $2^1/_2$ is a qualitative visual assessment of the cleanliness and quality of the resultant surface; it is not a measure of roughness, nor of potential adhesion *per se*. Cast (and wrought) iron surfaces may contain numerous cracks, fissures and voids. However, they respond well to grit-blasting and appear to provide similar or better adhesion characteristics to those associated with grit-blasted mild steel.

Fig. 3.6 shows the effect of three different surface treatments on the roughness of a mild steel surface; the left-hand picture shows a solvent-wiped surface, the centre picture shows a surface blasted with 180/220 mesh alumina, and the right-hand picture shows a surface blasted with coarser 60 mesh alumina. It is clear that the size of grit has a direct influence on the

Table 3.3 Strengths of lap shear joints made with mild steel adherends

Adhesive	Surface	Average shear strength (MPa)	
		Initial	After 60 weeks in water at 60°C
Cold cure epoxy	Degreased	13.5	Failed
	Grit-blast	23.5	10.5
	Grit-blast plus primer	24.0	15.7
Hot cure epoxy	Degreased	21.5	19.5
	Grit-blast	26.2	23.6
	Grit-blast plus primer	25.8	25.8
Acrylic	Degreased	11.1	Failed
	Grit-blast	18.4	8.3
	Grit-blast plus primer	22.0	13.4

Note: 20 mm × 10 mm single lap shear specimens with 1.2 mm adherend thickness used. All joints post-cured at 60°C prior to initial testing and water immersion

final surface profile. Harris and Beevers (1999) further discuss the effects of blasting on surface energy, surface roughness, and short- and long-term adhesion; they concluded that texture *per se* had little effect on adhesion. The effect of surface treatment on joint strength can be assessed using lap shear, pull-off and wedge cleavage tests. Some typical short- and long-term joint strength data for lap shear joints made with steel adherends and three types of adhesive are shown in Table 3.3; the benefit of the primer in maintaining bond durability is evident. Some short-term data for pull-off tests on cast iron surfaces for two different, but typical, construction epoxy adhesives and one epoxy laminating resin are shown in Table 3.4. The effect of two different types of primer and different surface treatments is shown. It is clear that grit-blasting represents a suitable surface preparation method and that the use of an epoxy primer provides satisfactory results. The data associated with cast iron primed with micaceous iron oxide highlight the fact that this material is cohesively weak and failure through the primer layer is inevitable. The strength realised may nevertheless be deemed acceptable for any particular design.

3.7 Surface preparation of timber

Historically, the adhesives used for the manufacture of timber products include: animal/casein glues, urea-formaldehyde, phenol-formaldehyde, resorcinol-formaldehyde, phenol-resorcinol-formaldehyde and melamine-

Table 3.4 Short-term pull-off data for epoxy adhesives and laminating cast iron surfaces

Adhesive	Surface preparation	Average pull-off strength (MPa)	Locus of failure
Epoxy 1	Grit-blast	37–40	90C, 10ACI
	Grit-blast plus primer[1]	20–23	95P, 5PCI
	Ground	39–40	70C, 30ACI
	Ground plus primer[1]	24–26	90P, 10PCI
	Needle gun	24–36	40C, 60ACI
	Needle gun plus primer[1]	20–26	90P, 10PCI
Epoxy 2	Grit-blast	30–35	100C
	Grit-blast plus primer[2]	24–34	20C, 30API, 50PCI
Epoxy laminating resin	Grit-blast plus primer[1]	15–18	100P

Key to loci of failure: ACI = adhesion failure at adhesive to cast iron interface. API = adhesion failure at adhesive to primer interface. C = cohesive failure within adhesive layer. P = failure within primer layer. PCI = adhesion failure at primer to cast iron interface
Note: Dolly diameter: 25 mm. Bondline thickness: 0.5 mm. Test speed 2 mm/min. Primer[1]: micaceous iron oxide. Primer[2]: epoxy primer

urea-formaldehyde. The formaldehyde-based adhesives found widespread application in glued-laminated timber beams and in hot-press applications such as plywood manufacture. The characteristic of such applications was a very thin bondline. The four main groups of alternative structural timber adhesives so far are epoxy, polyurethane, acrylic and emulsion polymer isocyanates. For structural repair work, gap-filling properties are required and two-part epoxy and polyurethane formulations are commonly used.

Timber provides a good substrate for adhesion, provided that a number of basic factors are observed. It is a cellular, natural, organic material whose porosity affects its characteristics as a substrate. There is a wide variety of different timber types, each with their own characteristics: they respond differently to treatments and to being bonded. Generally, hardwoods are more difficult to bond than softwoods because of their higher densities and the extractives that they contain (such as oils and tannins). Whatever the species, surface moisture contents at the time of bonding should be less than 20%; this can be achieved by localised drying (Wheeler and Hutchinson, 1998; Broughton and Hutchinson, 2001c).

The surface layers of timber become oxidised and degraded over time. Thus, it is essential to remove material and then bond the surfaces relatively quickly for optimum adhesion. There is general agreement that sound, freshly cut surfaces are ideal for painting and bonding. River *et al.* (1991) and Davis (1997) provide good accounts of the nature of timber surfaces and their treatment. The general treatments are:

- cutting with a plane, saw, auger, chisel or similar sharp tools;
- removal of dust;
- localised drying, if necessary;
- application of an adhesive-compatible primer, if necessary.

Surfaces that have been sanded or sawn should be cleaned carefully to remove dust which will be loosely bound or else clog up any cells which have been cut open. The use of adhesive-compatible primers is strongly recommended by many practitioners.

Compressive lap shear, pull-off and pull-out tests can be used to demonstrate the adequacy of prepared surfaces. It should be noted that the natural variability of timber substrates means that the scatter in data is generally rather large. Wheeler and Hutchinson (1998) have reported typical lap shear strengths of timber bonded with epoxy and polyurethane adhesives (see Table 3.5), and Broughton and Hutchinson (2001a, b) have reported on the pull-out behaviour of steel rods bonded into timber with epoxy adhesives. Wheeler and Hutchinson (1998) and Broughton and Hutchinson (2001c) report the effect of timber moisture content at the time of bonding. It was found that the polyurethane systems tested were very sensitive to the moisture content of the timber at the time of bonding (see Table 3.5). General guidance on resin-bonded repair systems for structural timber is available from Broughton and Hutchinson (2001a, b) and from TRADA (1995).

3.8 Surface preparation of FRP materials

Many strengthening applications involve the use of simple rectangular-section pultrusions, although bespoke laminations such as thick high modulus profiles are also common for metallic strengthening. Typically these are reinforced with carbon (CFRP), although aramid (AFRP) and glass (GFRP) may also be suitable. Some applications employ rods, either round or rectangular in section, for shear strengthening of concrete, for making connections between timber members, and for flexural strengthening of beams. GFRP rod comprising unidirectionally orientated fibres is satisfactory for timber, and may also be suitable for concrete if alkali-resistant glass fibres are used. Pre-moulded composite shells are of increasing interest for attaching to concrete columns, providing the potential for rapid on site joining. Such shells can be made by an open- or closed-moulding process using a variety of glass fibre forms. These shells are adhesively bonded both to themselves and to the column. In all such applications cur-

Table 3.5 Typical compressive shear stresses (MPa) at failure and mode of failure of lap joints made with timber adherends

Adhesive	Species	Timber moisture content at time of bonding (%)					
		10		18		22+	
		Average shear strength (MPa)	Locus of failure	Average shear strength (MPa)	Locus of failure	Average shear strength (MPa)	Locus of failure
Epoxy A	Douglas fir	10–14	N	11	N	8.3–8.9	N
	Oak	16	W	15.6	W	13–14	W
Epoxy B	Douglas fir	10–13	W	10	W	8.3–8.8	W
	Oak	8–10	W	9–12	W	9.4	W
Polyurethane A	Douglas fir	1.5–2	C	0.8–0.9	C	0.3–0.6	C
	Oak	1.3–2	C	0.10	C	0.01	C
Polyurethane B	Douglas fir	7.1	W	1.9	C	1.9	C
	Oak	7.3	W	1.4	C	0	C
Solid timber	Douglas fir	11	W	8.7	W	N	W
	Oak	14	W	11	W	N	W

Key to loci of failure: C = cohesive failure within adhesive layer. **N** = not recorded. **W** = failure within timber
Note: Overlap area was 45 × 45 mm. Bondline thickness = 1.0 mm. Test speed = 2 mm/min. Adherend thickness ≈ 10 mm

rently the matrix resin is likely to be a thermoset; epoxy, vinylester or polyester. However, phenolic and modified acrylic matrices may be used where fire resistance is required. Some future applications will see the introduction of thermoplastic matrix composites, and the adhesion characteristics of these materials are very different. Stitched or woven fabric preforms can, in principle, be used for almost any application, including column wrapping. In these cases the matrix is the laminating resin and no separate surface preparation step is required.

For strengthening applications involving pultrusions, factory-made profiles and shells, the material may be treated off site, enabling a variety of potential techniques to be used. However, relatively large areas of pultruded material will need to be treated in a reliable and consistent way.

The surface of a composite material may be contaminated with mould release agents, lubricants and fingerprints as a result of the production process. Further, the matrix resin may include waxes, flow agents and 'inter-

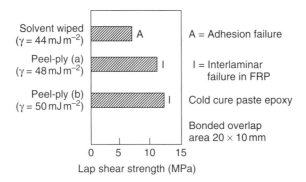

3.7 Surface treatments, energy values and lap shear strengths for joins made with 4 mm thick GFRP adherends (γ = surface free energy).

nal' mould release agents that can be left on the surface of a cured composite. Surface preparation not only serves to remove contamination such as fluorocarbon release agents, but may also increase the surface area for bonding, promote micro-mechanical interlocking and/or chemically modify a surface. Care must be taken to ensure that only the chemistry and morphology of a *thin* surface layer is modified. It is important not to break reinforcing fibres, nor to affect the bulk properties of the composite.

It is recommended in structural bonding that any random fibre mats and surface veils are removed from the surface to ensure stress transfer directly into the main reinforcement fibres. Such a task is rather difficult to control using mechanical removal methods such as grinding or grit-blasting. For this reason, careful consideration should be given to the fabrication of the composite material in the first place in order to make it ready for adhesive bonding with minimal disturbance. The main methods of surface preparation for composites are solvent degreasing, mechanical abrasion and use of the peel-ply technique, and these methods are often used in combination. A complete description of these and other techniques is given by ASTM D2093 (ASTM, 2003a), BSI (1995), Clarke (1996) and Hutchinson (1997). The effect of such treatments is to make the composite surface more 'wettable', as indicated in Table 3.1 and Fig. 3.7.

The basic sequence of steps in the process of surface preparation should be to *either*:

- remove grease and dust with a suitable solvent such as acetone or methyl ethyl ketone (MEK).
- remove release agents and resin-rich surface layers by abrasion. This can be accomplished by careful grinding, sanding (e.g. ASTM D2093, 2003a; Strongwell, 2008), light grit-blasting using very fine alumina (e.g. Bowditch and Stannard 1980), or cryo-blasting with solid carbon dioxide

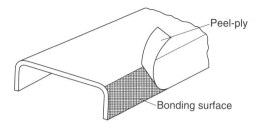

3.8 Incorporation of peel-ply layer on surface of FRP channel section.

pellets (ASTM D2093, 2003a; Strongwell, 2008). The resultant surface should be textured, dull and lustreless.
- remove dust and debris by solvent wiping.
- strip off a peel-ply layer (Fig. 3.8). A peel-ply is a sacrificial layer about 0.2 mm thick which is laid-up on the outermost surfaces of a composite material and co-cured with it. During cure, the peel-ply layer becomes consolidated into the surface of the composite.

The use of a peel-ply layer provides a uniquely practical surface preparation method such that unskilled operators can achieve a very clean and bondable surface on site. This technique was first advanced for use for such applications in construction by Hutchinson and Rahimi (1993). There are several special considerations associated with peel-ply technology, which are outside the scope of this chapter, but the subject is discussed more fully elsewhere (Wingfield, 1993; Hutchinson, 1997). It follows that combinations of composite resin matrix, peel-ply materials and composite processing conditions need to be assessed carefully. When the dry peel-ply material is pulled off, the top layer of resin on the FRP component is fractured and removed, leaving behind a clean, rough, surface for bonding. The resultant surface topography is essentially an imprint of the peel-ply fabric weave pattern (Fig. 3.9).

Finally, all FRP materials absorb small amounts of moisture due to ambient humidity levels or proximity to a wet environment. In CFRP it is the matrix resin that absorbs the moisture, but in GFRP and AFRP it is both the fibres and the matrix resin that can absorb moisture. For ambient temperature curing adhesive systems the presence of small amounts of moisture is unlikely to pose a problem, although it is recommended that the adherends be dried (Clarke, 1996); this may not, of course, be practical. However, with elevated temperature curing adhesive systems the FRP components *must* be dried as far as possible (to, say, less than 0.5% moisture content by weight). This is because the heat curing process draws moisture out of the adherend(s) and into the bondline, resulting in voiding and weakening of the adhesive itself (Hutchinson, 1997).

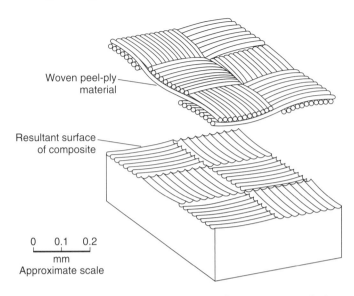

3.9 Typical surface topography arising from woven peel-ply material.

The suitability of surfaces for bonding following any treatment can be assessed using wettability tests and mechanical tests. Single lap shear joints, pull-off tests and wedge cleavage tests may be used to assess adhesion and joint strength. Short-term lap shear data for joints made with epoxy matrix woven CFRP adherends treated in many different ways are shown in Fig. 3.10. It is clear that the cold curing paste epoxy adhesive is very sensitive to surface conditions whereas the heat cured epoxy is relatively insensitive. This is a most important point in the context of externally bonded reinforcement.

3.9 Assessment of surface condition prior to bonding

It is fundamental to satisfactory adhesion that a substrate surface displays a number of characteristics or qualities. It is, therefore, important to identify and define the most important characteristics for promoting adhesion to particular substrate surfaces, and to be able to measure and verify those characteristics as part of a surface treatment production process.

Seven key surface qualities have been defined that can be measured and assessed prior to structural adhesive bonding (Hutchinson and Hurley, 2001):

- wettability;
- roughness (micro- and macro-);
- soundness;

Surface preparation of component materials 103

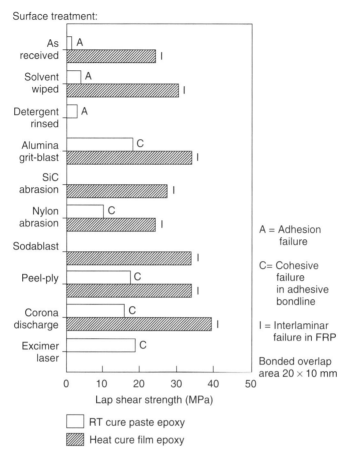

3.10 Effect of surface treatment of the short-term strength of single lap-shear joints made with 2 mm thick epoxy-matrix woven fabric CFRP (from Wingfield, J R J, *Int. J. Adhesion and Adhesives*, **13**, 151, 1993).

- contamination-free;
- stability in the operating environment;
- uniformity;
- adhesive compatibility.

A description of these qualities is given in Table 3.6. Many of these are linked to one or more other qualities.

The techniques for studying and assessing surfaces and surface quality can be classed as visual/optical, wettability, mechanical/adhesion, topographical/texture, electrical, laboratory analytical and commercially available. A best practice guide describes how users of adhesives in all industries can best assess surfaces prior to bonding (Lee *et al.*, 1998).

Table 3.6 Key surface qualities

Wettability	A measure of the attractiveness of a solid surface towards a liquid adhesive. The ideal adhesive should spread over a surface, displacing air and any contamination. To do this it should possess low surface tension and low viscosity.
Micro-roughness	The fine structure of a surface with dimensions 0.1 μm or less. On this scale, roughness increases mechanical keying and the effective area of contact with the adhesive, with consequent improvements in bond strength and durability. Chemical etches and mechanical abrasion will increase micro-roughness.
Soundness	Freedom from weak and loosely attached surface layers such as dust, dirt, mould releases, poorly adhering paint/primer and weak and crumbling oxide layers.
Stability	The stability of surface layers and oxides, towards water, organic compounds and elevated temperatures, as a function of time following treatment. The hydration of oxide layers on many metal surfaces is responsible for bond strength reduction over time. Pretreatments can be used to increase stability.
Contamination-free	Absence of foreign matter, both on a treated surface (i.e. cleanliness), or which could migrate through the bulk to a bonded interface with time (e.g. plasticisers in plastics and salts in masonry).
Uniformity	Visible or measurable consistency of the other characteristics and of the regularity of a treated surface area.
Adhesive compatibility	Particular adhesive material compatibility with particular substrate materials. If one component of the adhesive should react preferentially with a substrate, this can damage the latter and may result in failure of the adhesive to cure. On the other hand, some adhesives will cure only in the presence of metal surfaces.

The force required to break an adhesive bonded joint is resisted by a complex interaction of internally generated strains. Attempts to use short-term joint strengths as a measure of adhesion/surface preparation are misleading, because theoretical adhesive strengths will always be greater than experimentally measured joint strengths. It is widely acknowledged that most experimental assessments of the effects of adhesion and/or surface treatments are of limited value unless conducted after a period of exposure of joints to ageing in a wet environment (Mays and Hutchinson, 1992). Immersion of joints in deionised water at room temperature can be very useful. Suitable comparative mechanical/adhesion tests subject the bonded interface to tensile, peel or cleavage forces. Thus pull-off tests, lap shear tests, peel tests and fracture energy tests are used, depending on the adhesives and substrates involved. An externally applied constant or cyclic stress can be applied to joints to achieve a degree of acceleration. It is important to note, or monitor with time, the locus of failure since adhesion failures are of interest.

An example of the effect of different methods of surface preparation of GFRP on five key surface properties is given by Hutchinson (1997). Simple examples of assessment of surface condition include visual inspection, wettability tests using droplets of water and soundness tests using adhesive tape to detect the presence of dust. So-called 'Dyne pens' are available that can be used to establish whether or not the energy of a surface exceeds a value related to the surface tension of the ink in the pen. Such pens are used routinely in industry on smooth non-absorbent surfaces, such as thermoplastics, but their potential for use on concrete, metals and timber is doubtful.

There are generally established correlations between surface preparation methods (and characteristics) and in-service performance. However, the only direct link between the results generated and bond strength is provided by mechanical/adhesion test methods (Kinloch, 1983). Suitable procedures and protocols are recommended for the initial materials system *qualification* stage in construction (see Section 3.1) (www.compclass.org).

3.10 The bonding operation

A detailed specification should take care of the essential elements of workmanship and installation. Training of site operatives and their supervisors is essential because the quality and integrity of the bonded joints is largely in the hands of the operatives. Advice is available specific to the strengthening of concrete, metallic and timber structures (TRADA, 1995; Hollaway and Leeming, 1999; Cadei *et al.*, 2004; Concrete Society, 2004). The aspects that impinge directly on adhesion and joint quality are:

- evaluation of surface condition;
- surface preparation;
- protection of the working environment;
- mixing, application and curing of the adhesive;
- quality control samples.

The initial investigation of the condition of the structure should include a thorough inspection of the surface areas on which the bonding will be undertaken. Assessments of soundness should be undertaken and any requirements for defect repairs should be identified. Trials of the adhesion of candidate adhesive/primer systems should also be undertaken at this stage.

Table 3.7 Test methods recommended for the classification of adhesive and resin systems (Source: www.compclass.org)

Performance group	Performance characteristic	Test method	Description	Short-term	Long-term
Adhesion and Durability	Adhesion to FRP	ASTM D 3163[1] or EN 1542[2]	Tensile lap shear or pull-off	■	■
	Adhesion to metal	ASTM D3762[3] or EN 1542[2]	Crack propagation or pull-off	■	■
	Adhesion to concrete	EN 1542[2]	Pull-off	■	■
	Adhesion to timber	ASTM D3931[4] or EN 1542[2]	Compressive lap shear or pull-off	■	■

[1] ASTM, 2001b
[2] CEN, 1999
[3] ASTM, 2003b
[4] ASTM, 2005b

Precise details of the surface preparation procedures must be provided in the contract specification. These should be informed by laboratory trials used to establish control values for the materials system selected. Table 3.7 shows the test methods that are recommended for classifying and qualifying adhesive and resin systems for flexural strengthening of concrete, metals and timber (www.compclass.org); tests to establish adhesion to the FRP strengthening material are included. For example, a particular primer/adhesive combination may be established from tests on representative samples of substrate prepared in a way that mimics what is achievable on site. It is recommended that both short-term control values *and* long-term data be collected. The long-term data should be appropriate to the nature of the specific contract but, in the absence of guidance, immersion in deionised water for periods of at least 28 days is recommended. Particular surface qualities such as soundness, wettability and freedom from contamination can be measured on surfaces prepared on site and the results compared against recommended criteria.

The working environment should be dry and as clean as possible. Many adhesives will not cure below about 5 °C, so attention must be paid to controlling both the ambient temperature and the temperature of the component surfaces. For example, a warm adhesive applied to a very cold surface, even if properly prepared, may not adhere to it adequately because of the tendency to 'skin' due to thermal shock. Infra-red heaters, space heaters and heating blankets can be used both to warm the components prior to bonding and to assist with curing post-bonding. Dehumidifiers may also need to be used if the dew point is in excess of 10 °C. The surfaces of components to be bonded must be dry, and this can be achieved by proper attention to storage, an enclosed working environment and by physically drying or preheating the components.

Where two component adhesives are mixed together in a can or container, there is a finite usable life or pot-life. The viscosity, and therefore ease of application, of adhesive material can vary significantly with temperature. Adhesives with the correct characteristics at 20 °C may be impossibly viscous at 5–10 °C, or too fluid at 30 °C. It is important to spread the adhesive soon after mixing to dissipate the heat generated and extend its usable life. It is normal for the adhesive to be applied to both the parent substrate and the FRP. Curing and hardening can sometimes represent a significant process in scheduling and completing the bondline operation, giving rise to the need for enclosure and artificial heating of the working environment.

Quality control samples (Table 3.7) should be made carefully on site in accordance with the contract specification. It is vital that the quality of these samples is high, or else test data will be of limited use, and that they are tested by an experienced independent test house. These samples should include joints such as lap shear and pull-off specimens that can be tested to assess the adequacy of surface preparation achieved on site. The data can then be compared with the control data.

3.11 Quality assurance testing and acceptance criteria

The logic implicit in a *qualification scheme* is that the data obtained from QC testing can be compared directly with control data and with design input values. It is recommended that pull-off tests be conducted according to EN 1542 (CEN, 1999) for all substrate material; a minimum of three tests is suggested. For concrete, masonry and timber substrates the diameter of the steel dolly should be 50 mm; this should be reduced to 25 mm for metallic substrates. The acceptance criterion (or acceptable value) should be included in the specification produced by the designer. Two pieces of information should be recorded for bonded joints: strength and locus of failure. The acceptance criterion for strength is given by (www.compclass.org):

average of measured (strength) values > characteristic value

The characteristic value is defined as the mean value, derived from short-term laboratory control specimens, minus two standard deviations. If the average measured value lies between two and four standard deviations of the expected mean value then a decision on acceptability should be taken between the designer, client representative and contractor. If the average measured value is less than four standard deviations of the expected mean value, then the prepared surface should be rejected and the bonding operation repeated for the area affected.

The locus of failure should never be entirely at any substrate–adhesive, substrate–primer or adhesive–primer interface. The mechanical characteristics of different substrate materials mean that failure may be driven towards particular interfaces, but a clean substrate surface should never be exhibited. Control joints should always exhibit more than 75% cohesive failure of the adhesive layer; QC joints from site should exhibit a cohesive failure that remains above 50% (www.compclass.org).

3.12 Summary

Structural adhesives have a long history of use in construction, with epoxy adhesives having been exploited for externally-bonded steel plate reinforcement since the mid-1960s. Adhesive bonding represents the natural method for joining together dissimilar materials such as concrete, or metals or timber and polymer composites.

Control of the adhesive bonding operation is crucial to the satisfactory fabrication of all reliable and durable bonded joints. There are many aspects which must be considered, including the storage of materials, protection of the working environment, surface preparation of the components, adhesive mixing, dispensing and application, joint fit-up, bondline thickness control and curing. When bonding metals in particular, one of the most difficult aspects to control on site is surface preparation. However, it is essential to achieve a high standard of surface preparation in order to ensure long-term integrity and durability. This is facilitated by ensuring that operatives, and their supervisors, undergo training in adhesive bonding so that a clear understanding is developed of the standard of surface preparation that must be achieved.

3.13 References

ADAMS R D COMYN J and WAKE W C (1997) *Structural Adhesive Joints in Engineering*, London, UK, Chapman and Hall.

ASTM (2001a) *ASTM D2651-01. Standard Guide for Preparation of Metal Surfaces for Adhesive Bonding*, West Conshohocken, PA, USA, American Society of Testing and Materials.

ASTM (2001b) *ASTM D3163-01 Standard Test Method for Determining Strength of Adhesively Bonded Rigid Plastic Lap-Shear Joints in Shear by Tension Loading*, West Conshohocken, PA, USA, American Society for Testing and Materials.

ASTM (2003a) *ASTM D2093-03. Standard Practice for Preparation of Surfaces of Plastics Prior to Adhesive Bonding*, West Conshohocken, PA, USA, American Society for Testing and Materials.

ASTM (2003b) *ASTM D3762-03 Standard Test Method for Adhesive-Bonded Surface Durability of Aluminum (Wedge Test)*, West Conshohocken, PA, USA, American Society for Testing and Materials.

ASTM (2005a) *ASTM D4258-05. Standard Practice for Abrading Concrete*, West Conshohoken, PA, USA, American Society for Testing and Materials.

ASTM (2005b) *ASTM D3931-93a (2005) Standard Test Method for Determining Strength of Gap-Filling Adhesive Bonds in Shear by Compression Loading*, West Conshohocken, PA, USA, American Society for Testing and Materials.

BOWDITCH M R and STANNARD K J (1980) Bonding GRP with acrylic adhesives, in Allen K W (ed.) *Adhesion 5*, London, UK, Applied Science, 93–117.

BREWIS D M (ed.) (1982) *Surface Analysis and Pretreatment of Plastics and Metals*, London, UK, *Applied Science*.

BROUGHTON J G and HUTCHINSON A R (2001a) Pull-out behaviour of steel rods bonded into timber, *Materials and Structures*, **34**(236), 100–109.

BROUGHTON J G and HUTCHINSON A R (2001b) Adhesives systems for structural connections in timber, *International Journal of Adhesion and Adhesives*, **21**, 177–186.

BROUGHTON J G and HUTCHINSON A R (2001c) Effect of timber moisture content on bonded-in rods, *Construction and Building Materials*, **15**, 17–25.

BSI (1995) *BS EN 1840: 1995 Structural Adhesives – Guidelines for the Surface Preparation of Plastics*, London, UK, British Standards Institution.

BSI (1997) *BS EN 12768: 1997 Structural Adhesive – Guidelines for Surface Preparation of Metals*, London, UK, British Standards Institution.

CADEI J M C STRATFORD T J HOLLAWAY L C and DUCKETT W G (2004) *Strengthening Metallic Structures using Externally Bonded Fibre-Reinforced Polymers*, Report C595, London, UK, CIRIA.

CLARKE J L (ed.) (1996) *Structural Design of Polymer Composites: EUROCOMP Design Code and Handbook*, London, UK, E & F N Spon.

CONCRETE SOCIETY (2004) *Design Guidance for Strengthening Concrete Structures Using Fibre Composite Materials*, TR55, 2nd edn, Camberley, UK, The Concrete Society.

DAVIS G (1997) The performance of adhesive systems for structural timbers, *International Journal of Adhesion and Adhesives*, **17**, 247–255.

DTI (1993) *Review of Substrate Surface Treatments*, MTS Project 4, Measurement and Standards Programme on the Performance of Adhesive Joints, 1993–1996, Report No 2, London, UK, Department of Trade and Industry.

EDWARDS S C (1993). Surface coatings in Allen R T L Edwards S C and Shaw J D N (eds), *Repair of Concrete Structures*, 2nd edn, London, UK, Blackie Academic and Professional.

CEN (1999) *EN 1542: 1999 Products and systems for the protection and repair of concrete structures. Test Methods. Measurement of bond strength by pull-off*, Brussels, Belgium, European Committee for Standardisation.

GAUL R W (1984) Preparing concrete surfaces for coatings, *Concrete International*, **6**(7), 17–22.

HARRIS A F and BEEVERS A (1999) The effects of grit-blasting on surface properties for adhesion', *International Journal of Adhesion and Adhesives*, **19**, 445–452.

HOLLAWAY L C and LEEMING M B (eds) (1999) *Strengthening of Reinforced Concrete Structures*, Cambridge, UK, Woodhead.

HUTCHINSON A R (1993) Review of methods for the surface treatment of concrete, in DTI, *Characterisation of Surface Condition*, MTS Project 4, Report No 2 on the Performance of Adhesive Joints, London, UK, Department of Trade and Industry.

HUTCHINSON A R (1997) *Joining of Fibre-Reinforced Polymer Composite Materials*, Report 46, London, UK, CIRIA.

HUTCHINSON A R and RAHIMI H (1993) Behaviour of reinforced concrete beams with externally bonded fibre reinforced plastics, in Forde M C (ed.), *Proceedings Fifth International Conference on Structural Faults and Repair*, Edinburgh, UK, Engineering Technics Press, Vol. III, 221–228.

HUTCHINSON A R and HOLLAWAY L C (1999) Environmental durability, in Hollaway L and Leeming M B (eds), *Strengthening of Reinforced Concrete Structures*, Cambridge, UK, Woodhead, 156–182.

HUTCHINSON A R and HURLEY S A (2001) *Transfer of Adhesives Technology*, Project Report 84, London, UK, CIRIA.

ISO (2000) *ISO 8504-1: 2000 Preparation of steel substrates before application of paints and related products – Surface preparation methods – Part 1: General principles*, Geneva, Switzerland, International Organization for Standardization.

ISO (2007) *ISO 8501-1: 2007 Preparation of steel substrates before application of paints and related products – Visual assessment of surface cleanliness – Part 1: Rust grades and preparation grades of uncoated steel substrates and of steel substrates after overall removal of previous coatings*, Geneva, Switzerland, International Organization for Standardization.

KINLOCH A J (1983) *Durability of Structural Adhesives*, London, UK, Applied Science.

LEE R J, HANCOX N L and HUTCHINSON A R (1998) *The Assessment of Adherend Surface Quality: a Best Practice Guide*, London, UK, DTI Materials Metrology Programme.

MAYS G C and HUTCHINSON A R (1992) *Adhesives in Civil Engineering*, Cambridge, UK, Cambridge University Press.

MITTAL K L (ed.) (1992) *Silanes and Other Coupling Agents*, Utrecht, The Netherlands, VSP.

NPL (2000) *Surface Preparation for Coating: Guides to Good Practice in Corrosion Control*, London, UK, National Physical Laboratory.

PACKHAM D E (ed.) (2005) *Handbook of Adhesion*, 2nd edn, London, UK, John Wiley and Sons.

PLUEDDEMANN E P (1991) *Silane Coupling Agents*, 2nd edn, New York, USA, Plenum Press.

RAHIMI H and HUTCHINSON A R (2001). Behaviour of reinforced concrete beams with externally bonded fibre-reinforced plastic, *Journal of Composites for Construction, ASME*, **5**(1), 44–56.

RIVER B H, GILLESPIE R H and VICK C B (1991) Timber, in Minford J D (ed.) *Treatise on Adhesion and Adhesives*, Vol. 7, New York, USA, Marcel Dekker, 1–230.

SASSE H R (ed.) (1986) *Adhesion Between Polymers and Concrete*, London, UK, Chapman and Hall.

SASSE H R and FIEBRICH M (1983) Bonding of polymer materials to concrete, *RILEM Materials and Structures*, **16**(94), 293–301.

SIS (1967), *SIS 05 59 00, Pictorial surface preparation standards for painting steel surfaces*, Stockholm, Sweden, Swedish Standards Institution.

STRONGWELL DESIGN MANUAL (2008) (http://www.strongwell.com/designmanual), Strongwell, USA.

SYKES J M (1982) Surface treatments for steel, in Brewis D M (ed.) *Surface Analysis and Pre-treatment of Plastics and Metals* London, UK, Applied Science, 153–174.

TRADA (1995) *Resin-bonded Repair Systems for Structural Timber*, Wood Information Sheet 22, Section 4, High Wycombe, UK, Timber Research and Development Association.

WHEELER A S and HUTCHINSON A R (1998) Resin repairs to timber structures, *International Jaurnal of Adhesion and Adhesives*, **18**, 1–13.

WINFIELD P H, HARRIS A F and HUTCHINSON A R (2001) The use of flame ionisation technology to improve the wettability and adhesion properties of wood, *International Jaunral of Adhesion and Adhesives*, **21**, 107–114.

WINGFIELD J R J (1993) Treatment of composite surfaces for adhesive bonding, *International Journal of Adhesion and Adhesives*, **13**, 151–156.

4
Flexural strengthening of reinforced concrete (RC) beams with fibre-reinforced polymer (FRP) composites

J. G. TENG, The Hong Kong Polytechnic University, China;
S. T. SMITH, The University of Hong Kong, China;
J. F. CHEN, The University of Edinburgh, UK

4.1 Introduction

This chapter is concerned with the behaviour and design of fibre-reinforced polymer (FRP) plates (pultruded or wet lay-up plates) for the strengthening of reinforced concrete (RC) beams. The focus of the chapter is on RC beams strengthened with an FRP plate without prestressing or mechanical anchorage (Fig. 4.1). This method of strengthening RC beams was first researched in the mid-1980s at the Swiss Federal Laboratory for Materials Testing and Research (EMPA) (Meier *et al.*, 1993), but most of the research on FRP plate bonding for flexural strengthening has been carried out over the past 15 years. Detailed reviews of the extensive research conducted so far can be found in the open literature (e.g. Hollaway and Leeming, 1999; Teng *et al.*, 2002a; Täljsten, 2003; Oehlers and Seracino, 2004; Bank, 2006; ACI, 2007). As a result of extensive research to date, a reasonably detailed understanding has been obtained and comprehensive design theory developed. In practical applications, cautionary anchorage measures are often used at the plate ends, but they are generally not explicitly accounted for in design calculations. Prestressing the FRP plate does bring benefits such as the improvement of the serviceability of the beam, but difficulties still exist with *in situ* prestressing and anchoring operations, so the prestressing technique has not received wide acceptance in practice.

The chapter starts with a description of the different failure modes. The flexural strength equations for an FRP-plated section based on the plane section assumption are then presented. Interfacial stresses and bond behaviour between the FRP plate and the concrete substrate are next discussed, followed by a summary of strength models for debonding failures. Finally, a design procedure for the flexural strengthening of RC beams with FRP is given. All discussions in this chapter refer to a simply supported beam except at the end of the chapter where the application of the design equations presented in the chapter to indeterminate beams are explained.

Flexural strengthening of RC beams with FRP composites   113

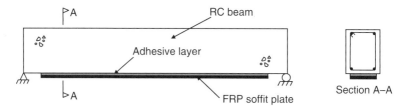

4.1 RC beam with an FRP plate bonded to its soffit.

This chapter and the three chapters that follow represent a succinct treatment of the behaviour and design of RC beams and columns strengthened with FRP composites. In line with the aim of the book as pointed out in the preface, these chapters have been written with the practitioners (including designers, engineers and contractors) in mind. That is, these chapters aim to present information of direct interest to practitioners based on the best understanding of the authors who have all worked together closely on the subject instead of providing a comprehensive or exhaustive summary of all published research. As a result, the material presented in this and the three subsequent chapters is, to a large extent, directly based on the research of the authors and their collaborators.

4.2   Failure modes

4.2.1 General

A number of distinct failure modes of RC beams bonded with an FRP soffit plate have been observed in numerous experimental studies (Teng *et al.*, 2002a; Oehlers and Seracino, 2004). A schematic representation of these failure modes is shown in Figs 4.2 and 4.3. Failure of an FRP-plated RC

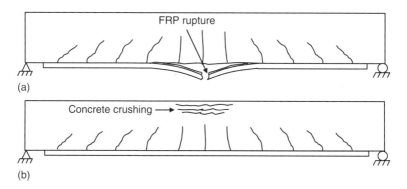

4.2 Conventional flexural failure modes of an FRP-plated RC beam: (a) FRP rupture; (b) crushing of compressive concrete.

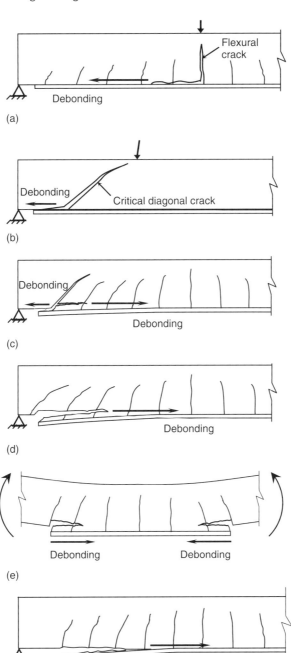

4.3 Debonding failure modes of an FRP-plated RC beam: (a) IC debonding; (b) CDC debonding; (c) CDC debonding with concrete cover separation; (d) concrete cover separation; (e) concrete cover separation under pure bending; (f) plate end interfacial debonding.

beam may be by the flexural failure of the critical section (Fig. 4.2) or by debonding of the FRP plate from the RC beam (Fig. 4.3). In the former type of failure, the composite action between the bonded plate and the RC beam is maintained up to the failure of the critical section, while the latter type of failure involves a loss of this composite action. Debonding failures generally occur in the concrete, which is also assumed in the design theory presented in this chapter. This is because, with the strong adhesives currently available and with appropriate surface preparation for the concrete substrate, debonding failures along the physical interfaces between the adhesive and the concrete and between the adhesive and the FRP plate are generally not critical.

Debonding may initiate at a flexural or flexural-shear crack in the high moment region and then propagate towards one of the plate ends (Fig. 4.3a). This debonding failure mode is commonly referred to as intermediate crack (IC) induced interfacial debonding (or simply IC debonding) (Teng *et al.*, 2002a, 2003; Lu *et al.*, 2007). Debonding may also occur at or near a plate end (i.e. plate end debonding failures) in four different modes: (i) critical diagonal crack (CDC) debonding (Fig. 4.3b) (Oehlers and Seracino, 2004); (ii) CDC debonding with concrete cover separation (Fig. 4.3c) (Yao and Teng, 2007); (iii) concrete cover separation (Figs 4.3d and 4.3e) (Teng *et al.*, 2002a); and (iv) plate end interfacial debonding (Fig. 4.3f) (Teng *et al.*, 2002a).

4.2.2 Flexural failure modes of an FRP-plated RC beam section

The flexural failure of an FRP-plated RC section can be in one of two modes: tensile rupture of the FRP plate (Figs 4.2a and 4.4) or compressive crushing of the concrete (Fig. 4.2b). These modes are very similar to the classical flexural failure modes of RC beams, except for small differences due to the brittleness of the bonded FRP plate. FRP rupture generally occurs following the yielding of the steel tension bars, although steel yielding may not have been reached if the steel bars are located far away from the tension face.

Figure 4.5 shows a typical load–deflection response of a simply-supported RC beam bonded with an FRP plate subjected to four point bending (Hau, 1999). In this particular beam, the plate was terminated very close to the supports and the beam failed in flexure by FRP rupture. Compared to the corresponding response of a control RC beam, the strengthened beam achieved a strength increase of 76%, but showed much reduced ductility. The strength increase and the ductility decrease are the two main consequences of flexural strengthening of RC beams using FRP plates. FRP-strengthened beams which fail by concrete crushing when a large amount

4.4 FRP-plated RC beam: FRP rupture.

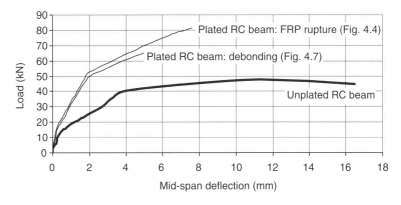

4.5 Typical load–deflection curves of plated and unplated RC beams.

of FRP is used also show much reduced ductility (Buyukozturk and Hearing, 1998).

4.2.3 Intermediate crack-induced interfacial debonding

When a major flexural or flexural-shear crack is formed in the concrete, the need to accommodate the large local strain concentration at the crack leads to immediate but very localised debonding of the FRP plate from the concrete in the close vicinity of the crack, but this localised debonding is not yet able to propagate. The tensile stresses released by the cracked concrete are transferred to the FRP plate and steel rebars, so high local interfacial stresses between the FRP plate and the concrete are induced near the crack. As the applied loading increases further, the tensile stresses in the plate and hence the interfacial stresses between the FRP plate and the concrete near the crack also increase. When these stresses reach critical values, debonding starts to propagate towards one of the plate ends, generally the nearer end.

Flexural strengthening of RC beams with FRP composites 117

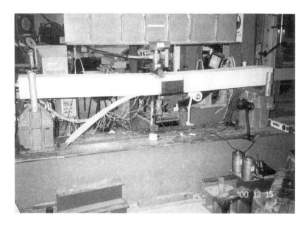

4.6 FRP-plated RC beam: intermediate flexural crack-induced interfacial debonding.

A typical picture of flexural crack-induced debonding is shown in Fig. 4.6. A thin layer of concrete remains attached to the plate (Fig. 4.6), which suggests that failure occurred in the concrete, adjacent to the adhesive-to-concrete interface. Intermediate crack-induced interfacial debonding failures are more likely in shallow beams and are, in general, more ductile than plate end debonding failures.

4.2.4 Concrete cover separation

Concrete cover separation involves crack propagation along the level of the steel tension reinforcement. Failure of the concrete cover is initiated by the formation of a crack near the plate end. The crack propagates to and then along the level of the steel tension reinforcement, resulting in the separation of the concrete cover. As the failure occurs away from the bondline, this is not a debonding failure mode in strict terms, although it is closely associated with stress concentration near the ends of the bonded plate. A typical picture of a cover separation failure is shown in Fig. 4.7a. Figure 4.7b shows a close-up view of the detached plate end, where the flexural tension reinforcement of the beam can be clearly seen. The cover separation failure mode is a rather brittle failure mode (Fig. 4.5).

4.2.5 Plate-end interfacial debonding

A debonding failure of this form is initiated by high interfacial shear and normal stresses near the end of the plate that exceed the strength of the weakest element, generally the concrete. Debonding initiates at the plate end and propagates towards the middle of the beam (Figs 4.2f and 4.8). This

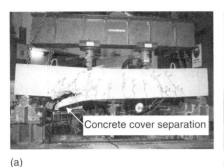
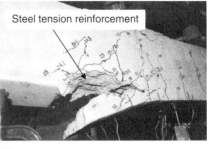

4.7 FRP-plated RC beam: concrete cover separation: (a) overall view; (b) close-up view.

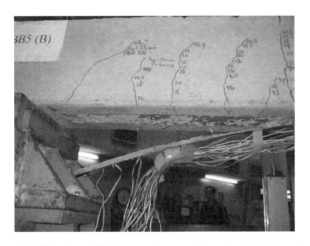

4.8 FRP-plated RC beam: plate-end interfacial debonding.

failure mode is only likely to occur when the plate is significantly narrower than the beam section as, otherwise, failure tends to be by concrete cover separation (i.e. the steel bars–concrete interface controls the failure instead).

4.2.6 Critical diagonal crack-induced interfacial debonding (CDC debonding)

This mode of debonding failure occurs in flexurally-strengthened beams where the plate end is located in a zone of high shear force but low moment (e.g. a plate end near the support of a simply-supported beam) and the amount of steel shear reinforcement is limited. In such beams, a major

Flexural strengthening of RC beams with FRP composites 119

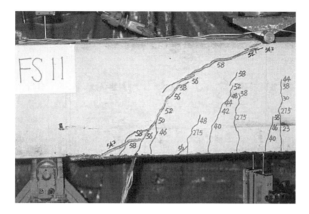

4.9 FRP-plated RC beam: CDC debonding.

diagonal shear crack (critical diagonal crack, or CDC) forms and intersects the FRP plate, generally near the plate end. As the CDC crack widens, high interfacial stresses between the plate and the concrete are induced, leading to the eventual failure of the beam by debonding of the plate from the concrete; the debonding crack propagates from the CDC crack towards the plate end (Fig. 4.9).

In a beam with a larger amount of steel shear reinforcement, multiple shear cracks of smaller widths instead of a single major shear crack dominate the behaviour, so CDC debonding is much less likely. Instead, cover separation takes over as the controlling debonding failure mode. In other cases, particularly when the plate end is very close to the zero-moment location, CDC debonding leads only to the local detachment of the plate end, but the beam is able to resist higher loads until cover separation occurs. The local detachment due to CDC debonding effectively moves the plate end to a new location with a larger moment, and cover separation then starts from this 'new end'. The CDC failure mode is thus related to the cover separation failure mode. It is surmised that, if a flexurally-strengthened beam is also shear-strengthened with U-jackets to ensure that the shear strength remains greater than the flexural strength, the CDC debonding failure mode may be suppressed.

4.2.7 Other aspects of debonding

The risk of debonding is increased by a number of factors associated with the quality of on-site application. These include poor workmanship and the use of inferior adhesives. The effects of these factors can be minimised if due care is exercised in the application process to ensure that debonding failure is controlled by concrete. In addition, small unevenness of the concrete surface may cause localised debonding of the FRP plate, but it is unlikely to cause the FRP plate to be completely detached.

4.3 Flexural strength of an FRP-plated section

4.3.1 General

The ultimate moment of an FRP-plated RC beam section can be calculated using the conventional RC beam design approach with appropriate modifications to account for the brittle nature of the externally bonded FRP plate (Teng *et al.*, 2002a). The following presentation is based on the work of Teng *et al.*, (2000) where the flexural strength equations were derived within the framework of BS 8110 (BSI, 1997) and using the design stress–strain curve for concrete, but similar equations using other stress–strain curves can be easily derived following the same general approach. The key assumption in developing the flexural strength design equations is that a plane section remains plane.

These design equations consider only flexural failure of the plated beam section by either FRP rupture (Fig. 4.2a) or concrete crushing (Fig. 4.2b) without premature debonding failure. The preferred modes of failure to be designed for are (i) concrete crushing following yielding of steel tension reinforcement and (ii) FRP rupture following yielding of steel reinforcement. In both modes, yielding of the steel tension reinforcement precedes failure by either concrete crushing or rupture of the FRP, which ensures that failure will occur after the formation of significant flexural cracks to give desirable warning of failure, despite the fact that these modes generally show limited ductility. Failure by concrete crushing or FRP rupture (particularly when the rupture strain is small as is the case for high-modulus CFRP) without yielding of steel reinforcement should be avoided as much as possible.

4.3.2 Design equations

Following BS 8110 (BSI, 1997), the ultimate strain at the extreme concrete compression fibre is taken to be 0.0035. The stress–strain curve for concrete adopted by BS 8110 (BSI, 1997) and more accurately defined in Kong and Evans (1987) is given in Fig. 4.10. As the behaviour of the FRP reinforcement is brittle, this ultimate compressive strain of concrete may not have been reached when the FRP reinforcement fails by rupture. Thus, the simplified rectangular stress block of the code (BS 8110, 1997) for the compression concrete is no longer valid. The strains and stresses over the depth of a plated beam are shown in Fig. 4.11.

Flexural strengthening of RC beams with FRP composites 121

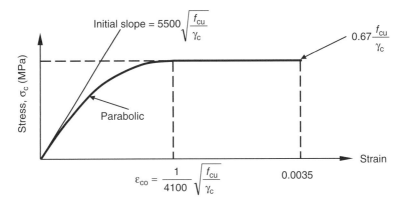

4.10 Stress–strain curve of concrete for design use.

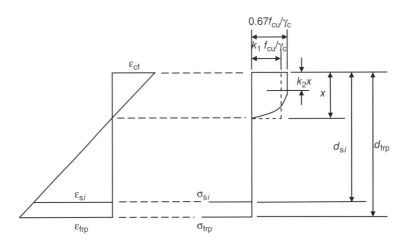

4.11 Strains and stresses over beam depth.

Compressive strains are taken to be positive here, so the strains in the compression concrete are positive, while the strains in the FRP are negative. The sign of strains in the reinforcing bars depends on their position in the beam. Based on the plane section assumption, the strains in the FRP ε_{frp} and those in the steel bars ε_{si} are related to the extreme compression fibre strain of concrete ε_{cf} as follows:

$$\varepsilon_{frp} = \varepsilon_{cf} \frac{x - d_{frp}}{x} \qquad [4.1]$$

and

$$\varepsilon_{si} = \varepsilon_{cf} \frac{x - d_{si}}{x} \qquad [4.2]$$

where x, d_{si} and d_{frp} are distances from the extreme concrete compression fibre to the neutral axis (i.e. neutral axis depth), the centroid of steel bars in layer i and the centroid of the FRP respectively. Similarly, when the strain in the FRP ε_{frp} is known, the value of ε_{cf} can be easily found as:

$$\varepsilon_{cf} = \varepsilon_{frp} \frac{x}{x - d_{frp}} \quad [4.3]$$

According to the stress–strain curve of Fig. 4.10, compressive stresses in the concrete are given by (Kong and Evans 1987):

$$\sigma_c = 5500 \left(\sqrt{\frac{f_{cu}}{\gamma_c}} \varepsilon_c - \frac{4100}{2} \varepsilon_c^2 \right) \quad \text{if } 0 \leq \varepsilon_c \leq \varepsilon_{co} \quad [4.4]$$

and

$$\sigma_c = 0.67 \frac{f_{cu}}{\gamma_c} \quad \text{if } \varepsilon_{co} \leq \varepsilon_c \leq 0.0035 \quad [4.5]$$

where σ_c is the compressive concrete stress, ε_c is the compressive concrete strain which has a value of $\varepsilon_{co} = \frac{1}{4100} \sqrt{\frac{f_{cu}}{\gamma_c}}$ at the attainment of the peak stress of concrete, f_{cu} is the cube compressive strength of concrete and γ_c is the partial safety factor for concrete.

For any given ε_{cf}, the total concrete compression force C is

$$C = k_1 \frac{f_{cu}}{\gamma_c} b_c x \quad [4.6]$$

where b_c is the beam width and k_1 is the mean stress factor defined by:

$$k_1 = \frac{\int_0^{\varepsilon_{cf}} \sigma_c d\varepsilon_c}{\frac{f_{cu}}{\gamma_c} \varepsilon_{cf}} \quad [4.7]$$

Substituting Eqs 4.4 and 4.5 into Eq. 4.7 yields:

$$k_1 = 0.67 \left(\frac{\varepsilon_{cf}}{\varepsilon_{co}} - \frac{\varepsilon_{cf}^2}{3\varepsilon_{co}^2} \right) \quad \text{if } 0 \leq \varepsilon_{cf} \leq \varepsilon_{co} \quad [4.8a]$$

and

$$k_1 = 0.67 \left(1 - \frac{\varepsilon_{co}}{3\varepsilon_{cf}} \right) \quad \text{if } \varepsilon_{co} \leq \varepsilon_{cf} \leq 0.0035 \quad [4.8b]$$

The depth of the neutral axis x can be determined by solving the following force equilibrium equation:

$$k_1 \frac{f_{cu}}{\gamma_c} b_c x + \sum_{i=1}^{n} \sigma_{si} A_{si} + \sigma_{frp} A_{frp} = 0 \quad [4.9]$$

where σ_{si} and σ_{frp} are the stresses in the steel bars and the FRP, respectively, A_{si} is the total area of steel in layer i, n is the total number of steel layers and A_{frp} is the area of the FRP. σ_{si} and σ_{frp} are given by:

$$\sigma_{si} = E_s \varepsilon_{si} \quad \text{if } |\varepsilon_{si}| < \frac{f_y}{\gamma_s E_s} \quad [4.10a]$$

$$\sigma_{si} = \text{sgn}\,\varepsilon_{si} \frac{f_y}{\gamma_s} \quad \text{if } |\varepsilon_{si}| \geq \frac{f_y}{\gamma_s E_s} \quad [4.10b]$$

and

$$\sigma_{frp} = E_{frp} \varepsilon_{frp} \geq \frac{-f_{frp}}{\gamma_{frp}} \quad [4.11]$$

where E_s and E_{frp} are moduli of elasticity of the steel bars and the FRP, respectively, f_y is the yield strength of the steel, f_{frp} is the tensile strength of the FRP and γ_s and γ_{frp} are the partial safety factors for steel and FRP, respectively. It should be noted that f_y and f_{frp}, being material properties, are always positive.

The position of the concrete compression force C is defined by D which is the distance from the extreme concrete compression fibre to the line of action of the concrete compression force. D can be related to the height of the compression zone through:

$$D = k_2 x \quad [4.12]$$

where the centroid factor k_2 of the compression force is given by:

$$k_2 = 1 - \frac{\int_0^{\varepsilon_{cf}} \varepsilon_c \sigma_c d\varepsilon_c}{\varepsilon_{cf} \int_0^{\varepsilon_{cf}} \sigma_c d\varepsilon_c} \quad [4.13]$$

Evaluating the right hand side of Eq. 4.13 results in:

$$k_2 = \frac{\frac{1}{3} - \frac{\varepsilon_{cf}}{12\varepsilon_{co}}}{1 - \frac{\varepsilon_{cf}}{3\varepsilon_{co}}} \quad \text{if } 0 \leq \varepsilon_{cf} \leq \varepsilon_{co} \quad [4.14a]$$

and

$$k_2 = \frac{\frac{\varepsilon_{cf}}{2} + \frac{\varepsilon_{co}^2}{12\varepsilon_{cf}} - \frac{\varepsilon_{co}}{3}}{\varepsilon_{cf} - \frac{\varepsilon_{co}}{3}} \quad \text{if } \varepsilon_{co} \leq \varepsilon_{cf} \leq 0.0035 \quad [4.14b]$$

For design use, the value of k_2 can be closely approximated by the following single expression for the entire strain range:

$$k_2 = 0.33 + 0.045 \frac{\varepsilon_{cf}}{\varepsilon_{co}} \quad [4.15]$$

The moment capacity of the beam M_u is finally determined by:

$$M_u = k_1 \frac{f_{cu}}{\gamma_c} b_c x \left(\frac{h}{2} - k_2 x\right) + \sum_{i=1}^{n} \sigma_{si} A_{si} \left(\frac{h}{2} - d_{si}\right)$$

$$+ \sigma_{frp} A_{frp} \left(\frac{h}{2} - d_{frp}\right) \quad [4.16]$$

where h is the depth of the RC beam. In Eq. 4.16, the three terms on the right-hand side represent the contributions of the compression concrete, the steel bars and the FRP, respectively.

The beam is deemed to have reached failure when either the concrete compression strain attains the maximum usable strain 0.0035 according to BS 8110 (BSI, 1997) and/or the FRP reaches its rupture strain $\varepsilon_{frp,rup} = f_{frp}/(\gamma_{frp} E_{frp})$. In practical design, the mode of failure should first be determined. Assuming that in a balanced design both the compression concrete and the FRP reach their respective failure states, then $\varepsilon_{cf} = 0.0035$ and $\varepsilon_{frp} = -\varepsilon_{frp,rup}$. The critical FRP ratio $\rho_{frp,cr}$ for a balanced section can be found from the equilibrium equation (Eq. 4.9) to be:

$$\rho_{frp,cr} = -\frac{k_1 \dfrac{f_{cu}}{\gamma_c} \dfrac{x_{cr}}{h} + \sum_{i=1}^{n} \sigma_{si} \rho_{si}}{\dfrac{f_{frp}}{\gamma_{frp}}} \quad [4.17]$$

where ρ_{si} is the steel reinforcement ratio of layer i, and x_{cr} is the critical depth of the neutral axis, given by:

$$x_{cr} = \frac{0.0035}{0.0035 + \dfrac{f_{frp}}{\gamma_{frp} E_{frp}}} d_{frp} \quad [4.18]$$

If $\rho_{frp} > \rho_{frp,cr}$, the beam fails by concrete crushing, but if $\rho_{frp} < \rho_{frp,cr}$ the beam fails by FRP rupture. It should be noted that both ρ_{si} and ρ_{frp} are defined here using the gross area of the cross-section.

In the case of concrete crushing, $\varepsilon_{cf} = 0.0035$ and ε_{frp} is found from Eq. 4.1. The mean stress factor k_1 is then directly obtained from Eq. 4.8b. The depth of the neutral axis x can be determined by solving Eq. 4.9 making use of Eqs 4.1, 4.2, and 4.10–4.12.

In the case of FRP rupture, $\varepsilon_{frp} = -\varepsilon_{frp,rup}$. The depth of the neutral axis x is first assumed a value so that Eq. 4.3 leads to an ε_{cf} value between ε_{co} and 0.0035. The mean stress factor k_1 is then obtained using Eq. 4.8b, which is then substituted into Eq. 4.9 to see if it is satisfied. The process of finding x is thus iterative and ends with an x value which satisfies Eq. 4.9. If Eq. 4.9

cannot be satisfied with $\varepsilon_{co} < \varepsilon_{cf} < 0.0035$, smaller values of x should be assumed to give $\varepsilon_{cf} \leq \varepsilon_{co}$. The mean stress factor k_1 is then calculated using Eq. 4.8a.

Hand calculations using the above design equations can be a lengthy process, so it is best to do these calculations using a computer spread sheet program such as EXCEL.

4.3.3 Presence of preloading

The preceding design equations are for beams without preloading, but preloading including self-weight is likely to exist in practical applications. Preloading leads to an initial tensile strain ε_{ini} at the tension face of the beam, where the FRP is to be bonded. Lam and Teng (2001) have examined the effect of preloading on slabs and their conclusions are equally applicable to RC beams. These conclusions, adapted for beams, are: (i) the effect of preloading due to self-weight and service loads is generally beneficial if the beam fails by FRP rupture, but this effect is generally insignificant, and (ii) the effect of preloading is more significant and detrimental if the beam fails by concrete crushing and should be investigated in design calculations. The effect of preloading should therefore be considered if the strengthened section is found to fail by concrete crushing. It should be noted that the failure mode of FRP-plated RC beams may change from FRP rupture to concrete crushing as a result of preloading. For such beams, the effect of preloading should also be considered in design.

The effect of preloading can be considered in strengthening design by modifying the strain value in the FRP as has been suggested by Saadatmanesh and Malek (1998). An alternative interpretation of their procedure is suggested here which employs a modified stress–strain curve of the FRP. This procedure involves two steps: (i) a cracked section analysis based on the assumptions of the design equations presented above is carried out to determine the strain ε_{ini} of the tension face of the beam without the bonded FRP at the critical section, and (ii) the strength of the beam with the bonded FRP is then evaluated using the design equations presented above, except that the modified stress–strain curve of FRP given in Fig. 4.12 should be used in calculations, which requires Eq. 4.11 to be replaced by:

$$\sigma_{frp} = 0 \quad \text{if } \varepsilon_{frp} \geq \varepsilon_{ini} \quad [4.19a]$$

and

$$\sigma_{frp} = E_{frp}(\varepsilon_{frp} - \varepsilon_{ini}) \geq \frac{-f_{frp}}{\gamma_{frp}} \quad [4.19b]$$

Tensile rupture of FRP now occurs at the following modified rupture strain:

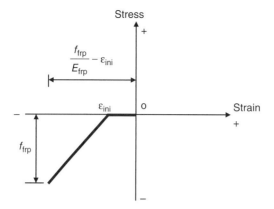

4.12 Modified FRP stress–strain curve for flexural design with preloading.

$$\varepsilon_{frp,rup} = \frac{f_{frp}}{\gamma_{frp} E_{frp}} - \varepsilon_{ini} \qquad [4.20]$$

which is greater than the original rupture strain as ε_{ini} is negative. The effect of preloading is therefore accounted for by a modified stress–strain curve for the FRP without affecting the rest of the design equations presented earlier. This procedure is conceptually very simple, as the design is basically the same as for a beam without preloading. Physically, the 'modified' beam is one which is bonded with FRP without any preloading and the FRP has no stiffness during the initial stage of straining up to a strain of ε_{ini} at the tension face of the beam.

4.4 Interfacial stresses

Many studies have shown that in an FRP-plated beam, high interfacial stresses exist between the FRP plate and the RC beam near the plate end. The two main components of interfacial stresses are the interfacial shear stress τ and the interfacial normal stress σ_y (Fig. 4.13). These high interfacial stresses play an important role in some of the debonding failure modes, including the modes of concrete cover separation and plate end interfacial debonding (Fig. 4.3). A simple analytical solution for these interfacial stresses has been presented by Smith and Teng (2001). Interfacial stresses predicted by finite element analysis show a much more complex picture (Teng *et al.*, 2002b), but results from the simple analytical solution of Smith and Teng (2001), as shown in Fig. 4.14 for a typical case, are sufficient to illustrate the stress concentration phenomenon in the vicinity of the plate end. Figure 4.14 shows that near the plate end, both the interfacial shear and normal stresses increase rapidly. For a given simply-supported beam

Flexural strengthening of RC beams with FRP composites   127

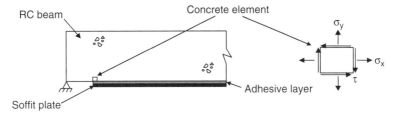

4.13 Stresses acting on a concrete element adjacent to the plate end.

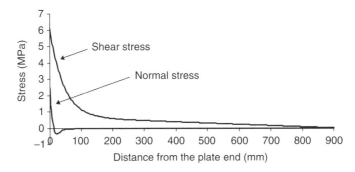

4.14 Typical interfacial shear and normal stress distributions.

under transverse loading, the magnitudes of these stresses increase with the distance between the support and the plate end, with both the elastic modulus and the thickness of the plate, and with the elastic modulus of the adhesive layer, but decrease with the thickness of the adhesive layer (Shen *et al.*, 2001; Teng *et al.*, 2002b).

4.5   Bond behaviour

4.5.1 General

A good understanding of the bond behaviour between the FRP plate and the substrate concrete is of great importance for understanding and predicting the debonding behaviour of FRP-plated RC beams. Bond behaviour between FRP and concrete has been widely studied experimentally using simple pull-off tests or using theoretical/finite element models (e.g. Chen and Teng, 2001; Wu *et al.*, 2002; Yuan *et al.*, 2004; Yao *et al.*, 2005). Figures 4.15 and 4.16 show, respectively, the schematic and a typical implementation (Yao *et al.*, 2005) of the widely used simple pull-off test. The discussions presented in this section use such a simple pull-off test as the reference case.

128   Strengthening and rehabilitation of civil infrastructures

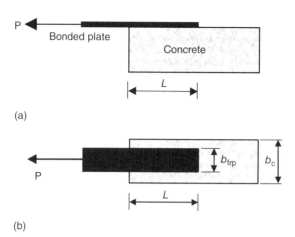

(a)

(b)

4.15 Schematic of a simple pull-off test: (a) elevation; (b) plan.

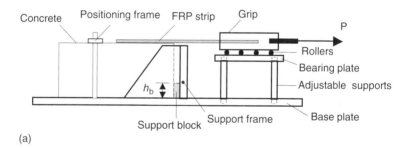

(a)

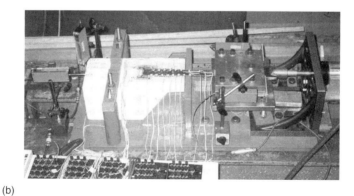

(b)

4.16 Set-up for pull-off tests: (a) elevation; (b) photograph.

4.5.2 Bond strength models

The ultimate tensile force that can be resisted by the FRP plate in a simple pull-off test before the FRP plate debonds from the concrete prism is referred to as the ultimate load or the bond strength. The bond strength is defined herein using the tensile force (or the tensile stress) in the plate instead of the average interfacial shear stress because the latter can be conceptually misleading. Existing research has shown conclusively (Chen and Teng, 2001; Yuan et al., 2004; Yao et al., 2005) that the ultimate load of a pull-off test initially increases as the bond length increases but, when the bond length reaches a threshold value, any further increase in the bond length does not lead to a further increase in the ultimate load. Therefore, when a long bond length is used, only part of the bond length is mobilised in resisting the ultimate load, so the use of an average interfacial stress referring to the entire bond length is inappropriate. This threshold value of the bond length is referred to as the effective bond length (Chen and Teng, 2001).

The fact that the bond strength cannot increase further once the bond length exceeds the effective bond length means that the ultimate tensile strength of an FRP plate may never be reached in a pull-off test, however long the bond length is. A longer bond length, however, can improve the ductility of the failure process. Most FRP-to-concrete bonded joints therefore fail by crack propagation in the concrete adjacent to the adhesive–to-concrete interface, starting from the loaded end of the plate. This phenomenon is substantially different from the bond behaviour of internal reinforcement, for which a bond length can always be designed for its full tensile strength if there is an adequate concrete cover.

Many theoretical models have been developed to predict the bond strength of FRP-to-concrete bonded joints (Chen and Teng, 2001; Lu et al., 2005). Among the existing bond strength models, the model developed by Chen and Teng (2001) has been found to provide the most accurate predictions of test results (Lu et al., 2005). Chen and Teng's (2001) bond strength model predicts that the stress in the bonded plate in MPa, to cause debonding failure in a simple pull-off test is given by:

$$\sigma_p = \alpha \beta_w \beta_L \sqrt{\frac{E_{frp} \sqrt{f'_c}}{t_{frp}}} \qquad [4.21]$$

where the width ratio factor

$$\beta_w = \sqrt{\frac{2 - b_{frp}/b_c}{1 + b_{frp}/b_c}} \qquad [4.22]$$

the bond length factor

$$\beta_L = \begin{cases} 1 & \text{if } L \geq L_e \\ \sin\left(\dfrac{\pi L}{2L_e}\right) & \text{if } L < L_e \end{cases} \quad [4.23]$$

and the effective bond length (mm)

$$L_e = \sqrt{\dfrac{E_{frp} t_{frp}}{\sqrt{f'_c}}} \quad [4.24]$$

in which E_{frp}, t_{frp} and b_{frp} are the elastic modulus (MPa), thickness (mm) and width (mm) of the FRP plate, respectively, f'_c and b_c are the concrete cylinder compressive strength (MPa) and width (mm) of the concrete block, respectively, and L is the bond length (mm). A value of 0.427 for α was found by Chen and Teng (2001) to provide a best fit of the test data gathered by them, while a value of 0.315 provides a 95 percentile lower bound which is suitable for use in ultimate limit state design.

4.5.3 Bond-slip models

An accurate bond-slip model for FRP-to-concrete interfaces is important for understanding and modelling the behaviour of FRP-strengthened RC structures. Lu et al., (2005) conducted a thorough review of bond-slip models and then proposed a set of three models of different levels of sophistication: the precise model, the simplified model and the bilinear model (Fig. 4.17). The bilinear model is the easiest to implement, without a significant loss of

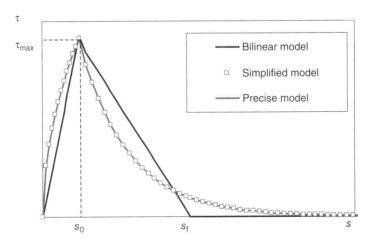

4.17 Lu et al.'s bilinear bond-slip models.

accuracy compared to the precise and simplified models and is defined by the following equations:

$$\tau = \tau_{max} \frac{s}{s_0} \quad \text{if } s \leq s_0 \qquad [4.25]$$

$$\tau = \tau_{max} \frac{s_f - s}{s_f - s_0} \quad \text{if } s_0 < s \leq s_f \qquad [4.26]$$

and

$$\tau = 0 \quad \text{if } s > s_f \qquad [4.27]$$

where

$$s_f = 2G_f/\tau_{max} \qquad [4.28]$$

In the above equations $\tau_{max} = 1.5\,\beta_w f_t$, $s_0 = 0.0195\,\beta_w f_t$, $G_f = 0.308\,\beta_w^2 \sqrt{f_t}$, s (mm) is the local slip, s_0 (mm) the local slip at the maximum local bond stress τ_{max} (MPa) and s_f (MPa) the local slip when the local bond stress τ (MPa) reduces to zero. The interfacial fracture energy is denoted by G_f (MPa.mm) and f_t is the concrete tensile strength (MPa). It should be noted that in Lu et al.'s (2005) original model, a slightly different expression was proposed for the width ratio factor β_w, but the expression given by Eq. 4.22 can be used in the above equations without any significant loss of accuracy.

4.6 Strength models for debonding failures

4.6.1 Plate end debonding

Many factors control the likely occurrence of a particular plate end debonding failure mode for a given plated RC beam. For example, for an RC beam with a relatively low level of internal steel shear reinforcement, each of the plate end debonding modes (Fig. 4.3) may become critical when the plate length or width is varied. When the distance between a plate end and the adjacent beam support (plate end distance) is very small, a CDC may form, causing a CDC debonding failure of the beam (Fig. 4.3b). If the plate end distance is increased, the CDC may fall outside the plated region, and only concrete cover separation is observed (Fig. 4.3d). Between these two modes, CDC debonding followed by concrete cover separation (Fig. 4.3c) may occur; this mode is critical if the CDC debonding failure load is lower than the shear resistance of the original RC beam as well as the cover separation failure load so that the load can still be increased following CDC debonding. As the plate end moves further away from the support, the cover separation mode remains the controlling mode, and the plate end crack that appears prior to crack propagation along the level of steel tension reinforcement becomes increasingly vertical (Smith and Teng, 2003). For the extreme case of a plate end in the pure bending region, the plate end crack is basically vertical (Fig. 4.3e). For any given plate end position, if the plate width is sufficiently small compared with that of the RC beam, the interface between the soffit plate and the RC beam becomes a more critical plane than the interface between the steel tension bars and the concrete, and plate end interfacial debonding (Fig. 4.3f) becomes the critical mode. However, this mode rarely occurs when the RC beam and the bonded plate have similar widths (unless the concrete surface is not properly prepared or inappropriate adhesive is used).

Given the larger variety of parameters that govern plate end debonding failures, the development of a reliable strength model is not a simple task. The recent model by Teng and Yao (2005, 2007) is the only model that appears to cover all the variations. The model caters for any combination of plate end moment and shear force via the following circular interaction curve:

$$\left(\frac{V_{db,end}}{0.85V_{db,s}}\right)^2 + \left(\frac{M_{db,end}}{0.85M_{db,f}}\right)^2 = 1.0 \qquad [4.29]$$

where $V_{db,end}$ and $M_{db,end}$ are the plate end shear force and the plate end moment at debonding, respectively, $M_{db,f}$ is the flexural debonding moment and $V_{db,s}$ is the shear debonding force. The use of the factor 0.85 in Eq. 4.29 is to ensure that the equation provides reasonable lower bound predictions of test results (Teng and Yao, 2007); for pure flexural debonding, this factor leads to 95 percentile lower bound predictions.

The flexural debonding moment, which is the bending moment that causes debonding of a plate end located in the pure bending zone of a beam, is found from:

$$M_{db,f} = \frac{0.488 M_{u,0}}{(\alpha_{flex} \alpha_{axial} \alpha_w)^{1/9}} \leq M_{u,0} \qquad [4.30]$$

where α_{flex}, α_{axial} and α_w are three dimensionless parameters defined by:

$$\alpha_{flex} = [(EI)_{c,frp} - (EI)_{c,0}]/(EI)_{c,0} \qquad [4.31]$$

$$\alpha_{axial} = E_{frp} t_{frp}/(E_c d_e) \qquad [4.32]$$

and

$$\alpha_w = b_c/b_{frp}, \quad b_c/b_{frp} \leq 3 \qquad [4.33]$$

where $(EI)_{c,frp}$ and $(EI)_{c,0}$ are the flexural rigidities of the cracked section with and without an FRP plate, respectively, $E_{frp} t_{frp}$ is the axial rigidity per unit width of the FRP plate, d is the effective depth of the beam and $M_{u,0}$

is the theoretical ultimate moment of the unplated section which is also the upper bound of the flexural debonding moment $M_{db,f}$.

The shear debonding force $V_{db,s}$, which is the shear force causing debonding of a plate end located in a region of (nearly) zero moment, can be found from:

$$V_{db,s} = V_c + \varepsilon_{v,e}\bar{V}_s \quad \text{with } \varepsilon_{v,e} \leq \varepsilon_y = \frac{f_y}{E_{sv}} \quad [4.34]$$

where V_c and $\varepsilon_{v,e}\bar{V}_s$ are the contributions of the concrete and the internal steel shear reinforcement to the shear capacity of the beam, respectively, and $\bar{V}_s$ is the shear force carried by the steel shear reinforcement per unit tensile strain, that is

$$\bar{V}_s = A_{sv} E_{sv} d_e / s_v \quad [4.35]$$

where A_{sv}, E_{sv} and s_v are the total cross-sectional area of the two legs of each stirrup, the elastic modulus and the longitudinal spacing of the stirrups. In Eq. 4.34, $\varepsilon_{v,e}$ is the tensile strain in the steel shear reinforcement, referred to here as the effective strain, and this effective strain may be well below the yield strain of the steel shear reinforcement. It should be noted that the bonded tension face plate also makes a small contribution to the shear debonding force $V_{db,s}$, but this contribution is small and is ignored in this debonding strength model.

Based on an analysis of test results obtained by Yao and Teng (2007), the best-fit expression for $\varepsilon_{v,e}$ is given by:

$$\varepsilon_{v,e} = \frac{10}{(\alpha_{flex}\alpha_E\alpha_t\alpha_w)^{1/2}} \quad [4.36]$$

where α_{flex} and α_w are given by Eqs 4.31 and 4.33, respectively, while the other two dimensionless parameters are defined by:

$$\alpha_E = E_{frp}/E_c \quad [4.37]$$

and

$$\alpha_t = (t_{frp}/d_e)^{1.3} \quad [4.38]$$

For the predictions of V_c in design, the design formula in any national code may be used.

4.6.2 IC debonding

Teng *et al.* (2003) and Lu *et al.* (2007) proposed two IC debonding strength models. The former is a simple modification of the bond strength model (Eq. 4.21). Both models predict a stress or strain value in the FRP plate at which IC debonding is expected to occur. According to Lu et al.'s (2007) model, this debonding stress is given by:

$$\sigma_{dbic} = 0.114(4.41 - \alpha)\tau_{max}\sqrt{\frac{E_{frp}}{t_{frp}}} \qquad [4.39]$$

$$\tau_{max} = 1.5\beta_w f_t \qquad [4.40]$$

and

$$\alpha = 3.41 L_{ee}/L_d \qquad [4.41]$$

where L_d(mm) is the distance from the loaded section to the end of the FRP plate while L_{ee}(mm) is given by:

$$L_{ee} = \sqrt{\frac{4E_{frp}t_{frp}}{\tau_{max}/s_0}} = 0.228\sqrt{E_{frp}t_{frp}} \qquad [4.42]$$

By substituting Eqs 4.40–4.42 into Eq. 4.39, the debonding stress can be rewritten as:

$$\sigma_{dbic} = \left(\frac{0.754}{\sqrt{E_{frp}t_{frp}}} - \frac{0.133}{L_d}\right) E_{frp}\beta_w f_t \qquad [4.43]$$

The following equation (Lu 2007), which provides a 95 percentile lower bound prediction of the ultimate loads of test beams given in Lu et al. (2007), should be used to ensure safe designs:

$$\sigma_{dbic} = \left(\frac{0.548}{\sqrt{E_{frp}t_{frp}}} - \frac{0.0967}{L_d}\right) E_{frp}\beta_w f_t \qquad [4.44]$$

4.7 Design procedure

4.7.1 Critical sections and plate end anchorage

For a strengthened RC beam, the moment capacity, taking into account the possibility of debonding, needs to be checked for two critical sections: (i) the maximum moment section, and (ii) the section just outside the effectively strengthened region, provided that the moment capacity of the original beam does not change along the beam length. If the moment capacity changes, it may be necessary to check the moment capacity for all sections at a change. Moment capacity changes can be due to changes in section size, the non-uniformity of the original steel reinforcement or the non-uniformity of the external FRP reinforcement. The design recommendations described herein are applicable to both situations, but specific reference

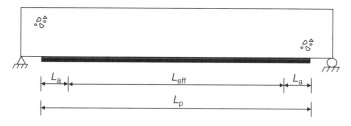

4.18 Definitions of L_a, L_{eff} and L_p.

is made only to beams of uniform original moment capacity for ease of presentation.

The effectively strengthened region L_{eff} is taken to be the plated region L_p minus an anchorage length L_a of the plate at each end (Fig. 4.18). The effective bond length determined from pull-off tests (Eq. 4.24), with an additional safety factor of 2, is recommended as the anchorage length. This ensures that in the case of IC debonding, the plate stress level at debonding is not adversely affected by an inadequate anchorage length. Following BS 8100 (BSI, 1997), this anchorage length (Eq. 4.24) is now given in terms of the concrete cube compressive strength as:

$$L_a = 2.11\sqrt{\frac{E_{frp}t_{frp}}{\sqrt{f_{cu}}}} \approx 2\sqrt{\frac{E_{frp}t_{frp}}{\sqrt{f_{cu}}}} \quad [4.45]$$

The capacity of the RC beam alone at the ends of the effectively strengthened region needs to be checked to ensure that it exceeds the applied moment here for the strengthened beam. This check can be done using any existing codes of practice such as BS 8110 (BSI, 1997). For checking the strength of the plated region, the following approach is recommended.

4.7.2 Strength check

Teng and Yao's (2005) model (Eq. 4.29) is recommended for the plate end debonding strength, while Lu et al.'s (2007) model (Eq. 4.44) is recommended for the intermediate crack induced debonding strength.

The strength check for the maximum moment section of the beam consists of the following four steps:

1. Determination of the ultimate plate stress σ_{dbic} for intermediate crack-induced debonding according to Lu et al.'s (2007) model (Eq. 4.44). The reduced tensile strength $f_{frp,r}$ of the FRP plate is worked out as the smaller of the ultimate tensile strength f_{frp} and σ_{dbic}, that is:

$$f_{frp,r} = \min(f_{frp}, \sigma_{dbic}) \quad [4.46]$$

2. The flexural strength M_u of the maximum moment section is then evaluated using the design equations presented in Section 4.3 with f_{frp} replaced by $f_{frp,r}$.
3. The plate end debonding strength $V_{db,end}$ is now determined using Teng and Yao's (2005, 2007) model (Eq. 4.29) incorporating appropriate partial safety factors for the concrete compressive strength and the steel yield stress.
4. The lower load-carrying capacity from steps 2 and 3 is the design value of the ultimate strength of the beam.

4.7.3 Detailing considerations and suppression of debonding failures

Whenever feasible, the plate end should be placed in a region where the moment is small (ideally zero), as this minimises the risk of plate end debonding. Furthermore, cautionary anchorage measures for the plate ends should always be adopted to improve the robustness of the strengthening system and to reduce the risk of a brittle debonding failure without prior warning, even when an RC beam strengthened with a bonded tension face plate satisfies all strength requirements. The provision of a U-jacket of suitable width and thickness at each plate end is an obvious choice for such a purpose. Tests by Smith and Teng (2003) have shown that, while the provision of a single U-jacket at each end improves the cover separation debonding load only by a limited extent, it does lead to a more ductile process.

Since the plate end debonding failure load can be as low as the shear resistance of the concrete alone in the RC beam (Oehlers, 1992; Smith and Teng, 2002a, b) (see also Eq. 4.34), plate end debonding controls the strength of RC beams bonded with a tension face plate in many cases. As a result, the high tensile strength of the FRP plate cannot be effectively utilised. The use of several U-jackets (or complete wraps if possible) in the high shear region near the plate end is believed to offer effective enhancement of the plate end debonding load. When flexural strengthening is applied to a beam, its shear strength often requires enhancement as well. In such situations, U-jackets or complete wraps distributed along the length of the beam (Fig. 4.19) can be used in conjunction with a tension face plate for combined flexural and shear strengthening (e.g. Al-Sulaimani et al., 1994; Kamiharako et al., 1997; Kachlakev and Barnes, 1999). Such U-jackets are believed to minimise plate end debonding risks as well. If properly used, such U-jacketing/wrapping also provides enhanced strength and ductility of the IC debonding mode.

If U-jacketing is not possible, the alternative of fibre anchors may be considered. Teng et al. (2000) and Lam and Teng (2001) used fibre anchors

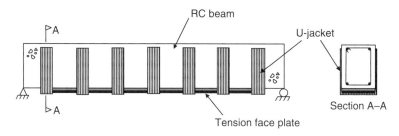

4.19 Combined use of tension face plate and U-jacketing to prevent plate end debonding.

to arrest the propagation of IC debonding in RC cantilever slabs successfully. Fibre anchors have also been used to arrest debonding in flexurally-strengthened RC beams (Oh and Sim, 2004; Eshwar *et al.*, 2005). Such fibre anchors are particularly useful for slabs where the use of U-jacketing is not possible. For slabs, the alterative U-anchorage proposed by Blaschko (2001) involving the insertion of a steel U-strip into the concrete may also be used, but there is a concern here with the possibility of corrosion of the steel strip.

An issue of concern in the use of anchorage measures is that no reliable design rule seems to exist. However, if U-jackets provided for shear strengthening ensure that the shear strength of a strengthened beam exceeds its flexural strength, plate end debonding failure may be assumed to have been completely suppressed, although IC debonding can still occur. For the design of U-jackets for shear strengthening purposes, readers should refer to Chapter 5. For the use of fibre anchors, little design guidance can be offered at present. The development of design methods for anchorage measures is an area where much research is needed. For the present, designs should be assisted by testing whenever necessary.

4.7.4 Partial safety factors

The reduced tensile strength of the FRP $f_{frp,r}$ is calculated as the minimum stress of either the tensile strength of the FRP or the stress determined from Lu *et al.*'s (2007) IC debonding strength model (Eq. 4.44). This reduced tensile strength is the characteristic strength of the FRP taking into account the effect of intermediate crack-induced debonding and should replace f_{frp} in the design equations presented in Section 4.3 where an FRP partial safety factor is included. Obviously, if the IC debonding stress is higher than the tensile strength of the FRP, no reduction in the FRP tensile strength is needed in the design calculations. It may be noted that this treatment does not lead to any inconsistency in the adopted safety margin in design if the partial safety factor for the tensile strength of FRP is the same as that for the bond strength.

4.8 Application of the design procedure to indeterminate beams

The design procedure presented in the preceding section is for simply-supported beams. Application of the design procedure to cantilever beams is straightforward as a cantilever beam can be taken as half of a simply-supported beam. An important issue for cantilever beams is that the FRP plate must be adequately anchored to the clamped support (Teng *et al.*, 2000; Lam and Teng, 2001).

This design procedure can also be applied to indeterminate beams, by simply treating each part of the beam between points of contraflexure as a simply-supported beam. In addition, if a clamped support exists, then the portion of the beam near the clamped support needs to be treated as a cantilever beam. One interesting and important situation of this kind arises in the strengthening of coupling beams in coupled shear walls. Here, the coupling beams are subject to double bending so, for effective strengthening, the beam needs to be dealt with as two cantilever beams. Of course, such coupling beams are subject to reversed cyclic loading from either wind or earthquake, so flexural strengthening is required for bending in both directions.

4.9 Acknowledgement

The authors are grateful for the financial support provided by The Hong Kong Polytechnic University (Project Code: BBZH).

4.10 References

ACI (2007) *Report on Fiber-reinforced Polymer (FRP) Reinforcement for Concrete Structures*, ACI 440R-07, Farmington Hills, MI, USA, American Concrete Institute.

AL-SULAIMANI G J, SHARIF A M, BASUNBUL I A, BALUCH M H and GHALEB B N (1994) Shear repair for reinforced concrete by fiberglass plate bonding, *ACI Structural Journal*, **91**(3), 458–464.

BANK L C (2006) *Composites in Construction: Structural Design with FRP Materials*, Hoboken, NJ, USA, John Wiley & Sons.

BSI (1997) *BS 8110*: 1997 *Structural Use of Concrete. Code of Practice for Design and Construction*, London, UK, British Standards Institution.

BLASCHKO M (2001) Anchoring device for FRP strips, in Teng J G (ed.), *Proceedings, International Conference on FRP Composites in Civil Engineering* – CICE 2001, Oxford, UK, Elsevier, 579–587.

BUYUKOZTURK O and HEARING B (1998) Failure behavior of precracked concrete beams retrofitted with FRP, *Journal of Composites for Construction*, ASCE, **2**(3), 138–144.

CHEN J F and TENG J G (2001) Anchorage strength models for FRP and steel plates bonded to concrete, *Journal of Structural Engineering*, ASCE, **127**(7), 784–791.

ESHWAR N, IBELL T J and NANNI A (2005) Effectiveness of CFRP strengthening on curved soffit RC beams, *Advances in Structural Engineering*, **8**(1), 55–68.

HAU K M (1999) *Experiments on Concrete Beams Strengthened by Bonding Fibre Reinforced Plastic Sheets*, Master of Science in Civil Engineering Thesis, The Hong Kong Polytechnic University, Hong Kong, China.

HOLLAWAY L C and LEEMING M B (eds) (1999) *Strengthening of Reinforced Concrete Structures Using Externally-Bonded FRP Composites in Structural and Civil Engineering*, Cambridge, UK, Woodhead.

KACHLAKEV D I and BARNES W A (1999) Flexural and shear performance of concrete beams strengthened with fiber reinforced polymer laminates, in Dolan C W, Rizkalla S H and Nanni A (eds), *Proceedings, 4th International Symposium on Fibre Reinforced Polymer Reinforcement for Concrete Structures – FRPRCS-4*, Farmington Hills, MI, USA, American Concrete Institute, 959–972.

KAMIHARAKO A, MARUYAMA K, TAKADA K and SHIMOMURA T (1997) Evaluation of shear contribution of FRP sheets attached to concrete beams, *Proceedings, 3rd International Symposium on Non-Metallic (FRP) Reinforcement for Concrete Structures – FRPRCS-3*, Sapporo, Japan, Japan Concrete Institute, **1**, 467–474.

KONG F K and EVANS R H (1987) *Reinforced and Prestressed Concrete*, 3rd edn, London, UK, Chapman & Hall.

LAM L and TENG J G (2001) Strength of cantilever RC slabs bonded with GFRP strips, *Journal of Composites for Construction*, ASCE, **5**(4), 221–227.

LU X Z, TENG J G, YE L P and JIANG J J (2005) Bond-slip models for FRP sheets/plates bonded to concrete, *Engineering Structures*, **27**, 920–937.

LU X Z (2007) Private communication.

LU X Z, TENG J G, YE L P and JIANG J J (2007) Intermediate crack debonding in FRP-strengthened RC beams: FE analysis and strength model, *Journal of Composites for Construction*, ASCE, **11**(2), 161–174.

MEIER U, DEURING M, MEIER H and SCHWEGLER G (1993) CFRP bonded sheets, in Nanni A (ed.), *Fibre-Reinforced-Plastic (FRP) Reinforcement for Concrete Structures: Properties and Applications*, Amsterdam. The Netherlands, Elsevier Science.

OEHLERS D J (1992) Reinforced concrete beams with plates glued to their soffits, *Journal of Structural Engineering*, ASCE, **118**(8), 2023–2038.

OEHLERS D J and SERACINO R (2004) *Design of FRP and Steel Plated RC Structures*, Oxford, UK, Elsevier.

OH H-S and SIM J (2004) Interface debonding failure in beams strengthened with externally bonded GFRP, *Composite Interfaces*, **11**, 25–42.

SAADATMANESH H and MALEK A M (1998) Design guidelines for flexural strengthening of RC beams with FRP plates, *Journal of Composites for Construction*, ASCE, **2**(4), 158–164.

SHEN H S, TENG J G and YANG J (2001) Interfacial stresses in beams and slabs bonded with a thin plate, *Journal of Engineering Mechanics*, ASCE, **127**(4), 399–406.

SMITH S T and TENG J G (2001) Interfacial stresses in plated beams, *Engineering Structures*, **23**(7), 857–871.

SMITH S T and TENG J G (2002a) FRP-strengthened RC beams-I: review of debonding strength models, *Engineering Structures*, **24**(4), 385–395.

SMITH S T and TENG J G (2002b) FRP-strengthened RC structures-II: assessment of debonding strength models, *Engineering Structures*, **24**(4), 397–417.

SMITH S T and TENG J G (2003) Shear-bending interaction in debonding failures of FRP-plated RC beams, *Advances in Structural Engineering*, **6**(3), 183–199.

TÄLJSTEN B (2003) *FRP Strengthening of Existing Concrete Structures. Design Guidelines*, 2nd edn, Division of Structural Engineering, Luleå University of Technology, Luleå, Sweden.

TENG J G and YAO J (2005) Plate end debonding failures of FRP- or steel-plated RC beams: A new strength model, in Chen J F and Teng J G (eds), *Proceedings, International Symposium on Bond Behaviour of FRP in Structures – BBFS 2005*, Hong Kong, China, Dec 7–9, 291–298.

TENG J G and YAO J (2007) Plate end debonding in FRP-plated RC beams-II: Strength model, *Engineering Structures*, **29**(10), 2472–2486.

TENG J G, LAM L, CHAN W and WANG J S (2000) Retrofitting of deficient RC cantilever slabs using GFRP strips, *Journal of Composites for Construction*, ASCE, **4**(2), 75–84.

TENG J G, CHEN J F, SMITH S T and LAM L (2002a) *FRP-Strengthened RC Structures*, Chichester, UK, John Wiley & Sons.

TENG J G, ZHANG J W and SMITH S T (2002b) Interfacial stresses in RC beams bonded with a soffit plate: a finite element study, *Construction and Building Materials*, **16**(1), 1–14.

TENG J G, SMITH S T, YAO J and CHEN J F (2003) Intermediate crack induced debonding in RC beams and slabs, *Construction and Building Materials*, **17**(6–7), 447–462.

YAO J and TENG J G (2007) Plate end debonding in FRP-plated RC beams-I: experiments, *Engineering Structures*, **29**(10), 2457–2471.

YAO J, TENG J G and CHEN J F (2005) Experimental study on FRP-to-concrete bonded joints, *Composites Part B: Engineering*, **36**(2), 99–113.

WU Z S, YUAN H and NIU H D (2002) Stress transfer and fracture propagation in different kinds of adhesive joints, *Journal of Engineering Mechanics*, ASCE, **128**(5), 562–573.

YUAN H, TENG J G, SERACINO R, WU Z S and YAO J (2004) Full-range behaviour of FRP-to-concrete bonded joints, *Engineering Structures*, **26**, 553–565.

5
Shear strengthening of reinforced concrete (RC) beams with fibre-reinforced polymer (FRP) composites

J. F. CHEN, The University of Edinburgh, UK;
J. G. TENG, The Hong Kong Polytechnic University, China

5.1 Introduction

This chapter is concerned with the design of externally bonded fibre-reinforced polymer (FRP) strips and plates for shear strengthening of reinforced concrete (RC) beams. Flexural and shear failures are the two main failure modes for normal unstrengthened RC beams. Flexural failure is generally preferred to shear failure as the strength-governing failure mode because the former is ductile which allows stress redistribution and provides warning to occupants, whilst the latter is brittle and catastrophic. Although the flexural failure mode in an RC beam strengthened with FRP in flexure shows much reduced ductility compared to a normal RC beam, as was discussed in Chapter 4, it is still a more ductile mode than shear failure.

When an RC member is deficient in shear, or when its shear capacity is less than the flexural capacity after flexural strengthening, shear strengthening must be considered. It is important to assess the shear capacity of RC beams which are intended to be strengthened in flexure, and to ensure that the shear capacity exceeds the flexural capacity.

Extensive research has been conducted on the strengthening of RC beams with externally bonded FRP composites in the last decade (e.g. Chajes et al., 1995; Chaallal et al., 1998; Khalifa et al., 1998; Maeda et al., 1998; Malek and Saadatmanesh, 1998; Triantafillou, 1998; Triantafillou and Antonopoulos, 2000; Chen and Teng, 2001a,b; 2003a,b, 2004; Täljsten, 2003; Adhikary et al., 2004; Bousselham and Chaallal, 2004; Concrete Society, 2004; Cao et al., 2005; Ali et al., 2006). This chapter first provides a brief discussion of possible shear strengthening schemes. A description of the shear failure modes then follows. Strength models for design use are next described, before a design procedure is presented.

5.2 Strengthening schemes

Many possibilities exist with the use of externally bonded FRP in shear strengthening. Common ways of attaching FRP shear reinforcement to a beam include side-bonding, in which the FRP is bonded to the sides only, U-jacketing, in which FRP U-jackets are bonded on both the sides and the tension face, and complete wrapping in which the FRP is wrapped around the entire cross-section. Both discrete strips and continuous sheets or plates may be used. For brevity, 'strips' are used in this chapter as a generic term to refer to FRP shear reinforcement including continuous sheets and plates, unless a clear differentiation is necessary for clarity. The fibres in the FRP may also be oriented at different angles to meet different strengthening requirements.

Because shear forces and bending moments in a beam may be reversed under conditions such as reversed cyclic loading and earthquake attacks, fibres may be arranged in two different directions to satisfy the requirement of shear strengthening in both directions. The use of fibres in two directions can obviously be beneficial with respect to shear resistance even if strengthening for reversed loading is not required, except for the unlikely case in which one of the fibre directions is exactly parallel to the shear cracks. In this sense, FRP plates with fibres in three (e.g. 0°/60°/120°) or more directions may also be used.

Figure 5.1 shows some typical examples of shear strengthening schemes where each scheme is represented by a notation as proposed by Teng *et al.* (2002) in which the first symbol represents the bonding configuration, the second symbol represents the fibre distribution and the last two sets of numbers represent the primary and secondary fibre orientations. For example, SS45/135 represents side-bonded FRP strips at 45 and 135°.

For a given design situation, the strengthening scheme needs to be selected based on such factors as the amount of increase required in the shear capacity, accessibility of the site (e.g. whether the whole perimeter of a beam can be accessed) and economic considerations.

As the FRP strips need to be bent around the corners of the beam in both U-jacketing and complete wrapping, the wet lay-up process (Teng *et al.*, 2002) is commonly used. One notable variation is the use of prefabricated L-shaped carbon fibre reinforced polymer (CFRP) plates (Basler *et al.*, 2001; Jensen and Poulsen, 2001; Meier *et al.*, 2001). In strengthening, the L-shaped plates are adhesively bonded to the web of the concrete beam, and are overlapped on the underneath of the beam over the full web width. The ends of the plate legs can be inserted into adhesive-filled openings in the slab for top-end anchorage (Meier *et al.*, 2001).

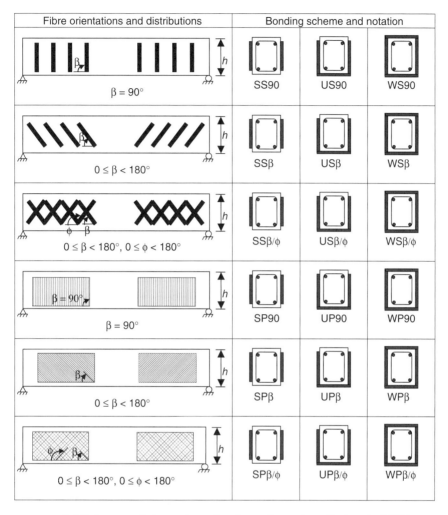

5.1 Schemes of shear strengthening using externally bonded FRP strips or plates.

5.3 Failure modes

The shear failure process of FRP-strengthened RC beams involves the development of either a single major diagonal shear crack or a number of diagonal shear cracks, similar to normal RC beams without FRP strengthening. For ease of description in this chapter, the existence of a single major diagonal shear crack (the critical shear crack) is assumed whenever necessary. Eventual failure of almost all test beams occurred in one of the two

5.2 Shear failure modes of FRP-strengthened RC beams: (a) rupture of complete FRP wraps; (b) debonding of FRP U-jackets.

main failure modes: tensile rupture of the FRP and debonding of the FRP from the concrete. Generally, both failure modes start with a debonding propagation process from the critical shear crack. Tensile rupture starts in the most highly-stressed FRP strip, followed rapidly by the rupture of other FRP strips intersected by the critical shear crack (Fig. 5.2a). In beams with complete FRP wraps, it is also common that many of the FRP strips intersected by the critical shear crack have debonded from the sides over the full height of the beam before tensile rupture failure occurs. In beams whose failure is by debonding of the FRP from the RC beam, failure involves a process of sequential debonding of FRP strips starting from the most vulnerable strip (Fig. 5.2b). The FRP rupture failure mode has been observed in almost all tests on beams with complete FRP wraps and in some tests on beams with FRP U-jackets, while the debonding failure mode has been observed in almost all tests on beams with FRP side strips and most tests on beams with FRP U-jackets.

Mechanical anchors can be used to prevent debonding and thus change the failure mode from debonding to rupture. However, care needs to be exercised to avoid local failure adjacent to the anchors such as those observed in tests conducted by Sato *et al.* (1997).

5.4 Shear strengths of FRP-strengthened reinforced concrete beams

5.4.1 Types of predictive models

Several different approaches have been used to predict the shear strength of FRP-strengthened RC beams. These include the modified shear friction method, the compression field theory, various truss models and the design code approach.

Deniaud and Cheng (2001a,b, 2003) proposed the modified shear friction method for predicting the shear strength of RC T-beams with U-jackets, which is a combination of Loov's (1998) shear friction method for normal beams and their own strip method for computing the contribution of FRP strips.

Apart from the design code approach and the modified shear friction method, a number of other approaches are available for normal RC beams in shear (ASCE-ACI 445, 1998), including the compression field theory (Collins, 1978), modified compression field theory (Vecchio and Collins, 1986) and various truss models such as the plasticity truss model and the rotating-angle softened truss model (Hsu, 1993). Broadly speaking, all these models belong to the truss model approach (Hsu, 1993). Attempts have been made to extend some of these models to FRP-strengthened RC beams, although the results are so far not very satisfactory overall (Deniaud and Cheng, 2001b; Colotti et al., 2001; Jensen and Poulsen, 2001). In these extensions, the FRP is assumed to contribute in a way similar to steel stirrups at an assumed average stress, so the strain distribution in the FRP is a critical issue. More recently, Aprile and Benedetti (2004) presented a model for the coupled flexural-shear design of RC beams strengthened with externally bonded FRP, where the compression field theory (Collins, 1978) with some modifications (ASCE-ACI 445, 1998) is used in predicting the shear strength.

However, the vast majority of existing research (e.g. Chajes et al., 1995; Chaallal et al., 1998; Khalifa et al., 1998; Maeda et al., 1998; Triantafillou, 1998; Triantafillou and Antonopoulos, 2000; Li et al., 2001; Tan, 2001, 2002; Pellegrino and Modena, 2002; Adhikary et al., 2003; Chen and Teng, 2003a,b; Täljsten, 2003; Deniaud and Cheng, 2004; Lu, 2004; Carolin and Täljsten, 2005; Monti and Liotta, 2007) and design guidelines (e.g. JSCE, 2000; fib, 2001; ISIS Canada, 2001; ACI, 2002; Concrete Society, 2004) has adopted the design code approach. The total shear resistance of FRP-strengthened RC beams in this approach is commonly assumed to be equal to the sum of the three components from concrete, internal steel shear reinforcement and external FRP shear reinforcement, respectively. Consequently, the shear strength of an FRP-strengthened beam V_n is given in the following form:

$$V_n = V_c + V_s + V_{frp} \quad [5.1]$$

where V_c is the contribution of concrete, V_s is the contribution of steel stirrups and bent-up bars and V_{frp} is the contribution of FRP. V_c and V_s may be calculated according to provisions in existing design codes. The contribution of FRP is found by truss analogy, similar to the determination of the contribution of steel shear reinforcement. Two parameters are important in determining the FRP contribution: the shear crack angle, which is generally assumed to be 45° for design use, and the average stress (or effective stress) in the FRP strips intersected by the critical shear crack. Different models differ mainly in the definition of this effective stress.

The most advanced models following the design code approach are probably those developed by Chen and Teng (2003a,b), in which the effective stress of the FRP strips is rationally defined instead of adopting empirical expressions based on curve-fitting of test results (Chen, 2003; Teng et al., 2004). They made a clear distinction between rupture failure and debonding failure, and developed two separate models for them. The key contribution of these two models is the realisation and the explicit account taken of the fact that the stress distribution in the FRP along the shear crack is likely to be strongly non-uniform at failure. Their debonding failure model has the additional advantage that an accurate bond strength model was employed, leading to accurate predictions.

The design code approach neglects the interactions between the external FRP and internal steel stirrups and concrete. The validity of this assumption has been questioned by several researchers (e.g. Teng et al., 2002; Denton et al., 2004; Qu et al., 2005; Mohamed Ali et al., 2006), but the approach is the least complex for design, the most mature and appears to be conservative for design in general. The design method presented below follows Chen and Teng's models (2003a,b) in this category.

5.4.2 FRP contribution to shear strength

For a general strengthening scheme with FRP strips of the same width bonded on both sides of the beam (Fig. 5.3) and an assumed critical shear crack inclined to the longitudinal axis of the beam by an angle $\theta = 45°$, the contribution of the FRP to the shear strength of the RC beam is given by:

$$V_{frp} = 2 f_{frp,e} t_{frp} w_{frp} \frac{h_{frp,e}(\sin\beta + \cos\beta)}{s_{frp}} \qquad [5.2]$$

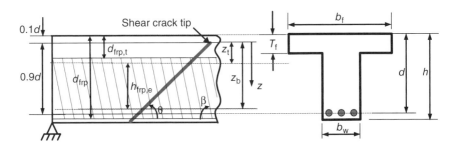

5.3 Notation for a general shear strengthening scheme.

where $f_{frp,e}$ is the average stress of the FRP intersected by the shear crack at the ultimate limit state, t_{frp} is the thickness of the FRP, w_{frp} is the width of each individual FRP strip (perpendicular to the fibre orientation), s_{frp} is the horizontal spacing of FRP strips (i.e. the centre-to-centre distance of FRP strips along the longitudinal axis of the beam), β is the angle of the inclination of fibres in the FRP to the longitudinal axis of the beam (measured clockwise for the left side of the beam as shown in Fig. 5.1) and $h_{frp,e}$ is the effective height of the FRP bonded on the web.

Figure 5.3 shows a general example where the side plates terminate slightly above the soffit of a T-section RC beam. Other strengthening schemes can be similarly represented by defining appropriate distances from the compression face to the top and the bottom of the FRP $d_{frp,t}$ and d_{frp}. A rectangular section can be treated as a special case of a T section by setting the flange thickness $T_f = 0$ (or the width of the flange b_f = the web thickness b_w). Assuming that:

- the shear crack ends at a distance of 0.1 d below the compression face of the beam, where d is the effective depth of the beam measured from the compression face to the centroid of steel tension reinforcement;
- the upper end of the effective FRP is 0.1 d below the actual upper end, which means that the effective upper end is at the crack tip when the actual upper end of the FRP is at the compression face of the beam;
- the lower end of the effective FRP is above the actual lower end by a distance of $(h - d)$, which means that the effective lower end of the FRP is at the centre of the steel tension reinforcement when the actual lower end of the FRP is at the tension face of the beam;

the effective height of the FRP $h_{frp,e}$ is given by:

$$h_{frp,e} = z_b - z_t \qquad [5.3]$$

where z_t and z_b are the coordinates of the top and the bottom ends of the effective FRP (Fig. 5.3):

$$z_t = d_{frp,t} \qquad [5.4a]$$

$$z_b = 0.9d - (h - d_{frp}) \qquad [5.4b]$$

in which $d_{frp,t}$ is the distance from the compression face to the top end of the FRP (thus $d_{frp,t} = 0$ for complete wrapping), h is the height of the beam and d_{frp} is the distance from the compression face to the lower end of the FRP. When FRP is bonded to the full height of the beam sides, Eq. 5.3 reduces to $h_{frp,e} = 0.9\ d$ as $z_t = 0$ and $z_b = 0.9$ d for such cases.

It should be noted that continuous FRP plates are treated as a special case of FRP strips (Fig. 5.4) with

148 Strengthening and rehabilitation of civil infrastructures

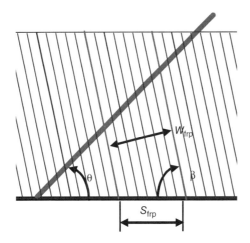

5.4 Relationship between w_{frp} and s_{frp} for continuous FRP plates.

$$s_{frp} = \frac{w_{frp}}{\sin \beta} \quad [5.5]$$

It is seen that $s_{frp} = w_{frp}$ only if $\beta = 90°$ (fibres vertically oriented). This is different from the situation for steel plates, arising from the fact that FRP composites can only bear a significant load in their main fibre direction.

The strain distribution in FRP strips intersected by the critical shear crack is closely related to the variation of the crack width at FRP rupture failure and, in general, the crack width is significantly non-uniform (Chen and Teng, 2003a). Similarly, the FRP stress distribution at debonding failure is also non-uniform, chiefly because the bond lengths of the FRP strips vary along the critical shear crack (Chen and Teng, 2003b). Therefore, the stress distribution in the FRP along the critical shear crack is non-uniform at the ultimate limit state for both modes of FRP rupture and FRP debonding failure. At the ultimate limit state, the average (or effective) stress in the FRP along the critical crack $f_{frp,e}$ is thus:

$$f_{frp,e} = D_{frp} \sigma_{frp,max} \quad [5.6]$$

in which $\sigma_{frp,max}$ is the maximum stress that can be reached in the FRP intersected by the critical shear crack and D_{frp} is the stress distribution factor defined by:

$$D_{frp} = \frac{f_{frp,e}}{\sigma_{frp,max}} \quad [5.7]$$

The values of $\sigma_{frp,max}$ and D_{frp} depend on whether the shear failure is controlled by FRP rupture or debonding.

5.4.3 FRP rupture failure

For FRP rupture failures, the following conservative and simple model was recommended by Chen and Teng (2003a) for practical use:

$$D_{frp} = \frac{1 + \dfrac{z_t}{z_b}}{2} \qquad [5.8]$$

which reduces to $D_{frp} = 0.5$ for strengthening schemes in which the bonded FRP strips cover the full height of the sides (e.g. complete wrapping) because $z_t = 0$.

Due to the detrimental effect of the beam corners on the tensile strength of the FRP, an FRP strip that is bent around a sharp corner is likely to rupture before its tensile strength is achieved. For design application, Chen and Teng (2003a) proposed a strength reduction factor of 0.8 to account for this effect. By including a partial safety factor for the FRP tensile strength γ_{frp}, the design value of the FRP tensile strength is given by:

$$\sigma_{frp,max} = \begin{cases} 0.8 \dfrac{f_{frp}}{\gamma_{frp}} & \text{if } \dfrac{f_{frp}}{E_{frp}} \leq \varepsilon_{max} \\ 0.8 \dfrac{\varepsilon_{max} E_{frp}}{\gamma_{frp}} & \text{if } \dfrac{f_{frp}}{E_{frp}} > \varepsilon_{max} \end{cases} \qquad [5.9]$$

where E_{frp} is the Young's modulus of the FRP. Equation 5.9 also includes a limit on the maximum usable strain of the FRP ε_{max} to control the width of shear cracks for design use. There is insufficient information in the literature to determine a suitable value for this maximum usable strain quantitatively. A value of $\varepsilon_{max} = 1.5\%$ may be used until a soundly based proposal is available.

5.4.4 Debonding failure

For debonding failure of FRP strips intersected by the critical shear crack, the maximum stress in the FRP occurs at the location where the FRP has the longest bond length. The maximum stress in the FRP is limited by either the bond strength or the tensile strength of the FRP. This maximum can be found from (Chen and Teng, 2003b):

$$\sigma_{frp,max} = \min \begin{cases} 0.8 \dfrac{f_{frp}}{\gamma_{frp}} \\ \dfrac{\alpha}{\gamma_b} \beta_w \beta_L \sqrt{\dfrac{E_{frp}}{t_{frp}}} \sqrt{f_c'} \end{cases} \qquad [5.10]$$

where the coefficient α has the best fit value of 0.427 and the 95 percentile characteristic value of 0.315 for design based on Chen and Teng's (2001c) bond strength model (also see Chapter 4), γ_b is the partial safety factor for debonding failures, β_L reflects the effect of bond length and β_w the effect of the FRP-to-concrete width ratio. The expressions for β_L and β_w are:

$$\beta_L = \begin{cases} 1 & \text{if } \lambda \geq 1 \\ \sin\dfrac{\pi\lambda}{2} & \text{if } \lambda < 1 \end{cases} \quad [5.11a]$$

and

$$\beta_w = \sqrt{\dfrac{2 - \dfrac{w_{frp}}{s_{frp}\sin\beta}}{1 + \dfrac{w_{frp}}{s_{frp}\sin\beta}}} \geq \dfrac{\sqrt{2}}{2} \quad [5.11b]$$

Note that $w_{frp}/(s_{frp}\sin\beta)$ is less than 1 for FRP strips with gaps. It becomes 1 when no gap exists between FRP strips and for continuous sheets or plates, yielding the lower limit value of $\sqrt{2}/2$ for β_w. The normalised maximum bond length λ, the maximum bond length L_{max} and the effective bond length L_e of the FRP strips are given by:

$$\lambda = \dfrac{L_{max}}{L_e} \quad [5.12a]$$

$$L_{max} = \begin{cases} \dfrac{h_{frp,e}}{\sin\beta} & \text{for U jackets} \\ \dfrac{h_{frp,e}}{2\sin\beta} & \text{for side plates} \end{cases} \quad [5.12b]$$

and

$$L_e = \sqrt{\dfrac{E_{frp}t_{frp}}{\sqrt{f'_c}}} \quad [5.12c]$$

The number 2 appears in the denominator for side plates in Eq. 5.12b because the FRP strip with the maximum bond length appears at the lower end of the critical shear crack for U-jacketing but at the middle for side plates.

Assuming that all the FRP strips intersected by the critical shear crack are able to develop their bond strength fully, the stress distribution factor for debonding failure D_{frp} can be derived as:

Shear strengthening of RC beams with FRP composites 151

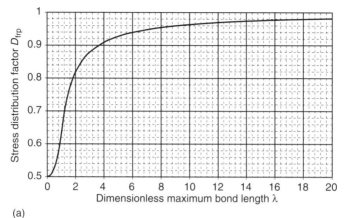

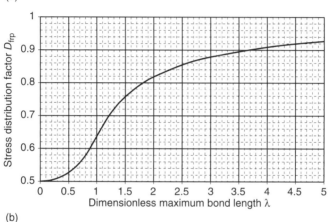

5.5 Stress distribution factor for U-jacketing and side bonding: (a) $0 < \lambda < 20$; (b) $0 < \lambda < 5$.

$$D_{frp} = \begin{cases} \dfrac{2}{\pi\lambda} \dfrac{1-\cos\dfrac{\pi\lambda}{2}}{\sin\dfrac{\pi\lambda}{2}} & \text{if } \lambda \leq 1 \\ 1 - \dfrac{\pi-2}{\pi\lambda} & \text{if } \lambda > 1 \end{cases} \qquad [5.13]$$

Equation 5.13 is applicable to both U-jackets and side strips. The actual calculated values are different for theses two cases even if the configuration of the bonded FRP is the same on the beam sides because the maximum bond length L_{max} for U-jackets is twice that for side strips (see Eq. 5.12b). The variation of D_{frp} with λ is shown in Fig. 5.5.

5.5 Design procedure

5.5.1 Shear capacity

Before designing a strengthening scheme, the beam should first be checked to ensure that diagonal compression failure will not occur after shear strengthening. This may be done according to existing design codes such as BS8110 (BSI, 1997) which requires the nominal shear stress v not to exceed $0.8\sqrt{f_{cn}}$ or $5\,\text{N/mm}^2$, i.e.

$$v = \frac{V}{bd} \le \text{lesser of } 0.8\sqrt{f_{cu}} \text{ and } 5\,\text{N/mm}^2 \quad [5.14]$$

in which V is the design shear force after strengthening. If such a requirement is not satisfied, other methods of strengthening such as enlargement of cross-section shall be considered instead. If it is satisfied, one may proceed to design an FRP strengthening scheme to ensure that the shear capacity of the beam after strengthening exceeds both the bending moment capacity and the design shear force ($V \le V_n$).

The shear capacity of an RC beam shear-strengthened with FRP can be calculated according to Eq. 5.1, where the contribution by concrete V_c and the contribution by steel shear reinforcement V_s may be obtained directly from formulae in existing codes of practice for RC structures such as BS 8110 (1997). The FRP contribution V_{frp} can be calculated from the equations presented in the preceding section.

For complete wrapping, the FRP contribution to the shear capacity V_{frp} shall be evaluated for the FRP rupture failure mode. For side bonding schemes, V_{frp} shall be evaluated for the debonding failure mode. It is generally unnecessary to check the shear strength for the FRP rupture mode for beams strengthened by side bonding, unless for very large beams strengthened with very thin FRP strips or for beams with FRP strips soundly anchored at strip ends using mechanical anchors. For U-jacketing, both failure modes shall be considered and the lower prediction of the two shall be used.

For the shear strengthening of RC beams using U-jacketed or side-bonded strips, an iterative procedure is required in design calculations because the coefficient β_w in Eq. 5.11b is related to the ratio of strip width to spacing $w_{frp}/(s_{frp}\sin\beta)$. An initial value of $\beta_w = 1$ may be used.

5.5.2 Spacing requirement for FRP strips

The design equations given in Section 5.4 were derived by treating strips as equivalent continuous plates. For a shear-strengthening scheme to be effective, it is necessary to ensure that the spacings between the strips are

Shear strengthening of RC beams with FRP composites

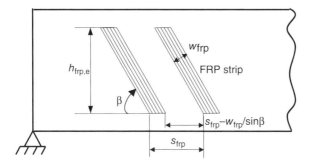

5.6 Spacing requirement of FRP strips.

sufficiently small as otherwise the critical shear crack may pass between two FRP strips without intersecting any of them, rendering the shear-strengthening scheme totally ineffective. Chen and Teng (2003b) suggested that the clear strip spacing between two FRP strips $s_{frp} - w_{frp}/\sin\beta$ (Fig. 5.6) should not exceed one half of the horizontal distance along the lower end of the effective FRP covered by the projection of the critical shear crack in the direction of fibres, which is given by $h_{frp,e}(1 + \cot\beta)/2$. As this calculated value can still be very large for large beams, it is necessary to limit it to an upper bound. The upper limit of 300 mm for internal steel links used in BS 8110 (1997) may be used. The expression for the maximum clear strip spacing is thus:

$$s_{frp} - \frac{w_{frp}}{\sin\beta} \leq \text{lesser of } \frac{h_{frp,e}(1+\cot\beta)}{2} \text{ and } 300\,\text{mm} \quad [5.15]$$

For continuous FRP plates, $s_{frp} - w_{frp}/\sin\beta = 0$ so Eq. 5.15 is automatically satisfied.

5.5.3 Strengthening design for indeterminate beams

The design equations and procedure presented in this chapter are applicable to both statically determinate and indeterminate beams. However, when designing shear strengthening measures for negative moment regions of continuous beams, it should be borne in mind that the upper face is in tension but the lower face is in compression. Shear cracks are then expected to start from the upper face of the beam in these regions. As a result, the dimensional quantities shown in Fig. 5.3 shall be measured from the lower face of the beam instead. Similarly, if U-jackets are used, the position of the U-jackets, if possible, must be reversed so that the U-jackets are bonded on the sides and the upper surface (instead of the sides and the lower face in positive moment regions); otherwise, the U-jackets may become only as effective as side plates/strips.

5.6 Acknowledgement

The authors are grateful for the financial support provided by The Hong Kong Polytechnic University (Project Code: BBZH).

5.7 References

ACI (2002) *Guide for the Design and Construction of Externally Bonded FRP Systems for Strengthening Concrete Structures*, ACI440.2R-02, Farmington Hills, MI, USA, American Concrete Institute.

ADHIKARY B B, MUTSUYOSHI H and ASHRAF M (2003) Effective shear strengthening of concrete beams using FRP sheets with bonded anchorage, Tan KH (ed.), *Proceedings, Fibre Reinforced Polymer Reinforcement for Concrete Structures – FRPRCS-6*, Singapore, World Scientific, 457–466.

ADHIKARY B B, MUTSUYOSHI H and ASHRAF M (2004) Shear strengthening of reinforced concrete beams using fiber-reinforced polymer sheets with bonded anchorage, *ACI Structural Journal*, **101**(2), 219–227.

ALI M S M, OEHLERS D J and SERACINO R (2006) Vertical shear interaction model between external FRP transverse plates and internal steel stirrups, *Engineering Structures*, **28**(3), 381–389.

APRILE A and BENEDETTI A (2004) Coupled flexural-shear design of R/C beams strengthened with FRP, *Composite Part B: Engineering*, **35**(1), 1–25.

ASCE-ACI 445 (1998) Recent approaches to shear design of structural concrete, *Journal of Structural Engineering*, ASCE, **124**(12), 1375–1417.

BASLER M, MUNGALL B and FAN S (2001) Strengthening of structures with the Sika CarboDur composite strengthening systems, Teng J G (ed.) *Proceedings, International Conference on FRP Composites in Civil Engineering – CICE 2001*, Oxford, UK, Elsevier Science, 253–262.

BOUSSELHAM A and CHAALLAL O (2004) Shear strengthening reinforced concrete beams with fiber-reinforced polymer: assessment of influencing parameters and required research, *ACI Structural Journal*, **101**(5), 660–668.

BSI (1997) *BS 8110-1: 1997 Structural Use of Concrete. Code of Practice for Design and Construction*, London, UK, British Standards Institution.

CAO S Y, CHEN J F, TENG J G, HAO Z and CHEN J (2005) Debonding in RC beams shear strengthened with complete FRP wraps, *Journal of Composites for Construction*, ASCE, **9**(5), 417–428.

CAROLIN A and TÄLJSTEN B (2005) Theoretical study of strengthening for increased shear bearing capacity, *Journal of Composites for Construction*, ASCE, **9**(6), 497–506.

CHAALLAL O, NOLLET M J and PERRATON D (1998) Strengthening of reinforced concrete beams with externally bonded fibre-reinforced-plastic plates: design guidelines for shear and flexure, *Canadian Journal of Civil Engineering*, **25**(4), 692–704.

CHAJES M J, JANUSZKA T F, MERTZ D R, THOMSON T A Jr and FINCH W W Jr (1995) Shear strengthening of reinforced concrete beams using externally applied composite fabrics, *ACI Structural Journal*, **92**(3), 295–303.

CHEN J F (2003) Design guidelines on FRP for shear strengthening of RC beams, in Forde M C (ed.), *Proceedings, Tenth International Conference on Structural Faults and Repair*, Edinburgh, UK, Technics Press (CD ROM).

CHEN J F and TENG J G (2001a) Shear strengthening of RC beams by external bonding of FRP composites: a new model for FRP debonding failure, in Forde M C (ed.), *Proceedings, Ninth International Conference on Structural Faults and Repair*, Edinburgh, UK, Technics Press (CD ROM).

CHEN J F and TENG J G (2001b) A shear strength model for FRP strengthened RC beams, in Burgoyne C J (ed.), *Proceedings, Fibre Reinforced Polymer Reinforcement for Concrete Structures – FRPRCS 5*, London, UK, Thomas Telford, 205–214.

CHEN J F and TENG J G (2001c) Anchorage strength models for FRP and steel plates bonded to concrete, *Journal of Structural Engineering*, ASCE, **127**(7), 784–791.

CHEN J F and TENG J G (2003a) Shear capacity of FRP strengthened RC beams: FRP rupture, *Journal of Structural Engineering*, ASCE, **129**(5), 615–625.

CHEN, J.F. and TENG, J.G. (2003b) Shear capacity of FRP strengthened RC beams: FRP debonding, *Construction and Building Materials*, **17**(1), 27- 41.

COLLINS M P (1978) Toward a rational theory for RC members in shear, *Journal of the Structural Division*, ASCE, **104**(ST4), 649–665.

COLOTTI V, SPADEA G and SWAMY R N (2001) Shear and flexural behaviour of RC beams externally reinforced with bonded FRP laminates: a truss approach, in Figueiras J, Ferreira A, Juvandes L, Faria R and Torres Marques A (eds), *Composites in Construction: Proceedings of the First International Conference*, Lisse, The Netherlands, Balkema, 517–522.

CONCRETE SOCIETY (2004) *Design Guidance for Strengthening Concrete Structures Using Fiber Composite Materials*, TR55, 2nd edn, Camberley, UK, The Concrete Society.

DENIAUD C and CHENG J J R (2001a) Shear behavior of reinforced concrete T-beams with externally bonded fiber-reinforced polymer sheets, *ACI Structural Journal*, **99**(3), 386–394.

DENIAUD C and CHENG J J R (2001b) Review of design methods for reinforced concrete beams strengthened with fibre reinforced polymer sheets, *Canadian Journal of Civil Engineering*, **28**, 271–281.

DENIAUD C and CHENG J J R (2003) Reinforced concrete T-beams, strengthened in shear with fiber reinforced polymer sheets, *Journal of Composites for Construction*, ASCE, **7**(4), 302–310.

DENIAUD C and CHENG J J R (2004). Simplified shear design method for concrete beams strengthened with fiber reinforced polymer sheets, *Journal of Composites for Construction*, ASCE, **8**(5), 425–433.

DENTON S R, SHAVE J D and PORTER A D (2004) Shear strengthening of reinforced concrete structures using FRP composites, in Hollaway LC, Chryssanthopoulos M K and Moy S S J (eds), *Advanced Polymer Composites for Structural Applications in Construction: ACIC 2004*, Cambridge, UK, Woodhead, 134–143.

fib (2001) *Externally bonded FRP reinforcement for RC structures*, Task Group 9.3, Bulletin No. 14, Lausanne, Switzerland, Federation Internationale du Beton.

HSU T T (1993) *Unified Theory of Reinforced Concrete*, Boca Raton, FL, USA, CRC Press.

ISIS CANADA (2001) *Strengthening Reinforced Concrete Structures with Externally-Bonded Fiber Reinforced Polymers*, The Canadian Network of Centres of Excellence on Intelligent Sensing for Innovative Structures, Winnipeg, Manitoba, Canada, ISIS Canada Corporation.

JENSEN A P and POULSEN E (2001) On the strengthening of RC beams by epoxy-bonded links of L-shaped CFRP strips against shear failure, in Figueiras J, Ferreira A, Juvandes L, Faria R and Torres Marques A (eds), *Composites in Construction: Proceedings of the First International Conference*, Lisse, The Netherlands, Balkema, 541–546.

JSCE (2001) *Recommendations for Upgrading Of Concrete Structures with Use of Continuous Fiber Sheets*, Concrete Engineering Series 41, Tokyo, Japan, Japan Society of Civil Engineers.

KHALIFA A, GOLD W J, NANNI A and AZIZ A (1998) Contribution of externally bonded FRP to shear capacity of RC flexural members, *Journal of Composites for Construction*, ASCE, **2**(4), 195–203.

LI A, ASSIH J and DELMAS Y (2001) Shear strengthening of RC beams with externally bonded CFRP sheets, *Journal of Structural Engineering*, ASCE, **127**(4), 374–380.

LOOV R E (1998) Review of A23.3.-94 simplified method of shear design and comparison with results using shear friction, *Canadian Journal of Civil Engineering*, **25**, 437–450.

LU X Z (2004) *Studies on FRP–Concrete Interface*, PhD Thesis, Tsinghua University, Beijing, China.

MAEDA T, ASANO Y, SATO Y, UEDA T and KAKUTA, Y. (1998) A study on bond mechanism of carbon fiber sheet, *Proceedings, 3rd International Symposium on Non-Metallic (FRP) Reinforcement for Concrete Structures*, Tokyo, Japan, Japan Concrete Institute, 279–285.

MALEK A M and SAADATMANESH H (1998) Analytical study of reinforced concrete beams strengthened with web-bonded fibre reinforced plates or fabrics, *ACI Structural Journal*, **95**(3), 343–351.

MEIER H, BASLER M and FAN S (2001) Structural strengthening with bonded CFRP L-shaped plates, in Teng J G (ed.), *Proceedings, International Conference on FRP Composites in Civil Engineering – CICE 2001*, Oxford, UK, Elsevier Science, 263–272.

MOHAMED ALI M S, OEHLERS D J and SERACINO R (2006) Vertical shear interaction model between external FRP transverse plates and internal steel stirrups, *Engineering Structures*, **28**(3), 381–389.

MONTI G and LIOTTA M (2007) Tests and design equations for FRP-strengthening in shear, *Construction and Building Materials*, **21**(4), 799–809.

PELLEGRINO C and MODENA C (2002) Fiber reinforced polymer sheet strengthening of reinforced concrete beams with transverse steel reinforcement, *Journal of Composites for Construction*, ASCE, **6**(2), 104–111.

QU Z, LU X Z and YE L P (2005) Size effect of shear contribution of externally bonded FRP U-Jackets for RC beams, in Chen J F and Teng J G (eds), *Proceedings, International Symposium on Bond Behaviour of FRP in Structures – BBFS 2005*, Hong Kong, China, Dec 7–9, 371–379.

SATO Y, KATSUMATA H and KOBATAKE Y (1997) Shear strengthening of existing reinforced concrete beams by CFRP sheet, *Non-Metallic (FRP) Reinforcement for Concrete Structures, Proceedings of the Third International Symposium*, Tokyo, Japan, Japan Concrete Institute, 507–513.

TÄLJSTEN B (2003) Strengthening concrete beams for shear with CFRP sheets, *Construction and Building Materials*, 17(1), 15–26.

TAN K H (2001) Shear strengthening of dapped beams using FRP, in Burgoyne C J (ed.), *Proceedings, Fibre Reinforced Polymer Reinforcement for Concrete Structures – FRPRCS 5*, London, UK, Thomas Telford, 249–258.

TAN Z (2002) *Experimental Research for RC Beam Strengthened with GFRP*, Masters Thesis, Tsinghua University, Beijing, China.

TENG J G, CHEN J F, SMITH S T and LAM L (2002) *FRP-Strengthened RC Structures*, Chichester, UK, John Wiley and Sons Ltd.

TENG J G, LAM L and CHEN J F (2004) Shear strengthening of RC beams using FRP composites, *Progress in Structural Engineering and Materials*, **6**, 173–184.

TRIANTAFILLOU T C (1998) Shear strengthening of reinforced concrete beams using epoxy-bonded FRP composites, *ACI Structural Journal*, **95**(2), 107–115.

TRIANTAFILLOU T C and ANTONOPOULOS C P (2000) Design of concrete flexural members strengthened in shear with FRP, *Journal of Composites for Construction*, ASCE, **4**(4), 198–205.

VECCHIO F J and COLLINS M P (1986) The modified compression-filed theory for reinforced concrete elements subjected to shear, *ACI Structural Journal*, **83**, 219–231.

6
Strengthening of reinforced concrete (RC) columns with fibre-reinforced polymer (FRP) composites

J. G. TENG and T. JIANG,
The Hong Kong Polytechnic University, China

6.1 Introduction

An important application of fibre-reinforced polymer composites is to provide confinement to reinforced concrete (RC) columns to enhance their load-carrying capacity and ductility. This method of strengthening is based on the well-known phenomenon that the axial compressive strength and ultimate axial compressive strain of concrete can be significantly increased through lateral confinement.

Various methods have been used to achieve confinement to columns using FRP composites (Teng *et al.*, 2002). *In situ* FRP wrapping has been the most commonly used technique, in which unidirectional fibre sheets or woven fabric sheets are impregnated with polymeric resins and wrapped around columns in a wet lay-up process, with the main fibres orientated in the hoop direction. In addition, filament winding and prefabricated FRP jackets have also been used. The filament winding technique uses continuous fibre strands instead of sheets/straps so that winding can be achieved automatically by means of a computer-controlled winding machine (ACI, 1996). When prefabricated FRP jackets are used, the jackets are fabricated in half circles or half rectangles (Nanni and Norris, 1995; Ohno *et al.*, 1997) and circles with a slit or in continuous rolls (Xiao and Ma, 1997), so that they can be opened up and placed around columns.

Regardless of the type of FRP jacket used, any vertical joint in the FRP jacket should include an adequate overlap to ensure that failure of the joint will not precede failure of the jacket away from the joint when subjected to hoop tension. In the strengthening of rectangular columns, the sharp corners of the columns should be rounded to reduce the detrimental effect of the sharp corners on the tensile strength of the FRP, and to enhance the effectiveness of confinement. Rectangular columns may also receive shape modifications before FRP jacketing to enhance the effectiveness of confinement. For example, a rectangular section may be modified to an elliptical section before FRP jacketing (Teng and Lam, 2002).

Strengthening of RC columns with FRP composites 159

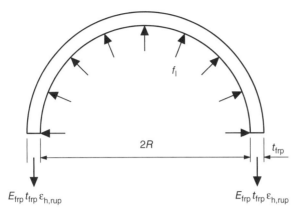

6.1 Confining action of FRP jacket.

A column under axial or eccentric compression is commonly covered with a FRP jacket that is continuous over the height of the column, but spiral wraps or discrete ring wraps may also be used. This chapter is concerned only with the behavior and design of columns confined by a continuous FRP jacket covering the column over the required height. Any confinement from transverse steel reinforcement in the original column is ignored in the design theory presented for columns later in the chapter, as the interaction between steel and FRP confinement is not yet well understood and those columns which require strengthening often have very limited transverse steel reinforcement that justifies this assumption.

6.2 Behaviour of FRP-confined concrete

6.2.1 General

The behaviour of an FRP-confined RC column differs from that of a conventional RC column in that the behaviour of concrete now depends on the amount of FRP confinement. The behaviour and modelling of FRP-confined concrete is thus fundamental to the prediction of the behaviour and the design of FRP-confined RC columns. This section therefore focuses on the behaviour of FRP-confined concrete in columns of different section forms.

6.2.2 FRP-confined concrete in circular columns

When an FRP-confined RC column is subject to axial compression, the concrete expands laterally and this expansion is confined by the FRP. The confining action of an FRP jacket for circular concrete columns is shown in Fig. 6.1. For circular columns, the concrete is subject to uniform

6.2 Failure of an FRP-confined circular concrete column by FRP rupture.

confinement. Eventual failure occurs once the hoop rupture strain of FRP $\varepsilon_{h,rup}$ is reached (Fig. 6.2), provided premature failure at the vertical lap joint is prevented by the provision of an adequate overlap length. The maximum confining pressure is reached when the FRP ruptures. This maximum confining pressure is given by

$$f_l = \frac{E_{frp} t_{frp} \varepsilon_{h,rup}}{R} = \frac{\rho_{frp}}{2} E_{frp} \varepsilon_{h,rup} \qquad [6.1]$$

where f_l is the maximum lateral confining pressure, E_{frp} is the elastic modulus of FRP in the hoop direction, t_{frp} is the total thickness of FRP jacket, R is the radius of the confined concrete core and ρ_{frp} is the FRP volumetric ratio and $= 2t_{frp}/R$.

A very important conclusion of existing research is that the FRP hoop rupture strain in an FRP-confined circular concrete column falls substantially below that from a flat coupon tensile test (Lam and Teng, 2004). This fact can be attributed to at least three factors (Lam and Teng 2004): (i) the curvature of the FRP jacket; (ii) the deformation localisation of the cracked concrete; and (iii) the existence of an overlapping zone. In addition, the biaxial stress state the jacket is subjected to is also believed to be a significant factor. A reasonable and accurate stress–strain model for FRP-confined concrete must be based on the actual FRP hoop rupture strain.

Strengthening of RC columns with FRP composites   161

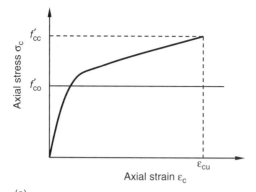

(a)

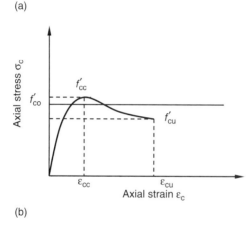

(b)

6.3 Typical stress–strain curves of FRP-confined concrete: (a) ascending type; (b) descending type.

The stress–strain curve of FRP-confined concrete features a monotonically ascending bilinear shape with sharp softening in a transition zone around the stress level of the unconfined concrete strength (Fig. 6.3a) if the amount of FRP exceeds a certain threshold value. This type of stress–strain curves (the ascending type) was observed in the vast majority of existing tests. With this type of stress–strain curves, both the compressive strength and the ultimate axial strain (or referred to simply as the ultimate strain for brevity) are reached simultaneously and are significantly enhanced. However, existing tests have also shown that in some cases such a bilinear stress–strain behaviour cannot be expected. Instead, the stress–strain curve features a post-peak descending branch and the compressive strength is reached before the tensile rupture of the FRP jacket (Fig. 6.3b) (the descending type). It should be noted that this type of stress–strain curve may end at a stress value (axial stress at ultimate axial strain) either larger or smaller than the compressive strength of unconfined concrete.

162　Strengthening and rehabilitation of civil infrastructures

(a)　　　　　　　　　　　　(b)

6.4 Failures of FRP-confined square and rectangular concrete columns with rounded corners by FRP rupture: (a) square column; (b) rectangular column.

6.2.3　FRP-confined concrete in rectangular columns

It has been well established that FRP confinement is much less effective for rectangular columns (including square columns as a special case) than for circular columns, even with the rounding of corners. This is because in the former, the confining pressure is non-uniformly distributed and only part of the concrete core is effectively confined. Failure generally occurs at the corners by FRP tensile rupture (Fig. 6.4). The stress–strain curves are more likely to feature a descending branch, but in such cases FRP confinement normally provides little strength enhancement. The effectiveness of confinement increases as the amount of FRP or the corner radius increases, and as the aspect ratio of the section (ratio between the longer and shorter sides of a rectangular section) reduces. Figure 6.5 shows two experimental stress–strain curves of FRP-confined concrete in square columns corresponding to two different amounts of FRP confinement. For ease of comparison, the axial stress and the strains are normalised by the compressive strength of unconfined concrete and its corresponding axial strain, respectively. Of the two curves, the one for a smaller corner radius and a single-ply jacket features a descending branch.

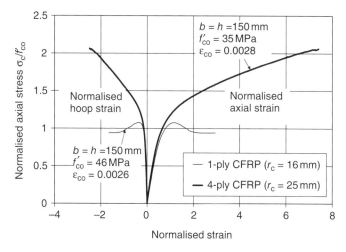

6.5 Typical stress–strain curves of FRP-confined concrete in square columns.

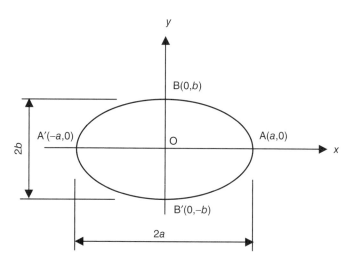

6.6 Definition of an ellipse.

6.2.4 FRP-confined concrete in elliptical columns

Elliptical sections are not commonly used, but they can arise as a result of shape modification of rectangular sections. Figure 6.6 shows an ellipse with a major axis (A–A′ axis) length of $2a$ and a minor axis (B–B′ axis) length of $2b$. A and A′ are referred to as the major vertices while B and B′ are referred to as the minor vertices of the ellipse. In the mathematical literature, the shape of an ellipse is determined by its eccentricity $e = \sqrt{a^2 - b^2}/a$.

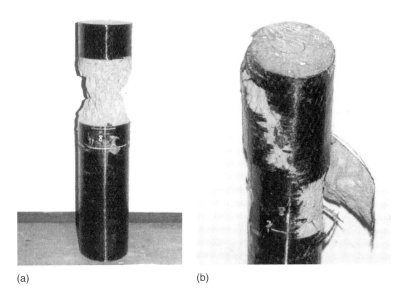

6.7 Failures of FRP-confined elliptical concrete columns: (a) $a/b = 5/4$; (b) $a/b = 5/3$.

In the present chapter, the simple major axis-to-minor axis length ratio a/b (or simply the aspect ratio) is used as the key shape parameter instead.

For the same FRP volumetric ratio, the amount of confinement provided by the FRP jacket to an elliptical column is between those of circular and rectangular columns. In more specific terms, if a rectangular column is confined by an FRP jacket with shape modification allowed, an elliptical jacket is less effective than a circular one but more effective than a rectangular one. The behaviour depends strongly on the aspect ratio. If this ratio is small and close to 1, the behaviour is similar to that of circular columns. As the aspect ratio increases, the effectiveness of FRP confinement reduces.

The location of FRP rupture is generally at a corner when the aspect ratio is large (Fig. 6.7) but, when this ratio is small, this location was found to vary in tests conducted by Teng and Lam (2002). Figure 6.8 shows the stress–strain curves of FRP-confined elliptical columns of three different aspect ratios. The bilinear stress–strain behaviour is clearly seen for the two aspect ratios of 5/4 and 5/3, but the other curve for an aspect ratio of 5/2 features a descending branch. These curves illustrate clearly the effect of aspect ratio on the behaviour of FRP-confined elliptical columns.

6.2.5 Size effect

It should be noted that the behaviour of FRP-confined concrete described above is mainly based on observations made on small-scale test specimens.

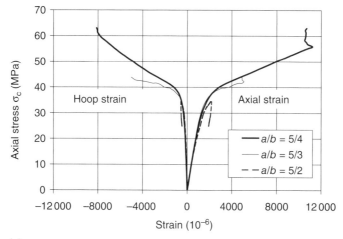

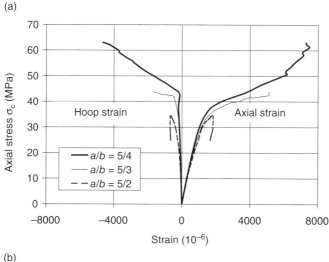

6.8 Typical stress–strain curves of FRP-confined concrete in elliptical columns: (a) at minor vertices; (b) at major vertices.

For circular columns, the limited number of existing tests on large-scale columns (Youssef, 2003; Carey and Harries, 2005; Mattys et al., 2005; Rocca et al., 2006; Yeh and Chang, 2007) have indicated that the size of columns does not have a significant effect on the observed behaviour. In other words, the behaviour of large-scale circular columns under concentric compression can be extrapolated from that of small-scale columns. For FRP-confined rectangular and elliptical columns, the effect of column size on the behaviour of the confined concrete is much more uncertain due to the very limited test data available (Youssef, 2003; Rocca et al., 2006; Pantelides and Yan,

2007). The stress–strain models for FRP-confined concrete presented in the following sections do not account for the possible size effect, so caution should be exercised when they are used to predict the behaviour of large non-circular columns.

6.3 Design-oriented stress–strain models for FRP-confined concrete

6.3.1 General

Many stress–strain models have been proposed for FRP-confined concrete. These models can be classified into two categories (Teng and Lam 2004): (i) design-oriented models (Fardis and Khalili, 1982; Karbhari and Gao, 1997; Samaan *et al.*, 1998; Miyauchi *et al.*, 1999; Saafi *et al.*, 1999; Toutanji, 1999; Lillistone and Jolly, 2000; Xiao and Wu, 2000, 2003; Lam and Teng, 2003a,b; Berthet *et al.*, 2006; Harajli, 2006; Saenz and Pantelides, 2007; Teng *et al.*, 2007a; Wu *et al.*, 2007; Youssef *et al.*, 2007), and (ii) analysis-oriented models (Mirmiran and Shahawy, 1997; Spoelstra and Monti, 1999; Fam and Rizkalla, 2001; Chun and Park, 2002; Harries and Kharel, 2002; Marques *et al.*, 2004; Binici, 2005; Teng *et al.*, 2007b). In models of the first category, the compressive strength, ultimate strain and stress–strain behaviour of FRP-confined concrete are predicted using closed-form equations based directly on the interpretation of experimental results. In models of the second category, the stress–strain curves of FRP-confined concrete are generated using an incremental numerical procedure which accounts for the interaction between the FRP jacket and the concrete core. The simple form of design-oriented models makes them convenient for design use while analysis-oriented models are more suitable for incorporation in computer-based numerical analysis such as non-linear finite element analysis. In this section, design-oriented models developed by Lam and Teng (2003a,b) are introduced. Analysis-oriented stress–strain models are discussed in Section 6.4.

6.3.2 FRP-confined concrete in circular columns

Concrete in circular columns receives uniform confinement from an FRP jacket; its behaviour has attracted a large number of studies and is now well understood. While different design-oriented stress–strain models have been published, Lam and Teng's model (Lam and Teng, 2003a; Teng *et al.*, 2007a) appears to be advantageous over other models due to its simplicity and accuracy. In particular, the model adopts a simple form which naturally reduces to that for unconfined concrete when no FRP confinement is provided. Moreover, the simple form of this model caters for the easy modification of the expressions for the ultimate condition of FRP-confined concrete (the compressive strength and the ultimate axial strain). The latest version of this model (Teng et al., 2007a) takes due account of the effect of the jacket stiffness on the ultimate condition of FRP-confined concrete, which is the key to the accurate prediction of stress–strain curves of FRP-confined concrete.

Lam and Teng's design-oriented stress–strain model is based on the following assumptions: (i) the stress–strain curve consists of a parabolic first portion and a linear second portion; (ii) the slope of the parabola at zero axial strain (the initial slope) is the same as the elastic modulus of unconfined concrete; (iii) the non-linear part of the first portion is affected to some degree by the presence of an FRP jacket; (iv) the parabolic first portion meets the linear second portion smoothly (i.e. there is no change in slope between the two portions where they meet); (v) the linear second portion terminates at a point where both the compressive strength and the ultimate axial strain of confined concrete are reached; and (vi) the stress axis is intercepted by the linear second portion at the compressive strength of unconfined concrete.

Lam and Teng's stress–strain model for FRP-confined concrete is described by the following expressions:

$$\sigma = E_c \varepsilon_c - \frac{(E_c - E_2)^2}{4 f'_{co}} \varepsilon_c^2 \qquad 0 \leq \varepsilon_c < \varepsilon_t \qquad [6.2a]$$

$$\sigma_c = f'_{co} + E_2 \varepsilon_c \qquad \varepsilon_t \leq \varepsilon_c \leq \varepsilon_{cu} \qquad [6.2b]$$

where σ_c and ε_c are the axial compressive stress and the axial compressive strain, respectively, E_c is the elastic modulus of unconfined concrete, E_2 is the slope of the linear second portion, f'_{co} is the compressive strength of unconfined concrete and ε_{cu} is the ultimate axial compressive strain of confined concrete.

Eq. 6.2 defines a stress–strain curve (Fig. 6.9) consisting of a parabolic first portion and a linear second portion with a smooth transition at ε_t which is given by:

$$\varepsilon_t = \frac{2 f'_{co}}{E_c - E_2} \qquad [6.3]$$

The slope of the linear second portion E_2 is given by

$$E_2 = \frac{f'_{cc} - f'_{co}}{\varepsilon_{cu}} \qquad [6.4]$$

With the ultimate point (compressive strength and ultimate axial strain) known, the stress–strain curve can be fully determined. The latest expressions (Teng et al., 2007a) for the definition of the ultimate point are given below.

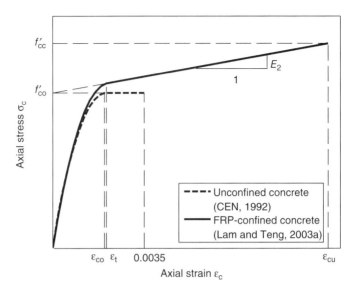

6.9 Lam and Teng's stress–strain model for FRP-confined concrete.

The compressive strength f'_{cc} is given by:

$$\frac{f'_{cc}}{f'_{co}} = 1 + 3.5(\rho_K - 0.01)\rho_\varepsilon \qquad \rho_K \geq 0.01 \qquad [6.5a]$$

$$\frac{f'_{cc}}{f'_{co}} = 1 \qquad \rho_K < 0.01 \qquad [6.5b]$$

and the ultimate axial compressive strain ε_{cu} by:

$$\frac{\varepsilon_{cu}}{\varepsilon_{co}} = 1.75 + 6.5\rho_K^{0.8}\rho_\varepsilon^{1.45} \qquad [6.6a]$$

where $\rho_K = E_{frp}t/(E_{sec,o}R)$ is the confinement stiffness ratio and $\rho_\varepsilon = \varepsilon_{h,rup}/\varepsilon_{co}$ is the strain ratio. It should be noted that the confinement ratio f_l/f'_{co} can be interpreted as a product of ρ_K and ρ_ε ($f_l/f'_{co} = \rho_K\rho_\varepsilon$). $E_{sec,o}$ and ε_{co} are the secant modulus and the axial strain at the compressive strength of unconfined concrete, respectively, with $E_{sec,o} = f'_{co}/\varepsilon_{co} \cdot \varepsilon_{co} = 0.002$ is recommended for design use.

In Eq. 6.5, a minimum confinement stiffness ratio of $\rho_K = 0.01$ is used to ensure an ascending type stress–strain curve. Otherwise, a descending type stress–strain curve is expected. In such cases, the limited strength enhancement is ignored and the post-peak descending branch is represented using a horizontal line which intercepts the stress axis at the compressive strength of unconfined concrete.

It should also be noted that in an earlier version of this model (Lam and Teng, 2003a), the compressive strength and the ultimate axial strain are defined as:

$$\frac{f'_{cc}}{f'_{co}} = 1 + 3.3 \frac{f_l}{f'_{co}} \qquad \frac{f_l}{f'_{co}} \geq 0.07 \qquad [6.5c]$$

$$\frac{f'_{cc}}{f'_{co}} = 1 \qquad \frac{f_l}{f'_{co}} < 0.07 \qquad [6.5d]$$

and

$$\frac{\varepsilon_{cu}}{\varepsilon_{co}} = 1.75 + 12 \frac{f_l}{f'_{co}} \left(\frac{\varepsilon_{h,rup}}{\varepsilon_{co}} \right)^{0.45} \qquad [6.6b]$$

In this old version, a minimum confinement ratio is used to ensure an ascending type stress–strain curve and the effect of the jacket stiffness on the compressive strength is not properly reflected. These expressions are included here for reference because the models presented in the following two subsections for FRP-confined concrete in rectangular and elliptical columns are based on this old version.

6.3.3 FRP-confined concrete in rectangular columns

Due to the non-uniformity of confinement, the axial stress of concrete in an FRP-confined rectangular column varies over the section, so in developing stress–strain models, the average axial stress (load divided by sectional area) is used. Lam and Teng (2003b) presented a stress–strain model for FRP-confined concrete in rectangular columns, which was modified from their earlier model for concrete uniformly confined by FRP (Lam and Teng, 2003a) as described above. This model is described below.

A rectangular column with rounded corners is shown in Fig. 6.10, where the width is denoted by b and is assumed to be smaller than the depth h for convenience. Square columns are considered as a special case of rectangular columns with $b = h$. The aspect ratio h/b defines the shape of a rectangular section. To improve the effectiveness of FRP confinement, corner rounding is generally recommended. Due to the presence of internal steel reinforcement, the corner radius r_c is generally limited to small values.

In Lam and Teng's (2003b) stress–strain model for concrete in rectangular columns, the shape of the stress–strain curve is still described by Eqs 6.2–6.4, but the ultimate point is redefined by introducing two shape factors and the concept of an equivalent circular column. The compressive strength is a simple modification of Eq. 6.5 by the introduction of a shape factor denoted by k_{s1}. Thus,

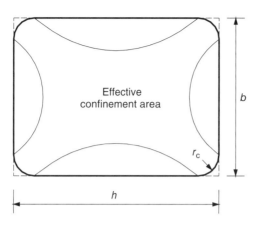

6.10 Effective confinement area of a rectangular section.

$$\frac{f'_{cc}}{f'_{co}} = 1 + 3.3k_{s1}\frac{f_l}{f'_{co}} \qquad k_{s1}\frac{f_l}{f'_{co}} \geq 0.07 \qquad [6.7a]$$

$$\frac{f'_{cc}}{f'_{co}} = 1 \qquad k_{s1}\frac{f_l}{f'_{co}} < 0.07 \qquad [6.7b]$$

where the term $k_{s1}f_l/f'_{co}$ is the effective confinement ratio. Similarly, the ultimate axial strain is given by the following equation in which a different shape factor, k_{s2}, is introduced:

$$\frac{\varepsilon_{cu}}{\varepsilon_{co}} = 1.75 + 12k_{s2}\frac{f_l}{f'_{co}}\left(\frac{\varepsilon_{h,rup}}{\varepsilon_{co}}\right)^{0.45} \qquad [6.8]$$

In Eqs 6.7 and 6.8, f_l is the confining pressure in an equivalent circular column given by Eq. 6.1. The equivalent circular column has a radius R being equal to half the diagonal distance of the section. That is

$$R = \frac{1}{2}\sqrt{h^2 + b^2} \qquad [6.9]$$

This means that the equivalent circular section circumscribes the rectangular section (Fig. 6.11).

The shape factors depend on two parameters, the effective confinement area and the aspect ratio. It is commonly accepted that in a rectangular section, only part of the concrete is effectively confined by transverse reinforcement through the arching action. In Lam and Teng's (2003b) model, the effective confinement area is contained by four parabolas as illustrated in Fig. 6.11, with the initial slopes of the parabolas being the same as the adjacent diagonal lines. The effective confinement area ratio A_e/A_c is therefore given by:

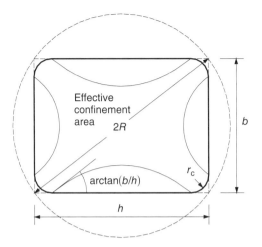

6.11 Lam and Teng's model for FRP-confined concrete in rectangular sections.

$$\frac{A_e}{A_c} = \frac{1 - \left(\dfrac{\dfrac{b}{h}(h - 2r_c)^2 + \dfrac{h}{b}(b - 2r_c)^2}{3A_g} \right) - \rho_{sc}}{1 - \rho_{sc}} \quad [6.10]$$

where A_e is the effective confinement area, A_c is the total cross-sectional area of concrete enclosed by the FRP jacket, A_g is the gross area of the column section with rounded corners and ρ_{sc} is the cross-sectional area ratio of the longitudinal steel reinforcement.

The two shape factors, one for strength enhancement k_{s1} and the other for strain enhancement k_{s2}, are then given by:

$$k_{s1} = \left(\frac{b}{h}\right)^2 \frac{A_e}{A_c} \quad [6.11]$$

and

$$k_{s2} = \left(\frac{h}{b}\right)^{0.5} \frac{A_e}{A_c} \quad [6.12]$$

6.3.4 FRP-confined concrete in elliptical columns

As mentioned in Section 6.2.4, a rectangular column can be transformed into an elliptical column for more effective FRP confinement. For concrete in FRP-confined elliptical columns, a compressive strength equation is available (Teng and Lam, 2002), but the corresponding equation for the ultimate strain has not been developed. Similar to a rectangular section, the axial stress and the confining pressure vary over an elliptical section, so the compressive strength equation is based on the average stress in the section.

The compressive strength of concrete in elliptical columns (Teng and Lam, 2002) is given by:

$$\frac{f'_{cc}}{f'_{co}} = 1 + 3.3 k_s \frac{f_1}{f'_{co}} \qquad k_s \frac{f_1}{f'_{co}} \geq 0.07 \qquad [6.13a]$$

$$\frac{f'_{cc}}{f'_{co}} = 1 + 3.3 k_s \frac{f_1}{f'_{co}} \qquad k_s \frac{f_1}{f'_{co}} < 0.07 \qquad [6.13b]$$

where f_1 is the confining pressure in an equivalent circular column and can be found from Eq. 6.1 once the radius of the equivalent circular column is known. The equivalent circular column is one with the same FRP volumetric ratio. The area of an ellipse is $\pi a b$, while its circumference is closely approximated by $\pi[1.5(a+b) - \sqrt{ab}]$. Therefore, the FRP volumetric ratio for an elliptical column confined by an FRP jacket with a thickness of t_{frp} is given by

$$\rho_{frp} = \frac{t_{frp}[1.5(a+b) - \sqrt{ab}]}{ab} \qquad [6.14]$$

The shape factor k_s is simply given by the following simple equation:

$$k_s = \left(\frac{b}{a}\right)^2 \qquad [6.15]$$

6.4 Analysis-oriented stress–strain models for FRP-confined concrete

6.4.1 General

Although design-oriented models are simple and convenient for use in practical design, they do not directly capture the interaction between the confining jacket and the concrete core. By contrast, this interaction is explicitly accounted for in analysis-oriented models through considerations of force equilibrium and displacement compatibility. Analysis-oriented models are more versatile and accurate in general, and are probably the preferred choice for use in accurate non-linear numerical analysis of concrete structures with FRP confinement. Conceptually, analysis-oriented models are only strictly valid for uniformly-confined concrete in circular columns.

In most existing analysis-oriented models for FRP-confined concrete, a theoretical model for actively-confined concrete (i.e. the confining pressure is externally applied and remains constant as the axial stress increases), which is referred to as an active confinement model for brevity hereafter, is employed as the base model. The stress–strain curve of FRP-confined concrete is generated through an incremental process, with the resulting stress–strain curve crossing a family of stress–strain curves for the same concrete under different levels of active confinement. The models of Mirmiran and Shahawy (1997), Spoelstra and Monti (1999), Fam and Rizkalla (2001), Chun and Park (2002), Harries and Kharel (2002), Marques et al., (2004), Binici (2005) and Teng et al., (2007b) are all of this type. This approach has been much more popular than some other approaches (e.g. Harmon et al., 1998; Becque et al., 2003) as it leads to conceptually simple yet effective models.

The stress–strain curve of FRP-confined concrete can be obtained through the following procedure: (i) for a given axial strain, find the corresponding lateral strain according to the lateral-to-axial strain relationship; (ii) based on radial displacement compatibility and force equilibrium between the concrete core and the FRP jacket, calculate the corresponding lateral confining pressure provided by the FRP jacket; (iii) use the axial strain and the confining pressure obtained from steps (i) and (ii) in conjunction with an active confinement model for concrete to evaluate the corresponding axial stress, leading to the identification of a point on the stress–strain curve of FRP-confined concrete; and (iv) repeat the above steps to generate the entire stress–strain curve. Figure 6.12 illustrates the concept of this incremental approach.

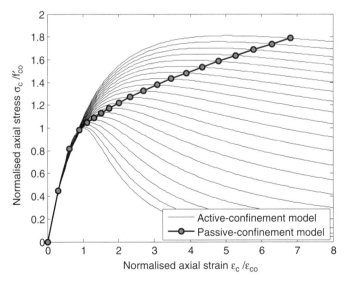

6.12 Generation of a stress–strain curve for FRP-confined concrete by an analysis-oriented model.

It is not difficult to realise that in the above procedure, the key elements determining the accuracy of the predictions are the active confinement model and the lateral-to-axial strain relationship. The performance of the active confinement model depends on two components: (i) the peak point of the stress–strain curve; and (ii) the stress–strain equation. All these models can be assessed in terms of the above three key aspects. For a discussion of the accuracy of most of these models, readers are referred to Teng and Lam (2004) and Jiang and Teng (2007). Teng et al.'s (2007b) model is one of the latest models and provides close predictions of test results. A refinement of Teng et al.'s (2007b) model has also recently been introduced by Jiang and Teng (2007) to improve its accuracy, particularly for weakly-confined concrete.

These models are all built on the assumption that the axial stress and the axial strain of concrete confined with FRP at a given lateral strain are the same as those of the same concrete actively confined with a constant confining pressure equal to that supplied by the FRP jacket. This assumption has been shown by Teng et al. (2007b) to be generally valid.

6.4.2 Teng *et al.*'s stress–strain model

Teng *et al.*'s (2007b) model is explained herein by following the process of generating stress–strain curves. The generation of any point on the stress–strain curve of FRP-confined concrete can start by specifying a value for the axial strain ε_c or for the lateral strain ε_l. Once the lateral strain is known, the corresponding axial strain ε_c can be easily found from the following axial-to-lateral strain equation that is applicable to unconfined, activelyconfined and FRP-confined concrete:

$$\frac{\varepsilon_c}{\varepsilon_{co}} = 0.85\left(1+8\frac{\sigma_l}{f'_{co}}\right)\left\{\left[1+0.75\left(\frac{\varepsilon_l}{\varepsilon_{co}}\right)\right]^{0.7} - \exp\left[-7\left(\frac{\varepsilon_l}{\varepsilon_{co}}\right)\right]\right\} \quad [6.16]$$

where the corresponding confining pressure σ_l is given by:

$$\sigma_l = \frac{E_{frp} t_{frp} \varepsilon_l}{R} \quad [6.17]$$

where ε_l is the lateral strain of the concrete core which is equal to the tensile hoop strain in the FRP jacket. If the axial strain is specified instead, the corresponding lateral strain needs to be found from Eqs 6.16 and 6.17 via an iterative process.

Since it is assumed that the axial stress and the axial strain of concrete confined with FRP at a given lateral strain are the same as those of the same concrete actively confined with a constant confining pressure equal to that supplied by the FRP jacket, the axial stress for a given axial strain can be found from the stress–strain curve of actively-confined concrete under a constant confining pressure corresponding to the known lateral strain. In Teng et al.'s (2007b) model, the following equation, which was originally proposed by Popovics (1973) and later employed in the model of Mander et al. (1988) for steel-confined concrete, is adopted to predict the stress–strain curves of actively-confined concrete:

$$\frac{\sigma_c}{f'^*_{cc}} = \frac{(\varepsilon_c/\varepsilon^*_{cc})m}{m-1+(\varepsilon_c/\varepsilon^*_{cc})^m} \quad [6.18]$$

where f'^*_{cc} and ε^*_{cc} are, respectively, the peak axial compressive stress and the corresponding axial compressive strain of concrete under a specific constant confining pressure. The constant m in Eq. 6.18 is defined by:

$$m = \frac{E_c}{E_c - f'^*_{cc}/\varepsilon^*_{cc}} \quad [6.19]$$

The peak stress on the stress–strain curve of actively-confined concrete is given by:

$$f'^*_{cc} = f'_{co} + 3.5\sigma_1 \quad [6.20]$$

while the axial strain at peak stress is given by:

$$\varepsilon^*_{cc} = \varepsilon_{co}\left[1 + 5\left(\frac{f'^*_{cc}}{f'_{co}} - 1\right)\right] = \varepsilon_{co}\left(1 + 17.5\frac{\sigma_1}{f'_{co}}\right) \quad [6.21a]$$

The application of the above steps each time leads to the determination of a single point on the stress–strain curve of FRP-confined concrete. To obtain an entire curve, a sufficiently large number of points should be generated.

Although Teng et al.'s (2007b) model was identified by Jiang and Teng (2007) to be the most accurate among the models examined by them, this model was also found to lead to some over-estimation of the axial stress of concrete confined with a weak FRP jacket, particularly when the confined concrete exhibits a descending type stress–strain curve (Jiang and Teng 2007). To improve the performance of this model for such concrete, Jiang and Teng (2007) proposed a refinement on a combined analytical and experimental basis. In the refined version, Eq. 6.21a is replaced by:

$$\varepsilon^*_{cc} = \varepsilon_{co}\left[1 + 17.5\left(\frac{\sigma_1}{f'_{co}}\right)^{1.2}\right] \quad [6.21b]$$

This simple modification improves the accuracy of the model for concrete with weak FRP confinement, but has little effect on its accuracy for concrete with stronger confinement.

6.5 Section analysis

6.5.1 General

Once the stress–strain models for FRP-confined concrete are defined, design of FRP-confined RC sections can be carried out using the conventional section analysis. Therefore, this section discusses the procedure of section analysis before presenting design recommendations for FRP-confined columns in the next section. It is necessary to note that the stress–strain models discussed in the two preceding sections are all for FRP-confined concrete under concentric compression; however, in a column under combined bending and axial compression, a strain gradient exists. For unconfined RC columns, the assumption that the stress–strain curve of concrete in an eccentrically-loaded column is the same as that of concrete under concentric compression is widely used. For FRP-confined RC columns, whether the same assumption is equally acceptable is not yet completely clear. The few existing studies on the behaviour of eccentrically-loaded circular columns with FRP confinement (Fam *et al.*, 2003; Hadi, 2006; Mosalam *et al.*, 2007) have not provided a clear understanding of the issue, although this assumption has been shown to lead to some over-estimations at large eccentricities (Fam *et al.*, 2003). Based on the information available to date, it seems reasonable at this stage to adopt this assumption in the analysis of FRP-confined circular RC sections as has been done by previous researchers (e.g. Yuan and Mirmiran, 2001; Cheng *et al.*, 2002; Teng *et al.*, 2002; Jiang and Teng, 2006).

The effect of eccentric loading on the stress–strain behaviour of FRP-confined concrete in rectangular and elliptical columns is much more uncertain. The very limited existing work (Parvin and Wang, 2001; Cao *et al.*, 2006) has been concerned with overall column behaviour and does not clarify the effect of eccentric loading on stress–strain behaviour of the confined concrete. In particular, a rectangular column or an elliptical column subjected to eccentric compression may be bent about either the strong axis or the weak axis. It is unlikely that the same stress–strain model developed from studies on FRP-confined concrete columns under concentric compression can be used for eccentric loading in both directions. Indeed, such a stress–strain model may not be applicable to eccentric loading in either direction. As a result, only the procedure of section analysis for circular columns is discussed in some detail in the following subsection; the section analysis of

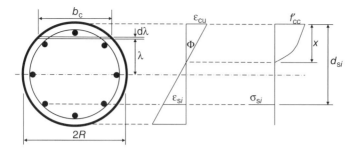

6.13 Strains and stresses over a circular column section at ultimate limit state.

rectangular or elliptical columns can be carried out in a similar manner once the stress–strain model is given.

6.5.2 Section analysis

When the stress–strain curve of FRP-confined concrete from concentric compression is directly used in the section analysis, it makes the analysis procedure similar to that for conventional RC columns, as may be found in numerous reinforced concrete textbooks (e.g. Park and Paulay, 1975; Kong and Evans, 1987). The only difference in the analysis procedure introduced by the presence of FRP confinement is the use of a suitable concrete stress–strain relationship that considers the confinement effect of the FRP. Numerical integration over the section can be carried out by the layer method in which the column section is divided into many small horizontal layers.

For the analysis of a circular section subjected to combined bending and axial compression as shown in Fig. 6.13, the latest version of Lam and Teng's design-oriented model for FRP-confined concrete in circular columns (Teng *et al.*, 2007a) is used. Complete composite action between concrete and FRP is assumed. Compressive stresses are taken to be positive and the tensile strength of concrete is ignored. Plane sections are assumed to remain plane. The axial load N and the bending moment M at any stage of loading carried by the section with the reference axis going through the centre of the section are found by integrating the stresses over the section:

$$N_u = \int_{\lambda=R-x}^{R} \sigma_c b_c d\lambda + \sum_{i=1}^{n} (\sigma_{si} - \sigma_c) A_{si} \qquad [6.22a]$$

and

$$M_u = \int_{\lambda=R-x}^{R} \sigma_c b_c \lambda \, d\lambda + \sum_{i=1}^{n} (\sigma_{si} - \sigma_c) A_{si} (R - d_{si})$$ [6.22b]

where b_c is the width of the section at a distance λ from the reference axis, σ_{si} is the stress in the i th layer of longitudinal steel reinforcement, A_{si} is the corresponding cross-sectional area of the longitudinal steel reinforcement. The stress of concrete σ_c in the compression zone can be determined from Eq. 6.2. σ_{si} can be calculated from:

$$\sigma_{si} = E_s \varepsilon_{si} \qquad |\varepsilon_{si}| < \frac{f_y}{E_s}$$ [6.23a]

$$\sigma_{si} = \frac{\varepsilon_{si}}{|\varepsilon_{si}|} f_y \qquad |\varepsilon_{si}| < \frac{f_y}{E_s}$$ [6.23b]

where E_s and f_y are the elastic modulus and yield strength of steel. Equation 6.22 is applicable at any stage of loading. The ultimate limit state of the column is reached when the strain at the extreme concrete compression fibre reaches the ultimate strain of FRP-confined concrete, signifying crushing of concrete due to FRP rupture. This ultimate strain is defined by Eq. 6.6a.

Typical interaction curves are shown in Fig. 6.14. These interaction curves are for a reference circular RC column with the following geometric and material properties: radius R = 300 mm, concrete cover = 30 mm, concrete strength f'_{co} = 30 MPa, bar diameter = 25 mm and steel yield stress = 460 MPa. Altogether 12 bars are distributed evenly around the circumference. The column is either unconfined or wrapped with a three-ply or six-ply carbon fibre reinforced polymer (CFRP) jacket, with a nominal ply thickness of 0.165 mm. The tensile strength and elastic modulus of the CFRP, based on a nominal ply thickness of 0.165 mm, are 3900 MPa and 230 000 MPa, respectively. The CFRP is assumed to rupture at an actual hoop rupture strain of $\varepsilon_{h,rup}$ = 0.01. These interaction curves are normalised by the axial load capacity N_{uo} (concentric compression) and moment capacity M_{uo} (pure bending) of the reference column when no FRP confinement is provided. It can be seen that the maximum benefit of FRP confinement occurs when the section fails in pure compression, but confinement is much less beneficial when the column fails in pure bending. The sharp slope change in the interaction curves at high axial loads for the wrapped columns (Fig. 6.14) starts when the neutral axis begins to fall outside the column section. As the neutral axis depth moves further away from the column section, the parabolic portion of the confined concrete stress–strain curve gradually moves outside the section, with stresses over the section being eventually governed entirely by the linear second portion of the stress–strain curve, at

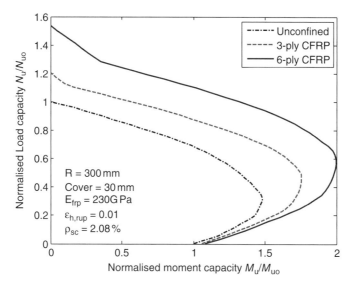

6.14 Bending moment-axial load interaction curves for a circular RC column.

the beginning of the linear portion of the interaction curve at high axial load.

6.5.3 Moment–curvature analysis

A moment–curvature analysis can also be carried out easily. For a given axial load, moment–curvature plots can be generated by specifying a sequence of suitable strain values for the extreme compression fibre of concrete ε_{cf} up to its ultimate value ε_{cu}. For each strain value, the neutral axis depth is varied until the resultant axial force acting on the section, calculated from Eq. 6.22a, equals the applied axial load. Once the neutral axis position has been determined, the moment can be evaluated using Eq. 6.22b and the curvature can be calculated by the following expression:

$$\Phi = \frac{\varepsilon_{cf}}{x} \qquad [6.24]$$

A typical moment–curvature curve is shown in Fig. 6.15. This curve is for the reference column defined in the preceding subsection with a 3-ply CFRP jacket whose interaction curve is shown in Fig. 6.14. The curve shown in Fig. 6.15 is for an axial load of 2000 kN.

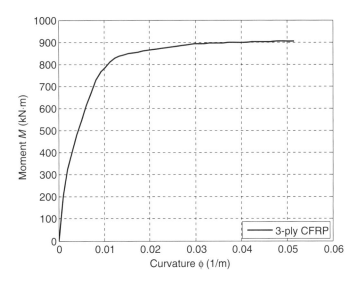

6.15 Moment–curvature curve of an FRP-confined circular RC column.

6.6 Design of FRP-confined reinforced concrete columns

6.6.1 General

On the basis of the section analysis described above, design equations can be easily established for short FRP-confined RC columns for which the effect of column slenderness is negligible. For slender FRP-confined RC columns, only limited research is available (e.g. Tao *et al.*, 2004; Fitzwilliam and Bisby, 2006; Jiang and Teng, 2006), so an appropriate procedure to deal with the effect of column slenderness is not yet available.

It is desirable for the design of FRP-confined RC columns to closely follow the established procedure for conventional RC columns that is familiar to engineers. In most design codes for RC structures [e.g. ENV 1992-1-1 (CEN, 1992), BS 8110 (BSI, 1997); ACI, 2005], a slenderness limit is given which should not be exceeded if an RC column is to be designed as a short column. This slenderness limit for short columns is also referred to simply as the slenderness limit for brevity hereafter. In design, a check of the slenderness of a column is the first step and determines whether the column is a short column or a slender column.

This section first presents Lam and Teng's stress–strain model with partial safety factors appropriately incorporated, followed by the definition of the slenderness limit to distinguish short FRP-confined RC columns from slender ones. Approximate design equations for short FRP-confined circular and rectangular RC columns subjected to either concentric or eccentric compression are then given. The authors are not aware of any existing design guidelines (e.g. ACI, 2002; Concrete Society, 2004; fib, 2001) where such a slenderness limit and similar design equations have been presented.

6.6.2 Lam and Teng's model for design use

For use in design following the general framework of BS 8110 (BSI, 1997), Lam and Teng's design-oriented model needs to be rewritten as follows by incorporating appropriate partial safety factors:

$$\sigma_c = E_c \varepsilon_c - \frac{(E_c - E_2)^2}{4(0.67 f_{cu}/\gamma_c)} \varepsilon_c^2 \qquad 0 \leq \varepsilon_c < \varepsilon_t \qquad [6.25a]$$

$$\sigma_c = 0.67 f_{cu}/\gamma_c + E_2 \varepsilon_c \qquad \varepsilon_t \leq \varepsilon_c \leq \varepsilon_{cu} \qquad [6.25b]$$

and

$$\varepsilon_t = \frac{2(0.67 f_{cu}/\gamma_c)}{E_c - E_2} \qquad [6.26]$$

with

$$E_2 = \frac{f'_{cc,d} - 0.67 f_{cu}/\gamma_c}{\varepsilon_{cu,d}} \qquad [6.27]$$

where $f'_{cc,d}$ and $\varepsilon_{cu,d}$ are the design values of the compressive strength and the ultimate strain. It is recommended that $\varepsilon_{co} = 0.002$ be adopted in design, thus:

$$E_c = \frac{2(0.67 f_{cu}/\gamma_c)}{\varepsilon_{cu}} = \frac{2(0.67 f_{cu}/\gamma_c)}{0.002} = 1000(0.67 f_{cu}/\gamma_c) \qquad [6.28]$$

where γ_c is the partial safety factor for concrete and f_{cu} is the characteristic cube compressive strength of concrete. The term $0.67 f_{cu}$ represents the axial compressive strength of the unconfined concrete in a real RC column in terms of the cube strength f_{cu}.

For circular columns, the design values of the compressive strength and the ultimate strain are given by:

$$f'_{cc,d} = \frac{0.67 f_{cu}}{\gamma_c} + 3.5 \frac{E_{frp} t_{frp}}{R} \left(1 - \frac{5}{\beta_K}\right) \frac{\varepsilon_{h,rup,k}}{\gamma_\varepsilon} \qquad \beta_K \geq 5 \qquad [6.29a]$$

$$f'_{cc,d} = \frac{0.67 f_{cu}}{\gamma_c} \qquad \beta_K < 5 \qquad [6.29b]$$

and

$$\varepsilon_{cu,d} = 0.0035 + 0.74\beta_K^{0.8}\left(\frac{\varepsilon_{h,rup,k}}{\gamma_\varepsilon}\right)^{1.45} \quad [6.30]$$

with

$$\beta_K = \frac{E_{frp}t_{frp}}{(0.67f_{cu})R} \quad [6.31]$$

where β_K is the confinement stiffness ratio for use in design, γ_ε is the partial safety factor for the FRP hoop rupture strain and $\varepsilon_{h,rup,k}$ is the characteristic value of the hoop rupture strain, which may be determined from a sufficiently large number of tests on FRP-confined concrete cylinders. Alternatively, this characteristic value may be taken as 50% and 70% of the characteristic values of the ultimate tensile strain from flat coupon tests for CFRP jackets and glass fibre reinforced polymer (GFRP) jackets, respectively, based on existing experimental evidence (e.g. Lam and Teng, 2003a; Lam and Teng, 2004).

For rectangular columns:

$$f'_{cc,d} = \frac{0.67f_{cu}}{\gamma_c} + 3.3k_{s1}\frac{E_{frp}t_{frp}}{R}\frac{\varepsilon_{h,rup,k}}{\gamma_\varepsilon} \quad k_{s1}\beta_K\varepsilon_{h,rup,k} \geq 0.07 \quad [6.32a]$$

$$f'_{cc,d} = \frac{0.67f_{cu}}{\gamma_c} \quad k_{s1}\beta_K\varepsilon_{h,rup,k} \geq 0.07 \quad [6.32b]$$

and

$$\varepsilon_{cu,d} = 0.0035 + 0.4k_{s2}\beta_K\left(\frac{\varepsilon_{h,rup,k}}{\gamma_\varepsilon}\right)^{1.45} \quad [6.33]$$

For elliptical columns, an expression for the design value of the ultimate strain is not available but the design value of the compressive strength is given by:

$$f'_{cc,d} = \frac{0.67f_{cu}}{\gamma_c} + 3.3k_s\frac{E_{frp}t_{frp}}{R}\frac{\varepsilon_{h,rup,k}}{\gamma_\varepsilon} \quad k_s\beta_K\varepsilon_{h,rup,k} \geq 0.07 \quad [6.34a]$$

$$f'_{cc,d} = \frac{0.67f_{cu}}{\gamma_c} \quad k_s\beta_K\varepsilon_{h,rup,k} < 0.07 \quad [6.34b]$$

It should be noted that in defining β_K (Eq. 6.31), R is the radius of the section for circular columns, but the radius of the equivalent circular column for rectangular columns $\left(=1/2\sqrt{b^2+h^2}\right)$ or elliptical columns $\left(=2ab/1.5(a+b)-\sqrt{ab}\right)$.

6.6.3 Slenderness limit for short FRP-confined reinforced concrete columns

The slenderness limit for short FRP-confined RC columns needs to be defined to ensure that the second-order effect leads to only a small amplification of the moment at the critical section or a small reduction (commonly 5% or 10%) of the axial load capacity, as is commonly adopted for RC columns. In Jiang and Teng's (2006) study, the latter option was adopted. The slenderness ratio of an FRP-confined RC column is herein defined as $\lambda = l_e/r$, where l_e is the effective length of the column and r is the radius of gyration. Based on the numerical results produced by the analytical model presented in Jiang and Teng (2006), the following expression is proposed for the slenderness limit:

$$\lambda_{crit} = \frac{60 \frac{e_2}{D}\left(1 - \frac{e_1}{e_2}\right) + 20}{\dfrac{f'_{cc,d}}{0.67 f_{cu}/\gamma_c}(1 + 30\varepsilon_{h,rup,k})} \qquad [6.35a]$$

where e_1 and e_2 are the load eccentricities at the two ends, respectively, with e_2 having the larger absolute value, and D is the diameter of the circular column under consideration. This expression has a clear physical meaning: the numerator defines the slenderness limit for short RC columns without FRP confinement, while the denominator accounts for the effect of FRP confinement. When no FRP confinement is provided and $e_2/D = 0.2$, Eq. 6.35a reduces to $\lambda_{crit} = 32 - 12e_1/e_2$, which is similar to but slightly more conservative than the expression adopted by ACI-318 (ACI, 2005). Besides, when $e_1/e_2 = 1$, Eq. 6.35a reduces to $\lambda_{crit} = 20$, which is the slenderness limit for short RC columns without FRP confinement given in GB-50010 (2002). It should be noted that when FRP confinement is provided, the denominator always has a value larger than unity. This implies that an RC column originally classified as a short column may need to be considered as a slender column when it is confined with FRP. This is because FRP confinement leads only to a very limited increase in the flexural rigidity of an RC section compared to the much larger increase in the strength of the same section.

Equation 6.35a is a general expression for all possible eccentricity combinations. When it is applied to concentrically-loaded columns ($e_1 = e_2 = 0$), it reduces to:

$$\lambda_{crit} = \frac{20}{\dfrac{f'_{cc,d}}{0.67 f_{cu}/\gamma_c}(1 + 30\varepsilon_{h,rup,k})} \qquad [6.35b]$$

Equation 6.35 is slightly unconservative in some cases when the slenderness limit is defined to correspond to a 5% axial load reduction. However, if a

10% loss of axial load capacity is acceptable, this expression provides a lower bound prediction for all cases. A 10% reduction in the axial load capacity has been adopted as the criterion for defining short columns in some of the existing literature (e.g. CEB-FIP, 1993).

It should be noted that Eq. 6.35 is also applicable to rectangular columns bent about either the weak or the strong axis if Lam and Teng's (2003b) stress–strain model is used in the section analysis, provided that the diameter of the circular section in Eq. 6.35 is replaced by the side dimension of the rectangular section in the bending direction. It should be further noted that when Eq. 6.35 is applied to rectangular columns bent in the stronger plane (about the strong axis if the effective length is the same in the two directions), an additional check of the slenderness of the column in the orthogonal plane with zero bending moment is necessary. This can easily be done using Eq. 6.35b. The slenderness limit needs to be satisfied in both planes for the column to be treated as a short column. Due to the current uncertainty in modeling the stress–strain behaviour of FRP-confined concrete in rectangular columns, readers are reminded to exercise due caution when using Eq. 6.35 in the strengthening design of rectangular columns.

6.6.4 Concentrically-loaded short columns

The design strength of an FRP-confined RC column under concentric compression is given by:

$$N_u = f'_{cc,d} A_c + \frac{f_{y,k}}{\gamma_s} A_{sc} \qquad [6.36]$$

where $f'_{cc,d}$ is the design value of compressive strength of FRP-confined concrete (Eqs 6.29, 6.32 and 6.34), $f_{y,k}$ is the characteristic yield strength of longitudinal steel, A_c and A_{sc} are cross-sectional areas of concrete and longitudinal steel, respectively, and γ_s is the partial safety factor for steel.

6.6.5 Eccentrically-loaded short columns

Circular columns

In the design of RC members, the distribution of compressive stresses in concrete is generally approximated using an equivalent rectangular stress block. Such an equivalent stress block provides the same axial force and the same moment contribution in a section analysis as the original stresses. The equivalent stress block can be described by two stress block factors, the mean stress factor a_1, which is defined as the ratio of the uniform stresses over the stress block to the compressive strength of concrete, and the block depth factor β_1, which is defined as the ratio of the depth of the stress block to that of the neutral axis. Due to FRP confinement, values of stress block factors in existing design codes for conventional RC members are no longer suitable for FRP-confined RC members. Appropriate stress block factors for FRP-confined circular RC columns are defined below.

Numerical results from section analyses using Lam and Teng's design-oriented stress–strain model (Teng et al., 2007a) showed that the value of β_1 varies slightly around 0.9 (Jiang and Teng, 2006). Therefore, the following equation is proposed for the block depth factor for simplicity:

$$\beta_1 = 0.9 \qquad [6.37a]$$

With the value of β_1 defined, the mean stress factor was found to be well approximated by the following simple linear equation (Jiang and Teng, 2006):

$$\alpha_1 = 1.17 - 0.2 \frac{f'_{cc,d}}{0.67 f_{cu}/\gamma_c} \qquad [6.37b]$$

Using the stress block factors given above, a simplified section analysis is presented herein, which represents a modification of the procedure given in GB-50010 (2002) for the design of circular RC columns. This method is only applicable to columns which have six or more evenly distributed longitudinal steel reinforcing bars. As shown in Fig. 6.16, the steel reinforcing bars are smeared into an equivalent steel cylinder of the same total cross-sectional area and with longitudinal strength only. The central angle corresponding to the depth of the equivalent stress block is equal to $2\pi\theta$. Based on the above assumption, the following design equations can be derived for short FRP-confined circular RC columns:

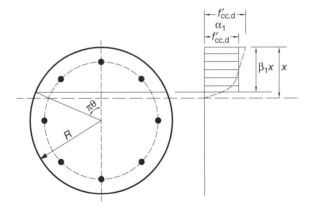

6.16 Equivalent stress block for an FRP-confined circular RC column.

$$N_u = \alpha_1 f'_{cc,d} A\theta \left(1 - \frac{\sin 2\pi\theta}{2\pi\theta}\right) + (\theta_c - \theta_t)\frac{f_{y,k}}{\gamma_s} A_{sc} \quad [6.38a]$$

and

$$M_u = \frac{2}{3}\alpha_1 f'_{cc,d} AR\frac{\sin^3 \pi\theta}{\pi} + \frac{f_{y,k}}{\gamma_s} A_{sc} R\frac{\sin \pi\theta_c + \sin \pi\theta_t}{\pi} \quad [6.38b]$$

with

$$0 \leq \theta_c = 1.25\theta - 0.125 \leq 1 \quad [6.38c]$$

$$0 \leq \theta_t = 1.125 - 1.5\theta \leq 1 \quad [6.38d]$$

where N_u and M_u are the axial load capacity, and the moment capacity, respectively, θ_c and θ_t are factors introduced to simplify the calculation of the contribution from the steel reinforcement and A is the area of the cross-section. A detailed derivation is available elsewhere (Jiang and Teng, 2006).

The predictions of Eq. 6.38 are compared with the results of accurate section analysis in Fig. 6.17 for a reference column which has the same geometric properties as the one used for comparison in Fig. 6.14. For simplicity, the characteristic values of material properties were used in the comparison shown in Fig. 6.17 by setting all partial safety factors to unity. The unconfined concrete was assumed to have $f_{cu} = 30$ MPa, which corresponds to a characteristic compressive strength of concrete in a real column ($=0.67 f_{cu}$) of 20.1 MPa and the steel was assumed to have $f_{y,k} = 460$ MPa. Figure 6.17a is for the case when the column is confined with a 3-ply CFRP jacket with an elastic modulus of 230 GPa and $\varepsilon_{h\,rup,k} = 0.01$ based on a nominal ply thickness of 0.165 mm, while Fig. 6.17b is for confinement by a 8-ply GFRP jacket with an elastic modulus of 80 GPa and $\varepsilon_{h,rup,k} = 0.015$ based on a nominal ply thickness of 0.17 mm. The approximate design equations are seen to provide accurate predictions. It should be noted that the interaction curves produced by the approximate design equations terminate at a high axial load level when the neutral axis starts to fall outside the cross-section. This is because the stress block factors are deduced on the assumption that the neutral axis stays within the cross-section, which means that the use of Eq. 6.37 may cause significant errors when the neutral axis falls outside the cross-section. Nevertheless, this limitation of the approximate design equations is insignificant as such cases are not normally encountered in design due to the inclusion of the minimum load eccentricity for all columns.

Strengthening of RC columns with FRP composites 187

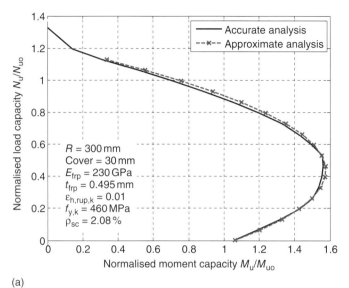

(a)

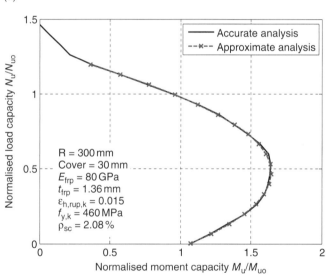

(b)

6.17 Comparison of interaction curves for circular columns: (a) 3-ply CFRP; (b) 8-ply GFRP.

Rectangular columns

The conventional section analysis can be employed in the FRP strengthening design of eccentrically-loaded short FRP-confined rectangular columns as long as the stress–strain model is defined. Similarly, simple approximate design equations are not difficult to develop with a given stress–strain model such as that given by Lam and Teng (2003b). Despite many existing studies, there is still much uncertainty in modelling the stress–strain behaviour of FRP-confined concrete in rectangular columns, particularly when the column is large and is subjected to combined axial loading and bending. This is an area where more research is still urgently needed.

If Lam and Teng's (2003b) model is adopted in the section analysis of short FRP-confined rectangular columns, it can be shown that the equivalent stress block is still closely approximated by Eq. 6.37. The accuracy of Eq. 6.37 will not change to any significant extent as a result of a future refinement of the ultimate condition equations for FRP-confined concrete in rectangular columns, as long as the shape of the stress–strain curve remains the same. Once the contribution of concrete to the load capacity is approximated using the equivalent stress block, the contribution of steel reinforcement can easily be evaluated using the conventional section analysis procedure.

Both the accurate section analysis using Lam and Teng's (2003b) model and the approximate section analysis using the equivalent stress block were employed to predict the behaviour of the same reference rectangular column as shown in Fig. 6.18. The column was assumed to have longitudinal steel reinforcing bars only at the four corners for simplicity. The material properties are the same as those used for comparison in Fig. 6.17a and the steel reinforcing bars were assumed to have a 30 mm diameter.

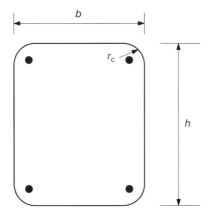

6.18 FRP-confined rectangular RC column section.

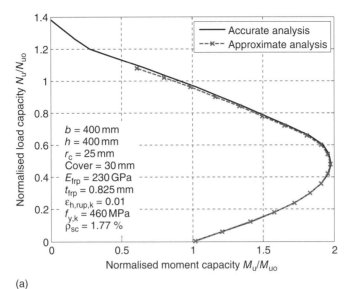

(a)

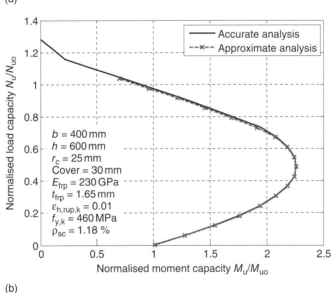

(b)

6.19 Comparison of interaction curves for rectangular columns: (a) square column; (b) rectangular column.

The predictions of the approximate analysis are compared with the results of the accurate section analysis in Fig. 6.19. Figure 6.19a is for a square column with a side dimension of 400 mm, a corner radius of 25 mm and a concrete cover of 30 mm, while Fig. 6.19b is for a rectangular column which has the same geometric properties as the square column except that the long side dimension is 600 mm. The interaction curves in Fig. 6.19b are for bending about the strong axis due to a load eccentricity in the direction of the longer side of the rectangular section. The approximate analysis is seen to provide accurate predictions.

Elliptical columns

For FRP-confined elliptical RC columns, no stress–strain model has been developed, so a great deal of further research is required before a reliable design procedure can be established. Such columns can be designed using the same type of section analysis, provided a reliable stress–strain model is available.

6.7 Acknowlegement

The authors are grateful for the financial support provided by the Research Grants Council of the Hong Kong SAR (Project No: PolyU 5059/02E) and The Hong Kong Polytechnic University (Project Codes: BBZH and RG88).

6.8 References

ACI (1996) *State-of-the-Art Report on Fiber Reinforced Plastic (FRP) Reinforcement for Concrete Structures*, ACI 440R-96, Farmington Hills, MI, USA, American Concrete Institute.

ACI (2002) *Guide for the Design and Construction of Externally Bonded FRP Systems for Strengthening Concrete Structures*, ACI 440.2R-02, Farmington Hills, MI, USA, American Concrete Institute.

ACI (2005) *Building Code Requirements for Structural Concrete and Commentary*, ACI 318-05, Farmington Hills, MI, USA, American Concrete Institute.

BECQUE J, PATNAIK A K and RIZKALLA S H (2003) Analytical models for concrete confined with FRP tubes, *Journal of Composites for Construction*, ASCE, **7**(1), 31–38.

BERTHET J F, FERRIER E and HAMELIN P (2006) Compressive behavior of concrete externally confined by composite jackets – Part B: modeling, *Construction and Building Materials*, **20**(5), 338–347.

BINICI B (2005) An analytical model for stress–strain behavior of confined concrete, *Engineering Structures*, **27**(7), 1040–1051.

BSI (1997) *BS 8110-1: 1997 Structural Use of Concrete. Code of Practice for Design and Construction*, London, UK, British Standards Institution.

CAO S, JING D and SUN N (2006) Experimental study on concrete columns strengthened with CFRP sheets confinement under eccentric loading, in Mirmiran A and Nanni A (eds), *Proceedings, Third International Conference on FRP Composites in Civil Engineering–CICE 2006*, Miami, FL, USA, Florida International University, 193–196.

CAREY S A and HARRIES K A (2005) Axial behavior and modeling of confined small-, medium-, and large-scale circular sections with carbon fiber-reinforced polymer jackets, *ACI Structural Journal*, **102**(4), 596–604.

CEB-FIP (1993) *Model Code 1990*, Bulletin No. 213/214, Lausanne, Switzerland, Comité Euro-International du Beton.

CEN (1992) ENV 1992-1-1: *Eurocode 2: Design of Concrete Structures – Part 1: General Rules and Rules for Buildings*, Brussels, Belgium, European Committee for Standardization.

CHENG H L, SOTELINO E D and CHEN W F (2002) Strength estimation for FRP wrapped reinforced concrete columns, *Steel and Composite Structures*, **2**(1), 1–20.

CHUN S S and PARK H C (2002) Load carrying capacity and ductility of RC columns confined by carbon fiber reinforced polymer. *Proceedings, 3rd International Conference on Composites in Infrastructure*–ICCI 2002, San Francisco, CA, USA, June 10–12 (CD ROM).

CONCRETE SOCIETY (2004) *Design Guidance for Strengthening Concrete Structures with Fibre Composite Materials*, TR 55, 2nd edn, Camberley, UK, The Concrete Society.

FAM A Z and RIZKALLA S H (2001) Confinement model for axially loaded concrete confined by circular fiber-reinforced polymer tubes, *ACI Structural Journal*, **98**(4), 451–461.

FAM A, FLISAK B and RIZKALLA S (2003) Experimental and analytical modeling of concrete-filled fiber-reinforced polymer tubes subjected to combined bending and axial loads, *ACI Structural Journal*, **100**(4), 499–509.

FARDIS M N and KHALILI H (1982) FRP-encased concrete as a structural material, *Magazine of Concrete Research*, **34**(122), 191–202.

FIB (2001) *Externally Bonded FRP Reinforcement for RC Structures*, Lausanne, Switzerland, Federation for Internationale du beton.

FITZWILLIAM J and BISBY L A (2006) Slenderness effects on circular FRP-wrapped reinforced concrete columns, *Proceedings Third International Conference on FRP Composites in Civil Engineering–CICE 2006*, Miami, FL, USA, Dec 13–15, 499–502.

GB-50010 (2002) *Code for Design of Concrete Structures*, Beijing, China, China Architecture and Building Press.

HADI M N S (2006) Behaviour of wrapped normal strength concrete columns under eccentric loading, *Composite Structures*, **72**(4), 503–511.

HARAJLI M H (2006) Axial stress-strain relationship for FRP confined circular and rectangular concrete columns, *Cement & Concrete Composites*, **28**(10), 938–948.

HARMON T G, RAMAKRISHNAN S and WANG E H (1998) Confined concrete subjected to uniaxial monotonic loading, *Journal of Engineering Mechanics*, ASCE, **124**(12), 1303–1308.

HARRIES K A and KHAREL G (2002) Behavior and modeling of concrete subject to variable confining pressure, *ACI Materials Journal*, **99**(2), 180–189.

JIANG T and TENG J G (2006) Strengthening of short circular RC columns with FRP jackets: a design proposal, *Proceedings Third International Conference on FRP Composites in Civil Engineering–CICE 2006*, Miami, FL, USA, Dec 13–15, 187–192.

JIANG T and TENG J G (2007) Analysis-oriented models for FRP-confined concrete, *Engineering Structures*, **29**(11), 2968–2986.

KARBHARI V M and GAO Y (1997) Composite jacketed concrete under uniaxial compression–verification of simple design equations, *Journal of Materials in Civil Engineering*, ASCE, **9**(4), 185–193.

KONG F K and EVANS R H (1987) *Reinforced and Prestressed Concrete*, 3rd edn, London, UK, Chapman & Hall.

LAM L and TENG J G (2003a) Design-oriented stress–strain model for FRP-confined concrete, *Construction and Building Materials*, **17**(6–7), 471–489.

LAM L and TENG J G (2003b) Design-oriented stress–strain model for FRP-confined concrete in rectangular columns, *Journal of Reinforced Plastics and Composites*, **22**(13), 1149–1186.

LAM L and TENG J G (2004) Ultimate condition of fiber reinforced polymer-confined concrete, *Journal of Composites for Construction*, ASCE, **8**(6), 539–548.

LILLISTONE D and JOLLY C K (2000) An innovative form of reinforcement for concrete columns using advanced composites, *The Structural Engineer*, **78**(23/24), 20–28.

MANDER J B, PRIESTLEY M J N and PARK R (1988) Theoretical stress–strain model for confined concrete, *Journal of Structural Engineering*, ASCE, **114**(8), 1804–1826.

MARQUES S P C MARQUES D C S C, DA SILVA J L and CAVALCANTE M A A (2004) Model for analysis of short columns of concrete confined by fiber-reinforced polymer, *Journal of Composites for Construction*, ASCE, **8**(4), 332–340.

MIRMIRAN A and SHAHAWY M (1997) Dilation characteristics of confined concrete, *Mechanics of Cohesive-Frictional Materials*, **2**(3), 237–249.

MIYAUCHI K, INOUE S, KURODA T and KOBAYASHI A (1999) Strengthening effects of concrete columns with carbon fiber sheet, *Transactions of the Japan Concrete Institute*, **21**, 143–150.

MATTYS S, TOUTANJI H, AUDENAERT K and TAERWE L (2005) Axial behavior of large-scale columns confined with fiber-reinforced polymer composites, *ACI Structural Journal*, **102**(2), 258–267.

MOSALAM K M, TALAAT M and BINICI B (2007) A computational model for reinforced concrete members confined with fiber reinforced polymer lamina: implementation and experimental validation, *Composites Part B: Engineering*, **38**, 598–613.

NANNI A and NORRIS M S (1995) FRP jacketed concrete under flexure and combined flexure-compression, *Construction and Building Materials*, **9**(5), 273–281.

OHNO S, MIYAUCHI Y, KEI T and HIGASHIBATA Y (1997) Bond properties of CFRP plate joint, *Proceedings, Third International Symposium on Non-Metallic (FRP) Reinforcement for Concrete Structures*, Sapporo, Japan, Japan Concrete Institute, 241–248.

PANTELIDES C P and YAN Z H (2007) Confinement model of concrete with externally bonded FRP jackets or posttensioned FRP shells, *Journal of Structural Engineering*, ASCE, **133**(9), 1288–1296.

PARK R and PAULAY T (1975) *Reinforced Concrete Structures*, NY, USA, John Wiley & Sons.

PARVIN A and WANG W (2001) Behavior of FRP jacketed concrete columns under eccentric loading, *Journal of Composites in Construction*, ASCE, **5**(3), 146–152.

POPOVICS S (1973) Numerical approach to the complete stress–strain relation for concrete, *Cement and Concrete Research*, **3**(5), 583–599.

ROCCA S, GALATI N and NANNI A (2006) Large-size reinforced concrete columns strengthened with carbon FRP: experimental evaluation, *Proceedings, Third International Conference on FRP Composites in Civil Engineering – CICE 2006*, Miami, FL, USA, Dec 13–15, 491–494.

SAAFI M, TOUTANJI H A and LI Z (1999) Behavior of concrete columns confined with fiber reinforced polymer tubes, *ACI Materials Journal*, **96**(4), 500–509.

SAMAAN M, MIRMIRAN A and SHAHAWY M (1998) Model of concrete confined by fiber composite, *Journal of Structural Engineering*, ASCE, **124**(9), 1025–1031.

SAENZ N and PANTELIDES C P (2007) Strain-based confinement model for FRP-confined concrete, *Journal of Structural Engineering*, ASCE, **133**(6), 825–833.

SPOELSTRA M R and MONTI G (1999) FRP-confined concrete model, *Journal of Composites for Construction*, ASCE, **3**(3), 143–150.

TAO Z, TENG J G, HAN L H and LAM L (2004) Experimental behaviour of FRP-confined slender RC columns under eccentric loading, in Hollaway L C, Chryssanthopoulos M K and Moy S S J (eds), *Proceedings, Second International Symposium on Advanced Polymer Composites for Structural Applications in Construction – ACIC 2004*, Cambridge, UK, Woodhead, 203–212.

TENG J G and LAM L (2002) Compressive Behavior of carbon fiber reinforced polymer-confined concrete in elliptical columns, *Journal of Structural Engineering*, ASCE, **128**(12), 1535–1543.

TENG J G and LAM L (2004) Behavior and modeling of fiber reinforced polymer-confined concrete, *Journal of Structural Engineering*, ASCE, **130**(11), 1713–1723.

TENG J G, CHEN J F, SMITH S T and LAM L (2002) *FRP-Strengthened RC Structures*, Chichester, UK, John Wiley and Sons.

TENG J G, JIANG T, LAM L and LUO Y Z (2007a) Refinement of Lam and Teng's design-oriented stress–strain model for FRP-confined concrete, *Proceedings, Third International Conference on Advanced Composites in Construction*, University of Bath, Bath, UK, Apr 2–4, 116–121.

TENG J G, HUANG Y L LAM L and YE L P (2007b) Theoretical model for fiber reinforced polymer-confined concrete, *Journal of Composites for Construction*, ASCE, **11**(2), 201–211.

TOUTANJI H A (1999) Stress–strain characteristics of concrete columns externally confined with advanced fiber composite sheets, *ACI Materials Journal*, **96**(3), 397–404.

WU G, WU Z S and LU Z T (2007) Design-oriented stress–strain model for concrete prisms confined with FRP composites, *Construction and Building Materials*, **21**(5), 1107–1121.

XIAO Y and MA R (1997) Seismic retrofit of RC circular columns using prefabricated composite jacketing, *Journal of Structural Engineering*, ASCE, **123**(10), 1357–1364.

XIAO Y and WU H (2000) Compressive behavior of concrete confined by carbon fiber composite jackets, *Journal of Materials in Civil Engineering*, ASCE, **12**(2), 139–146.

XIAO Y and WU H (2003) Compressive behavior of concrete confined by various types of FRP composite jackets, *Journal of Reinforced Plastics and Composites*, **22**(13), 1187–1201.

YEH F Y and CHANG K C (2007) Confinement efficiency and size effect of FRP confined circular concrete columns, *Structural Engineering and Mechanics*, **26**(2), 127–150.

YOUSSEF M N (2003) *Stress–strain Model for Concrete Confined by FRP Composites*, PhD Dissertation, University of California, Irvine, CA, USA.

YOUSSEF M N, FENG M Q and MOSALLAM A S (2007) Stress–strain model for concrete confined by FRP composites, *Composites Part B: Engineering*, **38**(5–6), 614–628.

YUAN W and MIRMIRAN A (2001) Buckling analysis of concrete-filled FRP tubes, *International Journal of Structural Stability and Dynamics*, **1**(3), 367–383.

7
Design guidelines for fibre-reinforced polymer (FRP)-strengthened reinforced concrete (RC) structures

J. G. TENG, The Hong Kong Polytechnic University, China;
S. T. SMITH, The University of Hong Kong, China;
J. F. CHEN, The University of Edinburgh, UK

7.1 Introduction

Several documents on the design and construction of externally bonded fibre reinforced polymer (FRP) systems for the strengthening of reinforced concrete (RC) structures have been published in recent years (fib, 2001; ISIS, 2001; JSCE, 2001; ACI, 2002; CSA, 2002; CECS, 2003; Concrete Society, 2004; CNR, 2005). In addition, the authors are aware of two other guidelines that will be published in the near future: the Australian guideline (Oehlers et al., 2008) and the Chinese national standard for the structural use of FRP composites in construction. These documents, herein referred to as *guidelines*, provide guidance for the selection, design and installation of FRP strengthening systems for RC structures. They represent the culmination of experimental investigations, theoretical studies and field implementations and monitoring of FRP strengthening systems up to the time of writing the respective guideline, and are therefore limited in one way or another, principally as a result of the limitation of the information then available.

Of the guidelines that currently exist around the world, those that have been more widely referred to by the international engineering community appear to be the fib (2001), JSCE (2001), ACI (2002) and Concrete Society (2004) guidelines. Systematic design recommendations have also been published by independent researchers (e.g. Teng et al., 2002; Täljsten, 2003; Oehlers and Seracino, 2004).

This chapter presents a critical review of the fib (2001), JSCE (2001), ACI (2002) and Concrete Society (2004) guidelines, with reference to the state-of-the-art knowledge presented in the three preceding chapters (Chapters 4–6). Therefore, the chapter is focussed on the design provisions for the flexural and shear strengthening of RC beams and the strengthening/retrofit of RC columns. Both the similarities and the differences between these guidelines are examined. The provisions in these guidelines are also contrasted with the latest research findings summarised in Chapters 4–6, which serves as a useful reference for future revisions of these guidelines. The information presented in this chapter is also intended to assist users of these guidelines in interpreting the design provisions contained in them and in looking for the best solutions within these guidelines and beyond.

7.2 General assumptions

In all four guidelines (*fib*, 2001; JSCE, 2001; ACI, 2002; Concrete Society, 2004), the design provisions are presented on the basis of the following general assumptions:

- The material properties and thus the condition of the structure to be strengthened have been obtained from a condition assessment of the structure.
- The application of the FRP system follows an appropriate quality assurance procedure, which should include proper preparation of the concrete surface prior to the installation of the FRP system, quality control of the installation process and regular inspection and assessment of the strengthening works after installation according to suitable maintenance and repair protocols put in place.
- The adhesive is sufficiently strong so that cohesive failure within the adhesive layer or adhesion failure between the adhesive layer and the concrete or the FRP reinforcement is prevented.

7.3 Limit states and reinforced concrete design

The *fib* (2001), JSCE (2001), ACI (2002) and Concrete Society (2004) guidelines all adopt the limit state design philosophy. The guidelines are based on traditional RC design principles and their corresponding codes for the design of RC structures. The ultimate limit state of strength is typically of main concern, followed by the serviceability limit state. If the strengthening is for the improvement of serviceability, then the serviceability limit state will obviously govern the design.

On the whole, the classifications in terms of ultimate and serviceability limit states in the different guidelines are largely similar. Naturally, strengths of members under bending, shear and compression are issues of ultimate limit state while deflections and cracking are serviceability issues. However, some differences do exist between the guidelines. For example, fatigue is classified as an ultimate limit state in ACI (2002) but as a serviceability limit state in Concrete Society (2004). The FRP system introduces some additional aspects to the limit states, such as the stress rupture of FRP systems which is treated as an ultimate limit state in ACI (2002) and as a serviceability limit state in Concrete Society (2004).

Table 7.1 Environmental reduction factors of ACI (2002)

Exposure conditions	Fibre and resin type	Environmental reduction factor C_E
Interior exposure	Carbon/epoxy	0.95
	Glass/epoxy	0.75
	Aramid/epoxy	0.85
Exterior exposure (bridges, piers and unenclosed parking garages)	Carbon/epoxy	0.85
	Glass/epoxy	0.65
	Aramid/epoxy	0.75
Aggressive environment (chemical plants and waste water treatment plants)	Carbon/epoxy	0.85
	Glass/epoxy	0.50
	Aramid/epoxy	0.70

7.4 Material properties: characteristic and design values

In all four guidelines, the calculations of characteristic and design strengths of concrete and steel reinforcement follow those specified in their corresponding codes for the design of RC structures. In addition, the characteristic strength of the FRP may be given by the manufacturer or via tension tests on coupon specimens. The test characteristic strength of FRP is determined from a statistical analysis of the mean and standard deviation of multiple coupon test results to ensure that the characteristic strength is exceeded by the test results in 95–99.9% of the cases.

The design strength of the FRP according to *fib* (2001), JSCE (2001) and Concrete Society (2004) is obtained by dividing the characteristic strength by a partial safety factor (otherwise known as a material safety factor or material factor). According to ACI (2002), the design strength of FRP is calculated by multiplying the characteristic strength by an environmental reduction factor which accounts for the effects of different environmental exposure conditions and fibre types (Table 7.1).

Partial safety factors for the tensile strength of FRP specified by *fib* (2001) and Concrete Society (2004) range from 1.2 to about 8.0 and they are generally dependent on the fibre type and the method of manufacture/application. These partial safety factors are intended to account for changes in material properties with time as well as other factors, which is the main reason for the dependency of partial safety factors on fibre types. In JSCE (2001), only FRPs formed via the wet lay-up process using carbon and aramid fibres are considered; the material factor is specified as a range (1.2–1.3) and does not depend on the fibre type or the method of application.

According to ACI (2002), the characteristic rupture strain is also multiplied by the Table 7.1 reduction factors so the elastic modulus is assumed not to be affected by the environment, which is consistent with existing experimental evidence. In *fib* (2001) the elastic modulus can either be based on a percentage of exceedence or the mean value. In JSCE (2001) the characteristic modulus of elasticity is taken as the mean of the tension test results. In the above three guidelines, the elastic modulus is assumed to be independent of environmental exposure. Concrete Society (2004), however, recommends a different approach where the characteristic values of both the elastic modulus and the rupture strain are divided by partial safety factors to obtain the design values.

None of the partial safety factors or environmental reduction factors for FRP appears to have a rigorous statistical basis. More research is therefore needed on this aspect.

7.5 Flexural strengthening

7.5.1 Failure modes and section moment capacity

The flexural capacity of RC flexural members such as beams and slabs can be increased by bonding FRP to their tension face as described in Chapter 4. All four guidelines primarily consider simply-supported beams although continuous beams can be analysed by considering them as a series of simply-supported beams that span between points of contraflexure.

Failure of an FRP-strengthened RC beam may occur by the compressive crushing of concrete, the tensile rupture of the FRP or debonding in one of several forms as fully described in Chapter 4. Preferably, the steel reinforcement should have sufficiently yielded at failure to ensure the formation of obvious cracks prior to failure by concrete crushing, FRP rupture or FRP debonding with the aim of ensuring some degree of ductility in these generally brittle modes of failure. Nevertheless, section failure with a low strain in the steel reinforcement is permitted in all four guidelines, and in three of them [except JSCE (2001)] the lack of ductility in such cases is compensated by requiring a significant strength reserve.

The evaluation of the ultimate moment of an FRP-strengthened RC section is based on fundamental principles of strain compatibility and internal force equilibrium as found in various reinforced concrete codes of practice [e.g. ENV 1992-1-1 (CEN, 1992); BS 8110 (BSI, 1997); ACI 318 (ACI, 1999); JSCE, 1996]. Plane sections are assumed to remain plane and the tensile strength of concrete is ignored. A set of design equations for the moment capacity of an FRP-strengthened section has been presented in Chapter 4.

For an FRP-strengthened RC section failing by concrete crushing, the use of the conventional stress block approach is perfectly valid. If the section fails by rupture of the FRP, due to the brittle nature of the FRP, the ultimate strain of concrete may not be reached and the conventional rectangular stress block approach, based on concrete crushing, is invalid. A modified rectangular stress block may be used as has been discussed in Chapter 4.

In *fib* (2001), JSCE (2002) and Concrete Society (2004), the design value of the strength of the section is calculated from the design values of the strengths of concrete, steel and FRP. ACI (2002) employs the resistance factor approach, where the nominal strength of the section is multiplied by a strength-reduction factor ϕ to arrive at the design strength. This strength reduction factor is dependent on the ductility of the section, based on the philosophy that a section with lower ductility should be compensated with a higher reserve of strength.

7.5.2 Debonding

Several distinct debonding modes have been observed in numerous experimental studies. They may be classified as (i) intermediate crack-induced interfacial debonding (IC debonding), (ii) critical diagonal crack debonding (CDC debonding), (iii) concrete cover separation, and (iv) plate end interfacial debonding. A detailed discussion of these failure modes is given in Chapter 4. The different design guidelines use different terminologies and approaches to categorise and design against debonding. In the following, these design approaches are summarised and discussed in accordance with the classification of and terminology for debonding failure modes given in Chapter 4.

Intermediate crack (IC) induced interfacial debonding

In *fib* (2001), IC debonding is referred to as 'peeling-off at flexural cracks'. It also discusses 'peeling-off at shear cracks' which is more likely to correspond to the mode of CDC debonding discussed in Chapter 4 than the mode of IC debonding at a flexural-shear crack. In *fib* (2001), three different approaches to design against IC debonding (i.e. peeling-off at flexural cracks) failures are discussed. In the first approach, the strain in the FRP is limited and the limiting value is discussed. *fib* (2001) admits that this is a crude simplification of the problem. The second approach follows the same principle as the JSCE (2001) method, but is far more complex than the latter and is thus unsuitable for use in practical design as admitted by *fib* (2001). In the third approach, the shear stress at the interface calculated based on simple equilibrium considerations is checked against a critical value which is a function of the concrete cylinder splitting tensile strength only.

In JSCE (2001), IC debonding is referred to as 'peeling failure'. According to JSCE (2001), IC debonding failure occurs when the maximum value of the axial stress change in the FRP over a typical crack spacing reaches a critical value which is a function of the interfacial fracture energy of the FRP-to-concrete interface. It recommends that this fracture energy should be determined through testing or taken as 0.5 N/mm when a test is not conducted. The crack spacing for use in this model is recommended to be 150~250 mm when the number of plies of FRP is below 3. These recommendations for the interfacial fracture energy and the crack spacing are obviously rather empirical.

The design recommendation in ACI (2002) does not distinguish between failure modes. To prevent debonding, a debonding strain limit, which is equal to the product of a reduction factor κ_m and the design rupture strain of the FRP, is placed on the tensile strain of the FRP, where κ_m is not to exceed 0.90. This model does not include such fundamental properties as concrete strength.

To design against debonding failure, Concrete Society (2004) imposes a limit of the strain in the FRP to 0.008 in conjunction with a limit of 0.8 N/mm^2 for the shear stress between the FRP and the concrete. Locations where this shear stress should be checked and formulas for calculating this shear stress are given. This model is very simplistic as it does not consider the geometric or material properties of the FRP and the concrete in the limiting values of strain and stress.

Yao et al. (2005) compared four IC debonding strength models (i.e. *fib*, 2001; JSCE, 2001; ACI, 2002; Teng et al., 2003 as referred to in Chapter 4) with the test results of four simply-supported FRP-strengthened RC beams and 18 FRP-strengthened cantilever slabs failing by IC debonding. Comparisons were made for both the IC debonding strain in the FRP and the IC debonding moment. The *fib* (2001) and ACI (2002) models greatly over-estimated debonding strains and most debonding moments while the JSCE (2001) model sometimes over-estimated and sometimes under-estimated the strains and moments. Teng et al.'s (2003) IC debonding strength model generally provided safe predictions of the experimental debonding strains as this model is a lower bound model for design use; the scatter of predictions was, however, large. It was therefore concluded that much further research needed to be carried out to develop a more accurate IC debonding strength model. The recent model by Lu et al. (2007), referred to in Chapter 4, arose from follow-on research and provides much closer predictions of IC debonding failures than Teng et al.'s (2003) model. Lu et al.'s (2007) model is herein recommended for use in design instead of the provisions in the four guidelines or Teng et al.'s (2003) model discussed above.

Critical diagonal crack (CDC) debonding

JSCE (2001) does not contain any specific discussions of the CDC debonding failure mode. In ACI (2002), cover delamination is mentioned, which is the same as cover separation defined in Chapter 4. It suggests that all debonding failure modes can be designed using the same set of strain limit formulas.

fib (2001) discusses two debonding failure modes which fall into the CDC debonding mode as classified in Chapter 4. The first is referred to as 'peeling-off at shear cracks' and the second is referred to as 'end shear failure'. It mentions the conclusion by Blaschko (1997) that peeling-off at shear cracks can be prevented if the shear force acting on the beam is smaller than the shear resistance contributed by the concrete of the RC beam alone. It further recommends that if this requirement is not met, then appropriate shear strengthening should be carried out. To design against end shear failure, Jansze's (1997) model based on the fictitious shear span concept is recommended.

In Concrete Society (2004), CDC debonding is referred to as 'shear-crack-induced FRP separation'. According to Concrete Society (2004), this debonding failure mode can be neglected if the shear capacity of the concrete in the RC beam alone can resist the applied shear force. If the applied shear force exceeds 67% of the ultimate shear capacity of the section then CDC debonding will occur. If the applied shear force lies between the shear capacity of the concrete alone and 67% of the ultimate shear capacity then 'careful consideration should be given to shear crack initiation', but no debonding design rules are given. The vertical displacement of opposite faces of a shear crack is assumed to be the main driver for this debonding mode.

Although no design equations are given for the CDC debonding failure mode in *fib* (2001) and Concrete Society (2004), the proposition that CDC debonding is not possible if the applied shear force is below the concrete component of the shear resistance of the RC beam has been supported by existing studies (e.g. Smith and Teng, 2002; Teng and Yao, 2005; 2007). The strength model of Teng and Yao (2005, 2007) for plate end debonding given in Chapter 4 provides a less conservative assessment of the CDC debonding failure load by including the contribution of the steel shear reinforcement. If the shear force exceeds the resistance evaluated by the strength model of Teng and Yao (2005, 2007), then the best approach is to provide shear strengthening to the beam as has been suggested in Chapter 4.

Concrete cover separation

In *fib* (2001), the end shear failure mode appears to cover the two possible modes of CDC debonding and cover separation. The proposed design model, based on Janzse's (1997) work, is discussed above.

ACI (2002) refers to concrete cover separation as 'concrete cover delamination' and provides detailing guidance for simply supported beams apart from the strain limit formulas mentioned earlier. It recommends that the FRP reinforcement should be extended a distance equal to the effective depth of the section past the position on the span corresponding to the cracking moment M_{cr} under factored loads. Guidance is also given in ACI (2002) for the termination of multiple plies of FRP. The FRP is to be anchored by transverse reinforcement (e.g. U-strips) if the shear force is greater than 2/3 of the shear capacity of the section due to concrete alone. For the case of continuous beams, the FRP is required to be terminated a distance of half the effective section depth or 150 mm minimum past the inflection point. These detailing rules appear to cover CDC debonding as a special case of 'concrete cover delamination', which is consistent with the observations made in Chapter 4.

No specific guidance to design against concrete cover separation is given in JSCE (2001) and Concrete Society (2004).

Smith and Teng (2002) assembled a database of 40 flexurally-strengthened RC beams failing by concrete cover separation and found models based on the shear capacity of the beam to be the most robust and promising. They found Jansze's model (1997) to give generally conservative predictions, but not in all cases.

The strength model of Teng and Yao (2005, 2007) given in Chapter 5 caters for all plate end debonding failure modes and is recommended for use in design against cover separation failures as well. Again, if the applied load exceeds the resistance evaluated by the strength model of Teng and Yao (2005, 2007), the best approach is to provide shear strengthening to the beam as has been suggested in Chapter 4.

Plate end interfacial debonding

ACI (2002) does not address this mode explicitly, except that its strain limit formulas can also be used to design against this debonding failure model. JSCE (2001) does not discuss this debonding failure mode either.

fib (2001) discuses 'peeling-off at end anchorage' which may be interpreted to be plate end interfacial debonding. It recommends that the end anchorage be checked on the basis of fracture mechanics and a bond-slip model. Holzenkämpfer's (1994) model, as modified by Neubauer and Rostásy (1997), is cited as an example. In Concrete Society (2004), design against plate end interfacial debonding is again based on the bond strength model of Neubauer and Rostásy (1997).

In Chapter 4, end anchorage requirement is recommended on the basis of the bond strength model of Chen and Teng (2001) because this model has been shown to be the most accurate available (Chen and Teng, 2001; Lu et al., 2005). Since plate end interfacial debonding is a plate end debonding failure mode covered by the strength model of Teng and Yao (2005, 2007) given in Chapter 4, this model is again recommended for use in the design against plate end interfacial debonding failures. Shear strengthening of the beam as suggested in Chapter 4 also needs to be considered should the resistance evaluated by the strength model of Teng and Yao (2005, 2007) fall below the applied load.

Surface irregularity

In *fib* (2001), unevenness or roughness of the surface is recognised to contribute to debonding at the FRP-to-concrete interface. *fib* (2001) notes that the effect of surface unevenness has not been studied sufficiently to recommend any procedure.

Concrete Society (2004) suggests that surface curvature may lead to the development of normal (i.e. peeling) stresses between the FRP and the concrete which will lead to debonding. The influence of surface curvature may be disregarded if over any 1 m length the concavity of the concrete or the installed FRP does not exceed 3 mm. If this concavity limit is exceeded, no design recommendation is given; however, in order to deal with the effect, specialist advice or reference to previous tests (e.g. Eshwar et al., 2003) should be sought. Since no general recommendations have been proposed, the best advice is that surface irregularity should be avoided as far as possible in practice.

7.5.3 Ductility

It is well known that RC beams strengthened with bonded FRP show reduced ductility, so it is important to ensure a minimum level of ductility in these beams. A simple method for ensuring ductility in the strengthened systems is to require the internal tension steel reinforcement to develop a sufficiently large strain. If this requirement cannot be met, strength reduction factors may be introduced so that the lack of ductility is compensated with a significant strength reserve. This approach is adopted by three of the guidelines, JSCE (2001) being the exception.

In *fib* (2001), geometric or material limits are placed on the concrete beam, FRP and steel reinforcement to ensure a ductile section based on ENV 1992-1-1 (CEN, 1992). In addition, it states that if the design value of resistance is at least 1.2 times the design value of action, then the ductility requirement no longer needs to be fulfilled. JSCE (2001) does not address the issue of ductility of RC beams flexurally strengthened with FRP. ACI (2002) adopts the same approach as ACI 318 (ACI, 1999) where sections with lower ductility are required to possess a higher reserve of strength (i.e. lower strength reduction factor ϕ). According to Concrete Society (2004), ductility is checked by ensuring that the internal tensile steel reinforcement strain is not less than the design value of yield strain of the steel reinforcement plus 0.002, unless the ultimate moment that the section can resist is greater than 1.15 times the demand.

7.5.4 Serviceability

All four guidelines have identified several serviceability issues that require checking, such as crack widths and deflections. Limitations on the stress level in the internal steel reinforcement and the FRP strengthening system are also imposed in order to prevent stress (creep) rupture of the FRP, excessive creep of the concrete, steel yielding and fatigue failure. ACI (2002) defines fatigue and stress rupture as ultimate limit states, however, they are summarised in this subsection on serviceability for ease of comparison with the recommendations of other guidelines.

Deflections, crack widths and stress limits

All four guidelines refer to their respective RC design codes for the calculation of deflections and crack widths, with some appropriate modifications, as well as the allowable limits.

To avoid undesirable damage to concrete and yielding of steel under service loads, various strain/stress limits are recommended by the guidelines. Both ACI (2002) and Concrete Society (2004) recommend that the stress in the internal steel reinforcement should not exceed 80% of its characteristic yield strength. In addition, Concrete Society (2004) limits the stress in the concrete to 60% of the characteristic compressive strength. *fib* (2001) recommends that the concrete stress should be no greater than 60% of the characteristic compressive strength under a rare load combination; however, this limit needs to be reduced under a quasi-permanent load combination. *fib* (2001) also limits the stress in the steel reinforcement to 80% of the characteristic yield strength under a rare load combination. The maximum stress levels in the steel and concrete according to JSCE (2001) are to be obtained from JSCE (1996). Note that these stress limits should be interpreted with due attention to the different definitions of design strengths of materials in the different guidelines.

Stress rupture of FRP

Stress limits are placed on the FRP to prevent stress rupture, otherwise known as creep rupture, of the FRP. Concrete Society (2004) recommends stress limits of 65%, 40% and 45% of the design tensile strength for carbon fibre reinforced polymer (CFRP), aramid fibre reinforced polymer (AFRP) and glass fibre reinforced polymer (GFRP) composites, respectively. The corresponding ACI (2002) limits are 55%, 30% and 20% of the design tensile strength, while the corresponding *fib* (2001) limits are 80%, 50% and 30% of the characteristic strength under a quasi-permanent load combination. No such limits are given in JSCE (2001).

Fatigue

The stress limits adopted by ACI (2002) to prevent fatigue failure of the FRP are the same as those employed to prevent stress rupture. Concrete Society (2004) recommends that the stress limits be 80%, 70% and 30% of the design tensile strength for CFRP, AFRP and GFRP, respectively. *fib* (2001) does not provide any such stress limits. The only specific provision in JSCE (2001) concerning the fatigue loading of FRP-strengthened RC beams is on the effect of fatigue loading on interfacial peeling.

7.6 Shear strengthening

All four design guidelines adopt the simple approach that the shear resistance of an FRP-strengthened RC beam is equal to the sum of the contributions of the concrete, internal steel shear reinforcement and external FRP shear reinforcement, as depicted by Eq. 5.1. The first two components can be easily evaluated according to existing RC design codes, while the contribution of the external FRP shear reinforcement V_{frp} is given in various forms of Eq. 5.2. The main differences between the four guidelines lie in the definition of the effective FRP stress (or strain) in evaluating V_{frp}, and in how the different strengthening schemes and different failure modes are dealt with. The common shear strengthening schemes include complete wraps (or complete wrapping or wrapping), U-jackets (or U-jacketing) and side-bonded strips (or side-bonding) as explained in Chapter 5.

JSCE (2001) employs a 'shear reinforcing efficiency' factor K which was obtained from regression of test results. No distinction is made between different strengthening schemes and different failure modes. The adopted K value represents the best fit of test data, and a member factor was determined from the test data to ensure a 95% confidence limit of the predictions. JSCE also specifies an alternative method in which the stress distribution of the continuous FRP sheets is evaluated based on a linear–brittle bond constitutive law (a linear path that ends abruptly at bond failure) to determine the shear contribution of the sheets, assuming a shear crack angle of 35°.

The *fib* (2001) provisions are based on the work of Triantafillou and Antonopoulos (2000). *fib* (2001) applies a reduction factor of 0.8 to Triantafillou and Antonopoulos' (2000) best-fit effective strain for design. In *fib*'s (2001) approach, different effective strain expressions are employed for the following three cases: (i) CFRP wrapping, (ii) CFRP U-jacketing and side-bonding, and (iii) AFRP wrapping. Therefore, in this approach, no distinction is made between CFRP U-jacketing and CFRP side-bonding, and GFRP is not covered at all. Another shortcoming of this approach is that the provisions are empirical in nature and material specific.

The ACI (2002) provisions are partially based on an approach originally developed by Khalifa *et al*. (1998). For complete wraps, it specifies an effective strain being the smaller of 0.004 or 75% of the ultimate FRP strain. For U-jacketing and side-bonding, the effective strain is determined using a bond mechanism approach based on the bond strength model proposed by Maeda *et al*. (1997), subjected to an upper strain limit of 0.004. A strength of the ACI guideline is that the different failure mechanisms are appropriately differentiated. Its weaknesses include the lack of a rational basis for the design effective strain for complete wraps and the use of a bond strength model which cannot correctly predict the effective bond length (Chen and Teng, 2001). Consequently, the design predictions are in poor agreement with test data (Chen, 2003).

The Concrete Society (2004) guideline adopts an approach based on Denton *et al*.'s (2004) work. For all the strengthening schemes considered, the effective strain in the FRP is taken to be the smallest of the following three values: (i) half of the FRP ultimate strain based on Chen and Teng's (2003a) research on FRP rupture failure in shear-strengthened beams, (ii) the debonding strain based on Neubauer and Rostásy's anchorage model (1997), and (iii) 0.004. The differences between complete wrapping, U-jacketing and side-bonding are reflected in the defined FRP depth $(d_{frp} - nl_{t,max}/3)$: $n = 0$ for complete wrapping, $n = 1$ for a U-jacketing, and $n = 2$ for side-bonding. Here, d_{frp} is the effective depth of the bonded FRP measured from the top of the FRP to the tension reinforcement and $l_{t,max}$ is the effective bond length based on Neubauer and Rostásy's (1997) model. It may be noted that the debonding of FRP in a completely wrapped beam is also considered as a design limit state in Concrete Society (2004), which is rational in many situations (Cao *et al*., 2005). The variation of the stress distribution in FRP along the shear crack is considered through a reduced

FRP depth here, but this approach is not as rigorous as that adopted in Chen and Teng's design proposal (2003a,b) (see Chapter 5). Combined with the fact that Concrete Society (2004) adopts a slightly inferior bond strength model (Chen and Teng, 2001), when compared with test data, its predictions are less accurate than those from Chen and Teng's design proposal (Chen, 2003, 2008).

7.7 Strengthening of columns with FRP wraps

7.7.1 General

Design guidance for the FRP strengthening of columns is given in *fib* (2001), ACI (2002) and Concrete Society (2004). In JSCE (2002), the effect of FRP confinement is briefly mentioned, but no quantitative method for evaluating this effect is provided. The behaviour and modelling of FRP-confined concrete is a key issue in the design of column strengthening measures, and has been discussed in detail in Chapter 6.

This section is limited to the strengthening of RC circular and rectangular (including square) columns with FRP wraps where the fibres are solely or predominantly oriented in the hoop direction. Such wraps provide confinement to the concrete to increase its axial compressive strength and the ultimate axial compressive strain. The latter is important in seismic upgrading as it often dictates the ductility of RC columns.

A number of issues of lesser significance are not discussed in this section, but readers can refer to the respective guidelines for details. The provision of FRP plates/strips with fibres oriented in the longitudinal direction for the flexural strengthening of columns is covered by Concrete Society (2004) but not by the other three guidelines. The design of such FRP plates/strips can follow the procedure for the flexural strengthening design of RC beams. The provision of a series of discrete wraps that do not cover the whole column height (partial wrapping) is possible but not common in practical applications. The effect of partial warping is covered by *fib* (2001) only and is not further discussed.

The provision of FRP wraps with hoop fibres is also an effective shear strengthening method for RC columns. Shear strengthening design is discussed in the preceding section. To achieve conservative designs, the thickness of FRP wraps required for shear strengthening should be added to that required for confinement.

7.7.2 Ultimate FRP jacket strain

As discussed in Chapter 6, the hoop rupture strain of an FRP jacket (i.e. the ultimate jacket strain) is lower than the ultimate tensile strain from tensile coupon tests. This aspect is noted, explicitly or implicitly, in all three guidelines covering column strengthening. *fib* (2001) provides three reasons why the ultimate jacket strain is lower than that from tensile tests but gives no specific recommendations on the ultimate jacket strain. ACI (2002) specifies the design effective strain for members subjected to combined compression and shear to be the smaller of 0.004 and 75% of the design rupture strain of FRP. Concrete Society (2004) does not address this issue explicitly as its design equations are directly based on tensile properties from flat coupon tests, but the adopted design equations include this effect implicitly.

7.7.3 Stress–strain model for FRP-confined concrete

fib (2001) recommends the analysis-oriented stress–strain model by Spoelstra and Monti (1999) for use in the section analysis of columns. This was an early analysis-oriented model proposed for FRP-confined concrete, based on a general approach that has also been employed by a number of other models (Teng and Lam, 2004; Teng *et al.*, 2007). However, this model significantly over-estimates the stress–strain response, particularly the ultimate axial strain, of FRP-confined concrete, as has been shown by Teng and Lam (2004). *fib* (2001) provides no corresponding stress–strain model for concrete in FRP-confined rectangular columns. ACI (2002) provides no stress–strain model for FRP-confined concrete.

Concrete Society (2004) recommends Lam and Teng's (2003a) model for FRP-confined concrete in circular columns with some modifications. Lam and Teng's (2003a) model has been presented in Chapter 6. The modifications include a different limit on the confinement level below which no strength gains should be assumed, which is based on the work of Xiao and Wu (2000), and a different equation for the compressive strength of FRP-confined concrete which was proposed by Lillistone and Jolly (2000). Both modified expressions are related to the jacket stiffness rather than the ultimate jacket strain. Concrete Society (2004) specifies that a stress–strain model for FRP-confined concrete under concentric compression can be used in section analysis of columns under combined compression and bending only when more than half of the section is in compression. Otherwise, any strength increases should be ignored and the stress–strain model for unconfined concrete should be used. Concrete Society (2004) does not provide a stress–strain model for FRP-confined concrete in rectangular columns. In addition, Concrete Society (2004) limits the design value of the ultimate axial compressive strain of FRP-confined concrete ε_{cc} to 0.01 to avoid reliance on concrete that has been crushed and has lost cohesion.

7.7.4 Compressive strength and ultimate strain of FRP-confined concrete

FRP-confined concrete in circular columns

fib (2001) adopted Spoelstra and Monti's (1999) simple equations for the stress at ultimate strain and the ultimate strain of FRP-confined concrete. It should be noted that the stress at ultimate strain defined in this guideline becomes equal to the compressive strength of FRP-confined concrete only when this ultimate stress exceeds the peak stress on the stress–strain curve. These equations were based on regression analysis of the predictions of Spoelstra and Monti's (1999) stress–strain model. A formula developed by Seible *et al.* (1995) is also given in this guideline.

According to ACI (2002), Mander *et al.*'s (1988) equation for the compressive strength of concrete confined by a constant confining pressure (active confinement) can be used to evaluate the compressive strength of FRP-confined concrete. Extensive research has shown that this assumption is inappropriate and conceptually incorrect (Teng and Lam, 2004) (see Chapter 6 for a detailed discussion).

In Concrete Society (2004), the confined concrete compressive strength is predicted by an equation proposed by Lillistone and Jolly (2000) while the ultimate axial strain is predicted by an equation proposed by Lam and Teng (2003a). The expression of Lillistone and Jolly (2000) can be easily shown to be unconservative by comparisons with available test data.

FRP-confined concrete in rectangular columns

The effect of FRP confinement is much less effective for rectangular (and square) columns than for circular columns as discussed in Chapter 6. *fib* (2001) provides definitions of lateral confining pressures for the x and y directions of a rectangular section, respectively, based on the effective confinement area concept. It is, however, not made clear how these effective confining pressures should be used. *fib* (2001) also suggests that for a rectangular section with ovalisation before FRP wrapping, the section can be replaced by an equivalent circular section with a radius equal to the average of the principal radii of the ellipse.

In ACI (2002), equations for the reinforcement ratio and the efficiency factor for square and rectangular sections are given, which can then be used to predict the compressive strength of FRP-confined concrete using the same equations as for circular columns. The confining effect of FRP is assumed to be negligible for sections with aspect ratios exceeding 1.5 or with face dimensions exceeding 900 mm unless their effectiveness is demonstrated by tests.

In Concrete Society (2004), the compressive strength of FRP-confined concrete in rectangular columns is predicted by an equation developed by Lam and Teng (2001) with a modified effective area of confinement. This Lam and Teng (2001) equation was developed as an earlier version of Eq. 6.7 which was presented in Lam and Teng (2003b). A shape factor and an equivalent circular column are defined for use in this equation. The use of this equation is limited by the guideline to small sections with aspect ratios not exceeding 1.5 and with the smaller face dimension not exceeding 200 mm.

7.7.5 Serviceability

fib (2001) does not contain specific serviceability requirements for column strengthening by FRP confinement. General discussions on serviceability-related issues such as stress rupture of FRP are given in Chapter 9 of this guideline.

According to ACI (2002), to avoid excessive radial cracking under service loads, the axial stress in the concrete should be kept below 65% of the compressive strength of unconfined concrete. With this stress limit, the FRP jacket will only be mobilised during temporary overloads. In addition, ACI (2002) includes the following requirements: (i) the axial stress in the internal steel longitudinal reinforcement should be limited to 60% of the yield stress to avoid plastic deformation when subjected to sustained or cyclic loading; (ii) the service load stresses in the FRP should never exceed its creep-rupture stress limit; and (iii) the effect of axial deformations of the column under service loads on the performance of the structure should be considered.

Concrete Society (2004) recommends that under service loads, the axial compressive strain of concrete should not exceed 0.0035. Furthermore, the stress level in the FRP jacket should be limited to 65%, 40% and 45% of the design rupture strength of CFRP, AFRP and GFRP, respectively (as previously described on pp 205), to avoid stress rupture failure of the FRP jacket. For bridge structures, Concrete Society (2004) recommends that the stress range in the FRP should be within 80%, 70% and 30% of the design rupture strength of CFRP, AFRP and GFRP, respectively (as previously described on pp 205), to avoid fatigue failure.

7.7.6 Ductility and seismic retrofit

The only provision ACI (2002) has on seismic retrofit is an equation proposed by Mander *et al.* (1988) for the ultimate strain of FRP-confined concrete. This equation was developed for actively-confined concrete and does not provide accurate predictions for FRP-confined concrete (Teng and

Lam, 2004). This guideline states that the confinement should enable the concrete to reach an ultimate axial compressive strain that meets the displacement demand. It also states that brittle shear failure should be suppressed. Both circular and rectangular sections are covered.

Concrete Society (2004) provides a very brief discussion of seismic upgrading using FRP composites, noting that it is not a major loading case for most structures in the UK.

fib (2001) discusses two approaches for the seismic upgrading of columns to meet specific ductility demands. The first of the approaches follows the same principle as that described in Chapter 7 of Teng *et al.* (2002). The second approach is that proposed by Japanese researchers (Mutsuyoshi *et al.*, 1999).

7.8 Acknowledgement

The authors are grateful for the financial support provided by The Hong Kong Polytechnic University (Project Code: BBZH).

7.9 References

ACI (1999) *Building Code Requirements for Structural Concrete (318–95) and Commentary*, ACI 318R-95, Farmington Hills, MI, USA, American Concrete Institute.

ACI (2002) *Guide for the Design and Construction of Externally Bonded FRP Systems for Strengthening Concrete Structures*, ACI 440.2R-02, Farmington Hills, MI, USA, American Concrete Institute.

BLASCHKO M (1997) *Strengthening with FRP*, Münchner Massivbau Seminar, TU München (in German).

BSI (1997) *BS 8110-1:1997. Structural Use of Concrete. Code of Practice for Design and Construction*, London, UK, British Standards Institution.

CAO S Y, CHEN J F, TENG J G, HAO Z and CHEN J (2005) Debonding in RC beams shear strengthened with complete FRP wraps, *Journal of Composites in Construction*, ASCE, **9**(5), 417–428.

CECS (2003) *CECS-146 Technical Specification for Strengthening Concrete Structures with Carbon Fiber Reinforced Polymer Laminates*, China Association for Engineering Construction Standardization, Beijing, China, China Planning Press.

CEN (1992) ENV 1992-1-1 *Eurocode 2: Design of Concrete Structures, Part 1: General Rules and Rules for Buildings*, Brussels, Belgium, European Committee for Standardization.

CHEN J F (2003) Design guidelines on FRP for shear strengthening of RC beams, in Forde M C (ed.), *Proceedings, Tenth International Conference on Structural Faults and Repair*, Edinburgh, UK, Technics Press (CD ROM).

CHEN J F (2008). Assessment of design guidelines for shear strengthening of RC beams with FRP, in preparation.

CHEN J F and TENG J G (2001). Anchorage strength models for FRP and steel plates bonded to concrete, *Journal of Structural Engineering*, ASCE, **127**(7), 784–791.

CHEN J F and TENG J G (2003a) Shear capacity of FRP strengthened RC beams: FRP rupture, *Journal of Structural Engineering*, ASCE, **129**(5), 615–625.

CHEN J F and TENG J G (2003b) Shear capacity of FRP strengthened RC beams: FRP debonding, *Construction and Building Materials*, **17**(1), 27–41.

CONCRETE SOCIETY (2004) *Design Guidance for Strengthening Concrete Structures with Fibre Composite Materials*, TR55, 2nd edn, Camberley, UK, The Concrete Society.

CNR (2005) *Instructions for the Design, Execution and Control of Strengthening Measures through Fibre-Reinforced Composites*, CNR-DT 200/04, Rome, Italy, Italian Research Council.

CSA (2002) *CSA S806-02 Design and Construction of Building Components with Fibre-Reinforced Polymers*, Ontario, Canada, Canadian Standards Association.

DENTON S R, SHAVE J D and PORTER A D (2004) Shear strengthening of reinforced concrete structures using FRP composite, in Hollaway, L C, Chryssanthopoulos M K and Moy S S J (eds), *Proceedings, Second International Conference on Advanced Polymer Composites for Structural Applications in Construction – ACIC 2004*, Cambridge, UK, Woodhead, 134–143.

ESHWAR N, IBELL T J and NANNI A (2003) CFRP strengthening of concrete bridge with curved soffits, in Forde M C (ed.), *Proceedings, Tenth International Conference on Structural Faults and Repair*, Edinburgh, UK, Technics Press (CD ROM).

fib (2001) *Externally Bonded FRP Reinforcement for RC Structures*, Technical Report, Task Group 9.3, Bulletin No. 14, Lausanne, Switzerland, The International Federation for Structural Concrete (*fib*).

ISIS (2001) *Strengthening Reinforced Concrete Structures with Externally-Bonded Fibre Reinforced Polymers*, Design Manual No. 4, Canadian Network of Centres of Excellence on Intelligent Sensing for Innovative Structures, Winnipeg, Manitoba, Canada, ISIS Canada Corporation.

HOLZENKÄMPFER P (1994) *Ingenieurmodelle des Verbundes Geklebter Bewehrung für Betonbauteile*, Dissertation, TU Braunschweig, Germany (in German).

JANSZE W (1997) *Strengthening of RC Members in Bending by Externally Bonded Steel Plates*, PhD Thesis, Delft University of Technology, Delft, The Netherlands.

JSCE (1996) *Standard Specifications for Concrete Structures – Design*, Tokyo, Japan, Japan Society of Civil Engineers (in Japanese).

JSCE (2001) *Recommendations for Upgrading of Concrete Structures with Use of Continuous Fiber Sheets*, Concrete Engineering Series 41, Tokyo, Japan, Japan Society of Civil Engineers.

KHALIFA A, GOLD W J, NANNI A and ABEL-AZIZ M (1998) Contribution of externally bonded FRP to shear capacity of RC flexural members, *Journal of Composites for Construction*, ASCE, **2**(4), 195–203.

LAM L and TENG J G (2001) Compressive strength of FRP-confined concrete in rectangular columns, in Teng J G (ed.), *Proceedings, International Conference on FRP Composites in Civil Engineering – CICE 2001*, Oxford, UK, Elsevier Science, 335–343.

LAM L and TENG J G (2003a) Design-oriented stress-strain model for FRP-confined concrete, *Construction and Building Materials*, **17**(6–7), 471–489.

LAM L and TENG J G (2003b) Design oriented stress-strain model for FRP-confined concrete in rectangular columns, *Journal of Reinforced Plastics and Composites*, **22**(13), 1149–1186.

LILLISTONE D and JOLLY C K (2000) An innovative form of reinforcement for concrete columns using advanced composites, *The Structural Engineer*, **78**(2), 20–28.

LU X Z, TENG J G, YE L P and JIANG J J (2005) Bond-slip models for FRP sheets/plates bonded to concrete, *Engineering Structures*, **27**, 920–937.

LU X Z, TENG J G, YE L P and JIANG J J (2007) Intermediate crack debonding in FRP-strengthened RC beams: FE analysis and strength model, *Journal of Composites for Construction*, ASCE, **11**(2), 161–174.

MAEDA T, ASANO Y, SATO Y, UEDA T and KAKUTA Y (1997) A study on bond mechanism of carbon fiber sheet, *Proceedings, Third International Symposium on Non-metallic (FRP) Reinforcement for Concrete Structures – FRPRCS-3*, Tokyo, Japan, Japan Concrete Institute, **1**, 279–285.

MANDER J B, PRIESTLEY M J N and PARK R (1988) Theoretical stress–strain model for confined concrete, *Journal of Structural Engineering*, ASCE, **114**(8), 1804–1826.

MUTSUYOSHI H, ISHBASHI T, OKANO M and KATSUKO F (1999) New design method for seismic retrofit of bridge columns with continuous fibre sheet – performance-based design, in Dolan C W, Rizkalla S H and Nanni A (eds), *Proceedings, Fourth International Symposium on Fibre Reinforced Polymer Reinforcement for Reinforced Concrete Structures – FRPRCS-4*, Farmington Hills, MI, USA, American Concrete Institute, 229–241.

NEUBAUER U and ROSTÁSY F S (1997) Design aspects of concrete structures strengthened with externally bonded CFRP plates, in Forde M C (ed.), *Proceedings, Seventh International Conference on Structural Faults and Repair*, Edinburgh, UK, Engineering Technics Press, Vol. II, 109–118.

OEHLERS D J and SERACINO R (2004) *Design of FRP and Steel Plated RC Structures*, Oxford, UK, Elsevier.

OEHLERS D J and SERACINO R and SMITH S T (2008) *Design Guideline for RC Structures Retrofitted with FRP and Metal Plates: Beams and Slabs*, Standards Australia, in press.

SEIBLE F, BURGUENO R, ABDALLAH M G and NUISMER R (1995) Advanced composite carbon shell systems for bridge columns under seismic loads, *Proceedings, National Seismic Conference on Bridges and Highways*, San Diego, CA, USA, Dec 10–13.

SMITH S T and TENG J G (2002) FRP-strengthened RC structures. II: Assessment of debonding strength models, *Engineering Structures*, **24**(4), 397–417.

SPOELSTRA M R and MONTI G (1999) FRP-confined concrete model, *Journal of Composites for Construction*, ASCE, **3**(3), 143–150.

TÄLJSTEN B (2003) *FRP Strengthening of Existing Concrete Structures. Design Guidelines*, 2nd edn, Division of Structural Engineering, Luleå University of Technology, Luleå, Sweden.

TENG J G and LAM L (2004) Behavior and modeling of fiber reinforced polymer-confined concrete, *Journal of Structural Engineering*, ASCE, **130**(11), 1713–1723.

TENG J G and YAO J (2005) Plate end debonding failures of FRP- or steel-plated RC beams: a new strength model, in Chen J F and Teng J G (eds), *Proceedings, International Symposium on Bond Behaviour of FRP in Structures – BBFS 2005*, Hong Kong, China, Dec 7–9, 291–298.

TENG J G and YAO J (2007) Plate end debonding in FRP-plated RC beams-II: strength model, *Engineering Structures*, **29**(10), 2472–2486.

TENG J G, CHEN J F, SMITH S T and LAM L (2002) *FRP-Strengthened RC Structures*, Chichester, UK, John Wiley & Sons.

TENG J G, SMITH S T, YAO J and CHEN J F (2003) Intermediate crack induced debonding in RC beams and slabs, *Construction and Building Materials*, **17**(6&7), 447–462.

TENG J G, HUANG Y L, LAM L and YE L (2007) Theoretical model for fiber reinforced polymer-confined concrete, *Journal of Composites for Construction*, ASCE, **11**(2), 201–210.

TRIANTAFILLOU T C and ANTONOPOULOS C P (2000) Design of concrete flexural members strengthened in shear with FRP, *Journal of Composites for Construction*, ASCE, **4**(4), 198–205.

XIAO Y and WU H (2000) Compressive behaviour of concrete confined by carbon fiber composite jackets, *Journal of Materials in Civil Engineering*, ASCE, **12**(2), 139–146.

YAO J, TENG J G and LAM L (2005) Experimental study on intermediate crack debonding in FRP-strengthened RC flexural members, *Advances in Structural Engineering*, **8**(4), 365–396.

8
Strengthening of metallic structures with fibre-reinforced polymer (FRP) composites

T. J. STRATFORD, University of Edinburgh, UK

8.1 Introduction

Fibre-reinforced polymer (FRP) composites can be used to address a variety of structural deficiencies in metallic infrastructure as described in Chapter 1:

- *increased load capacity* requirements can leave a structural member under-strength in flexure, shear or axial compression;
- the *connections* between metallic members may have insufficient strength;
- *corrosion* reduces the structural section available to carry load;
- metallic structures are susceptible to *fatigue failure*;
- brittle cast iron structures lack robustness and fail catastrophically under impact or thermal shock loading.

A superficial comparison of flexural strengthening for metallic and concrete structures suggests many similarities between the design methods; however, virtually every aspect of the design of metallic structures is different in detail. The structural failure modes, critical issues, degradation, and analysis techniques all differ significantly from concrete strengthening.

FRP strengthening for metallic structures is a younger technology than for concrete structures, and research into this method is still in progress. This chapter focuses on design guidance for strengthening flexural members. It is combined with a description of the current state-of-the-art for other forms of strengthening, so that the reader can take advantage of emerging applications of FRP strengthening to metallic structures.

8.2 Critical issues in the design of FRP strengthening for metallic structures

8.2.1 Failure modes for a strengthened member

There are a number of locations where failure can occur within externally bonded FRP strengthening applied to a metallic structure. These are

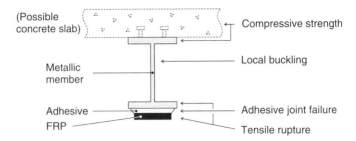

8.1 Possible locations of failure within externally-bonded FRP strengthening of metallic structures.

indicated in Fig. 8.1. The metallic member might be part of an existing structural composite section, such as steel beams acting compositely with a concrete slab or cast iron beams that are used to support brick jack arches.

The potential modes of failure for a metallic flexural member strengthened using an FRP include:

- *Adhesive joint failure.* 'Debonding' (separation of the FRP from the metallic substrate) frequently governs design. For metallic structures, failure occurs along the adhesive joint, unlike in concrete where failure occurs within the concrete substrate, along the flexural reinforcement.
- *Tensile rupture of the FRP.*
- *Tensile strength of the metallic member.* The addition of FRP-strengthening changes the stresses within the metallic member, possibly increasing the tensile stress. For brittle cast iron members, failure is based on the extreme fibre stresses; a cracked section is not usually allowed (unlike for the design of concrete strengthening).
- *Local buckling of the metallic member.* The metallic member should also be checked for local buckling in its strengthened state, for example, buckling of the compression flange or of the web in shear.
- *Compressive strength of the existing structure.* The maximum compressive stress in the section may increase, resulting in a compression failure. If a steel beam is topped by a composite concrete slab, increased compressive stresses could lead to failure within the concrete (Sen *et al.*, 2001; Tavakkolizadeh and Saadatmanesh, 2003a).
- *Compressive failure of the FRP.* FRP strengthening is not usually used as compressive strengthening, as its compressive strength is limited by localised micro-buckling of the fibres.
- *Global buckling of the strengthened member.*

8.2.2 The adhesive joint

Externally-bonded FRP strengthening relies crucially upon the adhesive joint to transfer load between the strengthening and the metallic member.

Failure occurs within the weakest link of the adhesive joint, which could be (i) within the metallic substrate, (ii) at the metal–adhesive interface, possibly due to surface contamination, (iii) within the adhesive, (iv) at the adhesive–FRP interface, or (v) within the FRP. A high-quality metal-to-FRP bonded joint will fail within the adhesive.

The use of structural adhesive joints for civil infrastructure follows their successful application in other industries. Most notably, bonded composite patches are regularly used to repair aluminium structures in the aerospace industry. Correctly designed and implemented bonded strengthening has superior performance over conventional mechanically fastened systems; however, numerous in-service failures of bonded joints in the aerospace industry have shown that a successful bonded joint requires (i) careful design, (ii) thorough and appropriate surface preparation and (iii) the correct conditions for adhesive cure (Davis and Bond, 1999). These lessons are equally critical to civil engineering applications.

Whilst the underlying requirements for a successful adhesive joint are similar for aerospace and infrastructure strengthening, the two applications require very different configurations and materials. Construction uses greater thicknesses of FRP, resulting in considerably higher peel stresses within the adhesive joint. Furthermore, the joint is formed on site (not under factory conditions), and it is rarely economic to undertake rigorous prototype testing, as each civil engineering project is unique (IStructE, 1999; Hutchinson and Hurley, 2001).

A designer must be aware of the following critical issues for the adhesive joint:

- *Mechanical design.* The strength of an adhesive joint is governed by the concentrated stresses at the ends of a strengthening plate (as described in more detail below).
- *Temperature effects.* Very significant stresses can develop within the adhesive layer due to differential thermal expansion between carbon fibre reinforced polymer (CFRP) strengthening and the metallic substrate (Denton, 2001). In addition, the mechanical properties of the adhesive vary with temperature; consequently, the strengthening materials must be suitable for the operating environment. Typical two-part ambient cure epoxy adhesives can have glass transition temperatures as low as 65 °C.
- *Sensitivity to defects.* Defects will always be present within an adhesive joint, and can reduce its strength, particularly if the defects coincide with highly stressed regions. The performance of bonded strengthening is hence very dependent on a high quality of surface preparation (see Chapters 3 and 12). The strengthening scheme should be designed for easy installation and the sensitivity of the strengthening to imperfect installation assessed (discussed below).

218   Strengthening and rehabilitation of civil infrastructures

- *Fatigue of the adhesive joint.* Various laboratory tests (Jones and Civjan, 2003; Tavakkolizadeh and Saadatmanesh, 2003b; Buyukozturk *et al.*, 2004) have shown that CFRP-to-steel adhesive joints have very good fatigue performance. However, only a small number of tests have been carried out to date, and further research is required to give sufficient fatigue life data for design purposes.
- *Environmental durability.* The environmental durability of an adhesive joint depends upon the adhesive used and the standard of surface preparation (Karbhari and Sulley, 1995; Hollaway and Cadei, 2002) (see also Chapter 3).
- *Bimetallic corrosion.* An electrochemical cell can form between carbon fibres and metal, leading to greatly accelerated corrosion of the substrate, which is hidden beneath the adhesive joint. Bimetallic (or 'galvanic') corrosion can be avoided by insulating the CFRP from the substrate, possibly using a glass or vinylester fabric within the adhesive to guarantee a minimum thickness of adhesive. This layer also reduces the stiffness of the adhesive joint and consequently reduces the maximum adhesive bond stress (Photiou *et al.*, 2006).

Report C595 (Cadei *et al.*, 2004) gives a more comprehensive treatment of the above issues than is possible here.

8.3   Selection of strengthening materials

FRP strengthening materials are usually provided as *systems*, comprising preformed FRP plates containing predominately longitudinal fibres and a compatible bonding adhesive. There are various options within these systems, such as the type of FRP used and the grade of adhesive. The designer should note that many of the systems on offer were originally developed for strengthening concrete structures and that suppliers can often provide alternative adhesives that might not be advertised in their civil engineering catalogue. As discussed in Chapter 1, any additional cost associated with (for example) a more expensive adhesive is generally more than offset by savings in the installation of the strengthening works.

8.3.1   FRP materials for different strengthening requirements

An efficient strengthening scheme will utilise a significant proportion of the FRP's tensile strength. This can be achieved in three different ways, according to the characteristics of the metallic substrate:

- by *maximising the modular ratio* of the FRP strengthening to the metal;

- by *transferring dead load stress* from the metal to the FRP;
- by allowing *yield* of the metallic substrate.

Maximising the FRP to metal modular ratio

A higher FRP-to-metal modular ratio allows the FRP to develop greater stress when the composite section is loaded, and results in a more efficient strengthened section. Whilst a thicker plate of a lower modulus FRP might be used, this increases the peel stresses within the adhesive (see below). Thicker plates are more expensive to fabricate and install.

UHM (ultra-high modulus) CFRP strengthening has the highest axial modulus and is consequently often most economic for strengthening metallic structures. Preformed UHM plates are available with Young's modulus $E = 360$ GPa (Cadei *et al.*, 2004), giving a modular ratio to cast iron of around 4. Failure of cast iron is governed by the tensile strength of brittle cast iron, making a high modular ratio strengthening material particularly beneficial.

As discussed below, a *ductile steel* member can be allowed to yield. A high modular FRP-to-metal ratio remains beneficial, but it may be possible to use a less expensive, lower modulus FRP strengthening material.

Dead load transfer

The FRP strengthening can only help carry loads applied *after* the adhesive joint has cured. Consequently, the FRP will not carry loads that were already present at the time of strengthening, such as the dead load of a bridge or floor slab. This is of particular importance for a *brittle* substrate (such as cast iron), as the tensile dead load stresses may be a large proportion of the metal's strength.

Dead load can be transferred from the existing structure to the FRP strengthening by either pre-tensioning the strengthening prior to bonding (e.g. Hythe Bridge, Cadei *et al.* (2004)) or relieving the structure of dead load stresses during the strengthening operation (e.g. by temporarily propping the structure on jacks, as used for Maunders Road Bridge, Cadei *et al.* (2004)). Pre-tensioning the strengthening allows a lower modulus CFRP to be used (UHM is not necessary), but requires tensioning jacks (or similar) and supplemental anchorages at the ends of the plates. Wherever the strengthening is required to carry permanent loads, creep within the adhesive must be considered.

Allowing yield of the substrate

Yield of a *ductile* substrate (such as steel) can be exploited to develop a large strain (and hence stress) in the elastic FRP strengthening at failure.

The strain capacity of the adhesive joint, however, may prevent the full strength of the FRP strengthening being exploited. Research work has shown that premature debonding of the FRP strengthening can be prevented using either additional mechanical restraints (Sen, 2001) or by overwrapping the strengthening with a glass fibre reinforced polymer (GFRP) U-wrap (Photiou *et al.*, 2006).

Other factors governing the choice of FRP strengthening materials

Carbon fibres are usually selected to strengthen metallic structures, due to their high stiffness, but other fibres might be chosen in specific circumstances. For example, carbon fibres conduct and may be inappropriate adjacent to electrical installations (such as overhead traction supply lines on railways), where aramid fibres could be used instead.

Glass or aramid fibres can also be used where the material stiffness is not critical. For example, member buckling can be prevented by an increase in the radius of gyration of a section, which can be achieved by positioning a low stiffness FRP remotely from the existing member (Liu *et al.*, 2005a).

The designer can choose the form of FRP strengthening. Preformed CFRP plates (manufactured by pultrusion or resin transfer moulding) are commonly used, due to the convenience with which they can be bonded to the existing structure. Forming the FRP *in situ* is more versatile, and can be applied to awkward shapes where preformed plates could not be used, such as curved beams and connection details (Garden, 2001). The FRP can be formed *in situ* using a combination of wet lay-up methods, pre-impregnated laminates, vacuum consolidation, resin transfer techniques and/or curing at elevated temperatures. The fabrication technique should be selected to suit the strengthening application.

8.3.2 The adhesive joint

Two-part ambient cure epoxy adhesives are usually used to bond the strengthening to the existing structure. These are the most easily applied adhesives for bonding large areas on a construction site. However, the designer should ensure that ambient cure epoxies are appropriate for the particular strengthening scheme. In particular, these adhesives soften at their glass transition temperature, which can be as low as 65 °C, a temperature which might easily be reached on a steel bridge exposed to the sun. Adhesives with higher glass transition temperatures generally require elevated cure temperatures, but this can be achieved using electric blankets or other heaters during the curing period.

Bonded FRP strengthening cannot currently be used where it is required to support loads during a fire. The bonding adhesive will first soften and then burn (producing toxic fumes) and the strengthening material will become detached from the structure. Fires in adjacent parts of a building can also cause failure through heat conducted to the adhesive joint along the metallic structure, making it very difficult to apply fire-proofing to protect the FRP strengthening. In many situations, however, the unstrengthened structure will be able to carry the reduced loading requirements during a fire, and it will not matter that the FRP strengthening does not survive.

As noted above, the adhesive joint fails within its weakest component. Wrought iron and early steel materials have a laminated structure, and their surfaces can delaminate under concentrated loads. Recent research work has shown that wrought iron is an inherently variable material, but can have sufficient shear strength for FRP strengthening (Moy, 2004a). The strength of the substrate should be established prior to design.

8.3.3 Factors of safety for limit state design

Material safety factors should be applied to the strength and stiffness of both the FRP and the adhesive. These material safety factors describe the variability of the material, variations due to the method of manufacture and long-term degradation of the materials. Moy (2001, 2004b), gives appropriate partial factors for the FRP materials. C595 (Cadei *et al.*, 2004) gives methods for determining partial factors for the adhesive, including environmental degradation and ageing effects. These safety factors assume that the metallic substrate has been prepared to give an adequate surface for reliable long-term bond.

8.4 Design of flexural strengthening

There are two principal stages to designing FRP flexural strengthening for metallic structures. The first of these is a *sectional analysis* to determine the amount of FRP material needed, and is described in Section 8.4.1. The second, a *bond analysis* to check the capacity of the adhesive joint, is described in Section 8.4.2.

8.4.1 Sectional analysis

A conventional plane sections analysis is used to size the reinforcement. This analysis must consider whether loads were applied before or after the adhesive cured (as shown in Fig. 8.2). Load applied prior to strengthening (moment M_0) is carried by the metallic section, and the resulting strains (σ) and stresses (ε) can be calculated using simple beam theory. Loads applied after strengthening (moment $M - M_0$) are carried by the composite (FRP

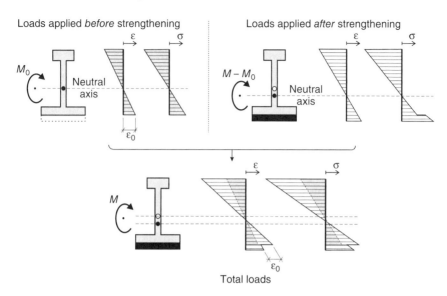

8.2 Determining the flexural stress distribution within a brittle metal beam (e.g. cast iron) strengthened with an FRP plate.

+ metal) section, which can be analysed using a transformed section approach. It is important to clearly define the position about which this increase in load acts. For zero axial load, $(M - M_0)$ must act about the neutral axis of the *strengthened* section, which is in a different position to the neutral axis of the unstrengthened section. As discussed above, differential thermal expansion can lead to very significant self-equilibrating stresses, and these must be considered after strengthening.

The total strain and stress distributions under the combined applied load are found by summation, and compared to allowable strain and stress limits. Checks should also be made for global or local buckling of the strengthened member. The amount of FRP strengthening is adjusted until all limits are satisfied.

For a *brittle* metal (such as cast iron), the stress distribution within the section will be elastic (Fig. 8.2) and the capacity of the section will usually be governed by the extreme section strains. Note that a cracked section is *not* allowed in metallic structures (unlike concrete), as the energy released by crack formation will usually lead to catastrophic failure of the adhesive joint.

A plane sections analysis can also be used to analyse strengthened *ductile* materials (such as steel), and pseudo-plastic materials (such as the concrete in a steel–concrete composite section). It is not possible to develop a fully-plastic section, as compatibility of the elastic FRP with the metallic section must be considered. Hence (as shown in Fig. 8.3), the central portion of the

Strengthening of metallic structures with FRP composites 223

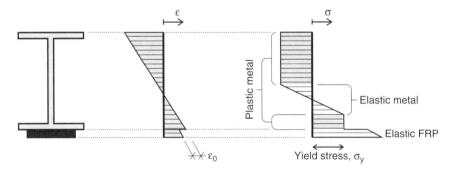

8.3 The flexural stress within a ductile metal beam strengthened with an FRP plate.

ductile member will remain elastic. Strengthening will change the stress distribution through the depth of the member, and this should be checked for local web buckling and compressive failures.

8.4.2 Bond analysis for the adhesive layer

The strength of the adhesive joint is governed by stress concentrations, which occur at geometric discontinuities, in particular, at the end of the strengthening plate. As indicated in Fig. 8.4, the misfit in deformations between the beam and strengthening plate leads to concentrations of both shear stress (τ) and peel stress (σ) across the adhesive joint. These concentrated stresses must be checked to ensure that they do not cause debonding failure. As for the sectional design process, the stresses in the adhesive joint are due to loads applied after the adhesive has cured, and the joint must accommodate stresses due to differential thermal expansion.

Two approaches can be used to analyse the capacity of an adhesive joint (both of which are used in the aerospace industry):

- a fracture mechanics approach;
- a stress-based analysis.

Fracture mechanics

Adhesive joint failure involves debonding by crack propagation along the adhesive joint, making a fracture mechanics approach attractive for modelling joint failure. Fracture mechanics assesses the energy required to propagate a crack along the adhesive joint; if the available energy is lower than this fracture energy, the adhesive joint will not fail. The energy required to drive fracture of the adhesive joint can be measured directly from coupon tests (such as the double cantilever beam test, BS 7991 (BSI, 2001) or the

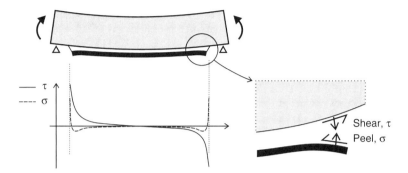

8.4 The longitudinal elevation of a flexural beam indicating the shear and peel stress concentrations due to misfit in deformations between beam and strengthening plate.

Boeing Wedge test, ASTM D3762 (ASTM, 2003)) and used in an analysis of the strengthened beam.

Fracture mechanics avoids the assumptions involved in a stress-based analysis (for example, linear-elasticity, lack of defects and no variation of stress through the thickness of the adhesive). However, whilst fracture mechanics techniques are widely used to design adhesive joints in the aerospace industry, they have yet to be developed for civil engineering purposes (Buyukozturk *et al.*, 2004). In particular, aerospace applications use much thinner strengthening materials, in which peel stresses are low. The mixed-mode (peel–shear) behaviour in typical civil engineering applications has not yet been reliably characterised using the fracture mechanics approach.

Stress analysis

Current best practice uses a linear-elastic stress analysis approach to check the capacity of the adhesive joint. The method determines the maximum stresses that are expected to occur in the adhesive layer, and compares these with the strength of adhesive obtained from material characterisation tests. The advantage of the stress-based analysis is that it is conceptually simple and produces visual results that are familiar to practising engineers. However, the designer must be aware of the assumptions behind the analysis, in particular:

- the adhesive is linear-elastic and brittle, whereas the real material behaviour will be non-linear, possibly exhibiting some plasticity;
- the adhesive layer is homogenous and defect free, whereas in reality the adhesive will contain defects such as voids and variable surface preparation;

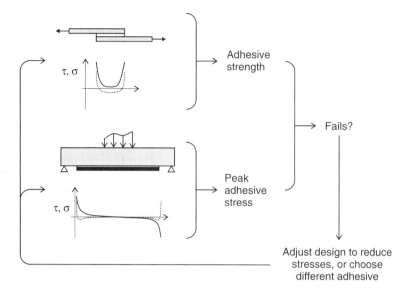

8.5 An outline of the elastic bond stress design process for checking the peak adhesive stress in a strengthened beam.

- the characterisation tests are assumed to be representative of the site-bonded joint;
- the method of analysis involves approximations (discussed in the next section).

Research has shown that an elasto-plastic adhesive analysis can be used to predict bond failure for steel sections strengthened using thin FRP laminates (Lam *et al.*, 2004; Liu *et al.*, 2005b). However, this work has yet to be extended to the thick FRP sections often required for metallic strengthening. Peel stresses are more significant for thicker FRP sections, leading to a more brittle mode of failure.

Figure 8.5 shows the elastic stress design process in outline. The strength of an adhesive joint is usually characterised using lap shear tests (e.g. BS 5350 (BSI, 2005) or ASTM D5868 (ASTM, 2001)), by either the single lap shear test shown or a double lap shear test. The lap shear test results are reported as average shear stresses at failure, but it is the stress concentration that is of interest in design (which is far greater than the average stress). The lap shear test results can be back-analysed using an elastic stress analysis that is consistent with the analysis used for the strengthened beam, to give the peak adhesive stress at failure, which can be used as the limiting stress for design. The adhesive supplier will usually quote lap shear test results performed on ideal specimens, with 'perfect' surface preparation and 'ideal' curing conditions, using two steel substrates. The strength of an adhesive joint made under site conditions with cast iron and/or FRP substrates may be lower, and the designer should consider whether additional representative tests are required.

The peak shear and peel stresses within the strengthened beam are determined using an elastic stress analysis. The design of the strengthening system is refined until the peak adhesive stress is less than the strength of the adhesive.

Methods of stress analysis

Various stress analyses have been proposed for designing strengthening plates bonded to beams, each involving a different set of approximations (Buyukozturk *et al.*, 2004). Low-order solutions assume constant shear and peel stresses through the thickness of the adhesive. Although global equilibrium requirements are satisfied, low-order solutions do not satisfy the local zero shear stress boundary condition at the end of the strengthening plate. For example report C595 (Cadei *et al.*, 2004) recommends the use of a low-order stress analysis, (Stratford and Cadei, 2006), to determine the peak stress within the adhesive joint between the plate and the beam, combined with Goland and Reissner's (1944) analysis of lap shear test results (which uses consistent assumptions to the beam analysis). Other low-order bond analyses are presented by (for example) Deng *et al.* (2004), Denton (2001) and Frost *et al.* (2003). Where the cross-section and material properties do not vary along a beam, closed-form solutions are available. For cases with more complex geometry, or where the metal is allowed to yield, finite difference solutions can be used (Deng *et al.*, 2004; Stratford and Cadei, 2006).

High-order bond analyses (such as Yang *et al.* (2004)) consider the variation in stress through the thickness of the adhesive joint and hence satisfy the zero shear stress boundary condition at the end of a strengthening plate. However, their complexity has so far prevented high-order solutions being adopted for design purposes. The difference between high- and low-order analyses is confined to a region approximately one adhesive layer thickness from the end of a strengthening plate (Buyukozturk *et al.*, 2004). Finite element models can be constructed for the adhesive layer, although these require high mesh densities in the vicinity of the stress concentrations.

Whichever method is used to determine the critical adhesive stresses, the following points should be noted:

- The sensitivity of the analysis to the adhesive thickness should be established and used to determine the allowable thickness of adhesive during installation.

- The adhesive stresses due to any plate curvature or other geometric features should be checked.
- The sensitivity of the adhesive joint to residual stresses within the existing section must be established. Sebastian (2005) has shown that the presence of residual stresses can increase the adhesive bond stresses, possibly resulting in premature bond failure.
- For continuous beams, migration of the points of contraflexure must be taken into account if the beam is allowed to yield during the sectional analysis. Sebastian (2003) has shown that, as the zones of plasticity spread in a strengthened steel, the adhesive stresses throughout the beam continue to increase non-linearly.

Reducing the peak adhesive stress to prevent premature debonding

If the predicted stresses exceed the strength of the adhesive, the designer has two options: to select a higher strength adhesive or to reduce the peak adhesive stresses. Finding a higher strength adhesive that is suitable for construction purposes is often impractical; hence the designer must usually investigate ways to reduce the peak stress within the adhesive joint. These include:

- increasing the width of the FRP strengthening, while maintaining the same cross-sectional area (this option is limited by the width of the member being strengthened);
- moving the end of the strengthening plate to a less highly stressed region of the beam;
- adding mechanical clamps (Sen *et al.*, 2001) either to provide an alternative load path to the adhesive joint, or to react the peel stresses in the adhesive and hence increase its capacity in shear;
- over-wrapping the FRP strengthening with a continuous GFRP U-wrap which is an effective method for preventing the spread of debonding adjacent to a zone of substrate plasticity, enabling the strengthening to stiffen the structure even after significant plastic damage in the beam (Photiou *et al.*, 2006);
- Adjusting the geometry of the strengthening plate and/or adhesive near the end of the plate, for example, by reducing the thickness of the plate (either in steps or by a gradual taper), or using a lower stiffness adhesive near the plate's end.

Figure 8.6 shows the predicted reduction in adhesive shear and peel stresses due to (a) stepping the plates and (b) tapering the plate end (calculated using an elastic bond stress analysis, Stratford and Cadei (2006)). For stepped plates, the peak shear stress is reduced by 30% and the maximum peel stress by 20% (compared to the unstepped case), whilst for the tapered

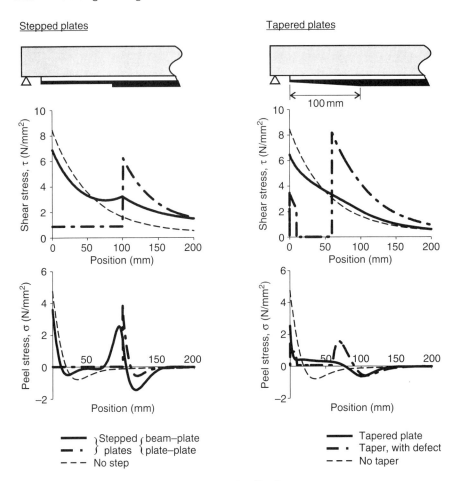

8.6 Methods of reducing the peak adhesive stress to prevent premature debonding, and the significance of a bond defect in the adhesive joint: (a) stepping the plate; (b) tapering the end of the plate (reprinted from Stratford and Cadei (2006), with permission from Elsevier).

plate the maximum shear stress is reduced by 25% and the maximum peel stress by 50%.

Whether or not the bond analysis predicts that stress reduction measures are required, it is generally considered good practice to taper the end of the plate, and also to provide a spew fillet to the adhesive. A spew fillet of adhesive around the edge of a strengthening plate reduces the stress concentration (Frostig et al., 1999); however, this reduction in stress should not be relied upon in design, as the spew fillet adhesive is prone to damage and environmental degradation.

Strengthening of metallic structures with FRP composites 229

Bond defects

The sensitivity of the adhesive joint to defects should be established and used to write the installation specification and to inform periodic inspection requirements. Highly-stressed regions of the adhesive joint (such as adjacent to the end of a strengthening plate) will be most susceptible to defects.

For example, the elastic bond analysis in Fig. 8.6b has been used to determine the effect of a 50 mm long defect, starting 10 mm from the end of a tapered plate. This might be due to an area of poor surface preparation. This defect gives adhesive shear stresses close to the values in a plate without a taper.

The size of defects considered in the analysis should be determined by considering the size of defect that can be detected. The adhesive joint is very difficult to inspect (see Chapter 12). Voids can be detected by tap-testing, but this will not detect touching debonds. Various non-destructive testing (NDT) methods are under development, although none can detect all types of bond defect. For example, ultrasonic NDT has been used to detect defects with a minimum diameter of 15 mm and thickness of 0.8 mm (Bastianini *et al.*, 2004).

Bond durability

There are few data available on the durability of adhesive joints between metallic structures and FRP. The most comprehensive study suggests that the joints have good environmental properties, but there are insufficient data to cover the various materials involved and the damage and deterioration mechanisms are not thoroughly understood (Karbhari and Sulley, 1995). As noted above, report C595 (Cadei *et al.*, 2004) suggests a partial safety factor for environmental degradation, although this is based upon FRP-to-FRP bonded joints.

Environmental attack will take place from the exposed edge of the adhesive joint, or any other exposed area. Moisture ingress can be particularly detrimental, and the long-term performance of the plate can be improved by barriers to water ingress. These include a spew fillet around the adhesive joint, and blocking other water ingress paths to the adhesive joint (for example, by ensuring that water cannot reach the adhesive joint via holes in a riveted wrought iron beam).

Fatigue of the adhesive joint

There has been a small amount of work on the fatigue properties of FRP-to-metal adhesive joints (Jones and Civjan, 2003; Tavakkolizadeh and

Saadatmanesh, 2003b; Buyukozturk *et al.*, 2004). These tests could be used as the basis of a fatigue life prediction for the adhesive joint. Further research work is being carried out on this subject.

8.5 Design in cases other than flexural strengthening

The previous section focused upon the design of FRP strengthening for metallic members in flexure because this has been the most common application to date, and consequentially the most fully developed. The sectional analysis and bond analysis concepts also apply if FRP strengthening is used to address other deficiencies in metallic structures.

8.5.1 Damaged or degraded beams

The metallic cross-section can be reduced by corrosion or localised damage (for example, due to impact). FRP strengthening can be used to replace the lost cross-section area, in much the same way as it can be used to increase the flexural strength of a beam.

Tests on full-scale steel girders with composite concrete slabs have shown that damaged beams can be repaired using CFRP laminates (Tavakkolizadeh and Saadatmanesh, 2003c). The tensile flange of each beam was partially or completely cut to simulate damage, and then strengthened using CFRP. For girders in which a large portion of the tensile flange had been damaged, failure was characterised by debonding of the FRP along the adhesive joint from mid-span. This was due to unequal deformation of the steel on either side of the damaged area, resulting in peel stresses. Other research has shown how a GFRP over-wrap can be applied to prevent debonding of FRP strengthening from the metallic beam at mid-span (Photiou *et al.*, 2006).

8.5.2 Extending the fatigue life of a metallic structure

Two methods are available for increasing the fatigue life of a metallic structure using FRP strengthening.

- *Prestressed* FRP strengthening has the advantage that it applies compression to the bottom flange of a beam, which will slow fatigue crack propagation at all positions within the member. This method is particularly appropriate for riveted structures, due to the large number of stress raisers within them, and has been successfully demonstrated in laboratory tests by Bassetti *et al.* (2000).
- *Unstressed* FRP patches can also be applied, so as to bridge fatigue cracks. These patches reduce the crack tip stress concentration by local

stiffening; carbon fibres are used to provide the required stiffness. Steel tension coupon specimens (initially cracked and uncracked) have been strengthened using CFRP and tested in fatigue (Jones and Civjan, 2003; Buyukozturk *et al.*, 2004). These tests showed that the fatigue life of steel can be more than doubled using CFRP, but this must be applied to both sides of a tensile member; single-sided strengthening induces bending deformation and does not result in an increase in fatigue life. Such bending effects are not present in beams, however, and the CFRP is required only on the underside of the tensile flange. Fatigue tests on steel beams have shown that the fatigue life can be increased around three times. Growth of the fatigue crack in the steel was accompanied by gradual debonding of the CFRP away from the steel crack.

Further work is required on fatigue strengthening before a general design model can be given; however, the above research can be used to inform design.

8.5.3 Buckling

The lateral buckling capacity of a member is dependent upon the member's flexural stiffness. Applying FRP directly to the surface of a member makes little difference to its stiffness. (However, the FRP will provide tensile strength to a cast iron member and prevent collapse during buckling-induced deflection.) To increase a member's buckling capacity, it is most efficient to increase the section's radius of gyration by positioning the FRP material a long way from the neutral axis. For example, metallic compression members can be encapsulated within an oversized FRP tube, which is filled with concrete to provide bond between the existing structure and the FRP (Liu *et al.*, 2005a).

The local buckling capacity of a member can be increased using bonded FRP strengthening. Research has focused upon applying the FRP directly to the surface of the member. For example, the local buckling strength of square hollow steel sections was increased by 18% using carbon fibres wrapped circumferentially, and the section stiffness increased by 28% with the fibres arranged longitudinally (Shaat and Fam, 2006). The technique has also been applied to increase the web capacity of steel I-beams (Patnaik and Bauer, 2004), although the beams tested had thin webs. The thickness of CFRP required for normally proportioned steel beams may be uneconomic, unless the CFRP is positioned away from the web of the beam.

8.5.4 Local strengthening of connections

Connection details can be strengthened by over-wrapping them with fibre strengthening (Garden, 2001). Wet lay-up methods (combined with vacuum consolidation techniques) allow the strengthening to be tailored to suit the connection configuration. Connection strengthening will require design on a case-by-case basis, in which the fibre arrangement is chosen to maximise the connection strength in the required direction.

8.5.5 Structural integrity

FRP strengthening can be used to improve the robustness of brittle cast iron structures, which crack under impact loading or thermal shock. FRP can be bonded to restrain the cast iron after it has cracked, so that the original structural form is not lost and overall collapse does not occur (Bastianini *et al.*, 2004).

8.6 References

ASTM (2001) *ASTM D5868-01. Standard Test Method for Lap Shear Adhesion for Fiber Reinforced Plastic (FRP) Bonding*, West Conshoshocken, PA, USA, American Society for Testing and Materials.

ASTM (2003) *ASTM D3762-03. Standard Test Method for Adhesive-Bonded Surface Durability of Aluminium (Wedge Test)*, West Conshohocken, PA, USA, American Society for Testing and Materials.

BASSETTI A, NUSSBAUMER A and HIRT M (2000) Crack repair and fatigue life extension of riveted bridge members using composite materials, in Hosny A-N (ed.), *Proceedings Bridge Engineering Conference 2000*, ESE-IABSE-FIB, Sharm El-Sheikh, Cairo, Egypt, Egyptian Society of Engineers, Vol. **1**, 227–238.

BASTIANINI F, CERIOLO L, DI TOMMASO A and ZAFFARONI G (2004) Mechanical and nondestructive testing to verify the effectiveness of composite strengthening on historical cast iron bridge in Venice Italy, *Journal of Materials in Civil Engineering*, **16**(5), 407–413.

BSI (2001) BS 7991; 2001 *Determination of the mode I adhesive fracture energy-G_{IC}, of structural adhesives using the double cantilever beam (DCB) and tapered cantilever beam (TDCB) specimens*, London, UK, British Standards Institute.

BSI (2005) *BS 5350-C5: 2002 Methods of test for adhesives. Determination of bond strength in longitudinal shear for rigid adherends*, London, UK, British Standards Institute.

BUYUKOZTURK O, GUNES O and KARACA E (2004) Progress on understanding debonding problems in reinforced concrete and steel members strengthened using FRP composites, *Construction and Building Materials*, **18**, 9–19.

CADEI J M C, STRATFORD T J, HOLLAWAY L C and DUCKETT W G (2004) *Strengthening Metallic Structures using Externally Bonded Fibre-Reinforced Polymers*, Report C595, London, UK, CIRIA.

DAVIS M and BOND D (1999) Principles and practices of adhesive bonded structural joints and repairs, *International Journal of Adhesion and Adhesives*, **19**, 91–105.

DENTON S R (2001) Analysis of stresses developed in FRP plated beams due to thermal effects, in Teng (ed.), *Proceedings International Conference on FRP Com-*

posites in Civil Engineering – CICE 2001, Oxford, UK, Elsevier Science, 527–536.

DENG J, LEE M M K and MOY S S J (2004) Stress analysis of steel beams reinforced with a bonded CFRP plate, *Composite Structures*, **65**, 205–215.

FROST S, LEE R J and THOMPSON V K (2003) Structural integrity of beams strengthened with FRP plates – analysis of the adhesive layer, in Forde M C (ed.), *Proceedings Tenth International Conference on Structural Faults and Repair*, Edinburgh, UK, Technics Press (CD ROM).

FROSTIG Y, THOMSEN O T and MORTENSEN F (1999) Analysis of adhesive-bonded joints, square-end, and spew-fillet – high-order theory approach, *Journal of Engineering Mechanics*, **125**(11), 1298–1307.

GARDEN H N (2001) Use of composites in civil engineering infrastructure, *Reinforced Plastics*, **45**(7), 44–50.

GOLAND M and REISSNER E (1944) The stresses in cemented joints, *Journal of Applied Mechanics*, **11**, 17–27.

HOLLAWAY L C and CADEI J M C (2002) Progress in the technique of upgrading metallic structuresv with advanced polymer composites, *Progress in Structural Engineering and Materials*, **3**, 131–148.

HUTCHINSON A R and HURLEY S A (2001) *Transfer of Adhesives Technology – Feasibility Study*, Project Report 84, London, UK, CIRIA.

ISTRUCTE (1999) *Guide to the Structural Use of Adhesives*, London, UK, SETO.

JONES S C and CIVJAN S A (2003) Application of fiber reinforced polymer overlays to extend steel fatigue life, *Journal of Composites for Construction*, **7**(4), 331–338.

KARBHARI V M and SULLEY S B (1995) Use of composites for rehabilitation of steel structures – determination of bond durability, *Journal of Materials in Civil Engineering*, **7**(4), 239–245.

LAM C C A, CHENG J J R and YAM C H M (2004) Study of the tensile strength of CFRP/steel double Lap Joints, *Proceedings Advanced Composite Materials in Building and Structures – ACMBS IV*, Calgary, Alberta, Canada, July 20–23 (CD ROM).

LIU X, NANNI A and SILVA P F (2005a) Rehabilitation of compression steel members using FRP pipes filled with non-expansive and expansive light-weight concrete, *Advances in Structural Engineering*, **8**(2), 129–142.

LIU H, ZHAO X-L, AL-MAHAIDI R and RIZKALLA S (2005b) Analytical bond models between steel and normal modulus CFRP, in Shen Z Y, Li G Q and Chan S L (eds), *Proceedings 4th International Conference on Advances in Steel Structures*, Amsterdam, The Netherlands, Balkema, 1545–1552.

MOY S S J (ed.) (2001) *FRP Composites – Life Extension and Strengthening of Metallic Structures*, London, UK, Thomas Telford.

MOY S S J, CLARK J and CLARKE H (2004a) The strengthening of wrought iron using carbon fibre reinforced polymer composites, in Hollaway L C, Chryssanthopoulos M K and Moy S S J (eds), *Advanced Polymer Composites for Structural Applications in Construction: ACIC 2004*, Cambridge, UK, Woodhead, 258–265.

MOY S S J (2004b) Design guidelines for the strengthening of metallic structures using fibre reinforced composites, *Proceedings Advanced Composite Materials in Building and Structures – ACMBS IV*, Calgary, Alberta, Canada, July 20–23 (CD ROM).

PATNAIK A K and BAUER C L (2004) Strengthening of steel beams with carbon FRP laminates, *Proceedings Advanced Composite Materials in Building and Structures – ACMBS IV*, Calgary, Alberta, Canada, July 20–23 (CD ROM).

PHOTIOU N K, HOLLAWAY L C and CHRYSSANTHOPOULOS M K (2006) Strengthening of an artificially degraded steel beam utilising a carbon/glass composite system, *Construction and Building Materials*, **20**, 11–21.

SEBASTIAN W M (2003) Nonlinear influence of contraflexure migration on near-curtailment stresses in hyperstatic FRP-laminated steel members, *Computers and Structures*, **81**, 1619–1632.

SEBASTIAN W M (2005) Path dependency in hybrid structures with simultaneous ductile and brittle connections and materials, *International Journal of Solids and Structures*, **42**, 4859–4879.

SEN R, LIBY L and MULLINS G (2001) Strengthening steel bridge sections using CFRP laminates, *Composites: Part B*, **32**, 309–322.

SHAAT A and FAM A (2006) Axial loading tests on short and long hollow structural steel columns retrofitted using carbon fibre reinforced polymer, *Canadian Journal of Civil Engineering*, **33**(4), 458–470.

STRATFORD T J and CADEI J M C (2006) Elastic analysis of adhesion stresses for the design of a strengthening plate bonded to a beam, *Construction and Building Materials*, **20**, 34–45.

TAVAKKOLIZADEH M and SAADATMANESH H (2003a) Strengthening of steel–concrete composite girders using carbon fiber reinforced polymers sheets, *Journal of Structural Engineering*, **129**(1), 30–40.

TAVAKKOLIZADEH and SAADATMANESH (2003b) Fatigue strength of steel girders strengthened with carbon fiber reinforced polymer patch, *Journal of Structural Engineering*, **129**(2), 186–196.

TAVAKKOLIZADEH and SAADATMANESH (2003c) Repair of damaged steel-concrete composite girders using carbon fibre-reinforced polymer sheets, *Journal of Composites for Construction*, **7**(4), 311–322.

YANG J, TENG J G and CHEN J F (2004) Interfacial stresses in soffit-plated reinforced concrete beams. *Proc Inst civil engineers: Structures and buildings*, **157**(1), 2004, 77–89.

9
Strengthening of masonry structures with fibre-reinforced polymer (FRP) composites

L. DE LORENZIS, University of Salento, Italy

9.1 Introduction

Masonry is a construction technique where a large number of small modular units, either natural or artificial, are assembled together, typically with mortar, to form a structure or a component. It can be considered the great protagonist in the history of construction, having been the primary construction technique for almost ten millennia all over the world. Despite the decline in use in the 19th and 20th centuries due to the advent of modern construction materials, masonry structures still represent a vast portion of the international built inventory, including most of the world's historical constructions. In recent times, masonry is once again becoming competitive due to ease of realisation, versatility and ease of substitution of the modular units, good mechanical and insulation properties, implying the possibility to couple load-bearing and insulation functions in the same components, and good durability and fire resistance. A revitalised interest in this type of construction is due to the concept of reinforced masonry, where steel bars are used to introduce tension resistance. This section however discusses only unreinforced masonry (URM).

The international and historical development of masonry has resulted in a wide variety of material and bonding patterns. Artificial modular units are typically clay bricks or concrete blocks, natural modular units are normally cut stones or, less commonly, raw clay bricks. Assemblage is made with mortar of different types, or with no mortar in particular cases. A large variety exists in bonding patterns for masonry walls, both in-plane and through the thickness. Ancient walls are often double- or even multiple-leaf walls and have different degrees of transverse connection and types of fill between the leaves.

Masonry elements are typically load-bearing or infill walls, which may be designed to resist lateral and/or gravity loads. Structural arches, vaults and domes are common in historic masonry structures. Other masonry elements are retaining walls, columns, platbands and lintels. Old masonry buildings often have unsafe characteristics, such as unbraced parapets, inadequate connections between the walls and to the roof and unbalanced thrusts from curved members. URM structures have shown their vulnerability during major earthquakes throughout the world. These and other factors, such as extreme loading, material deterioration or imposed displacements, prompt the need for rehabilitation and strengthening. Chapter 1 gave an overview of the most frequent causes of structural deficiency in masonry structures.

Given the importance of the masonry building stock and its vulnerability, a great deal of research has been devoted in the last decades to the development of structurally effective and affordable techniques for repair and strengthening of masonry structures. The evolution from research findings to design procedures is particularly challenging, as the large variability of masonry materials and systems implies a notable difficulty in generalising results obtained on one system of masonry to another. Moreover, strengthening historic structures involves additional challenges such as reversibility, minimal intrusion and the expectation of durability over a century timescale.

Traditionally, masonry strengthening has been accomplished using conventional materials and construction techniques, such as externally bonded steel plates, reinforced concrete overlays, grouted cell reinforcements and external post-tensioning among many others. Over these methods, the recently emerged use of fibre-reinforced polymer (FRP) composites presents a number of advantages (see also Chapter 1). FRPs do not corrode. Their size and weight allow the dynamic response of the structure to remain practically unaltered. Their low thickness minimises aesthetic impact. Recently, transparent FRP laminates have been developed for applications on historic masonry structures, with the result that the intervention is macroscopically invisible (Triantafillou, 2001). The application of FRP laminates has been proved reversible, by raising the temperature of the FRP above the glass transition temperature of the resin. However, the limited information available on long-term durability and compatibility of FRP with the masonry substrate in various moisture and temperature conditions means that considerable caution is still called for. This is a major obstacle preventing the widespread use of FRP, especially for strengthening historic structures. Also, there is still a lack of established design procedures specific to masonry structures. Masonry elements strengthened with FRP materials are frequently treated as reinforced concrete elements because of lack of specific knowledge. Further experimental and theoretical research is still needed to provide the designer with reliable design procedures in all strengthening cases.

This chapter focuses on strengthening of URM structures with externally bonded FRP laminates and near-surface mounted (NSM) FRP reinforcement. Details on these two systems can be found in Chapter 1. Very recently, the use of glass fibre grids bonded to the masonry substrate with sprayed polyurea has been proposed; for details see Galati *et al.* (2005). This chapter addresses the best-established applications of FRP in masonry structures and, when available, outlines the relevant methods of analysis and design. Other less common applications are listed in Chapter 1. At the present stage, the only design guidelines available on FRP strengthening of masonry structures are those recently issued by the National Research Council in Italy (CNR, 2004). Other organisations such as the American Concrete Institute are, at the time of writing, in the drafting phase of their masonry-related documents.

For FRP strengthening of historic masonry structures, the design of the intervention should comply with the theories of restoration (ICOMOS, 2001). For an extensive treatment of the problems associated with the use of FRP in historic masonry structures, see e.g. De Lorenzis and Nanni (2004). Durability-related aspects of masonry structures strengthened with FRP composites are addressed in Chapter 11.

9.2 General aspects of FRP strengthening for masonry structures

The use of FRP systems in masonry structures, as with other types of structures, is normally part of a global intervention of structural strengthening. The basic role of FRP systems is to transfer tensile stresses both *within* a structural member and *between* different members of the structure. The introduction of tension resistance radically modifies the way the structures react to external loads. In particular, FRP systems in masonry structures can be used to:

- increase the load-carrying capacity of load-bearing panels;
- transform non-structural elements (e.g. infill panels) in load-carrying elements by enhancing their strength and stiffness;
- increase the load-carrying capacity and reduce the lateral thrust of thrusting elements (arches and vaults);
- strengthen and stiffen non-thrusting flooring systems to enable their functioning as a rigid diaphragm (see Chapter 1, p 35);
- create a connection between different elements of a structural assembly, in order to obtain a three-dimensional response to external loads (e.g. by connecting the perimeter elements of an entire building at the height of the floor slabs);
- limit the opening of cracks;
- confine columns to enhance their strength and/or ductility.

An overview of possible FRP applications in masonry structures has been given in Chapter 1 (for a general summary see Fig. 1.3).

The strengthening design should aim at obtaining a state of tensile stress in the FRP system. FRP systems subjected to compressive stresses do not offer any significant benefit to the masonry structure, as their cross-sectional area is negligible compared to that of a masonry element and they are prone to local buckling leading to premature debonding.

Structural modelling of a masonry structure is a complex task. The distribution of forces and moments within a structural assembly or the distribution of stresses on a single element of a masonry structure can be computed with linear-elastic analysis, or with non-linear analyses accounting for the inelastic behaviour and limited tensile strength of the masonry material. A possible alternative which simplifies design consists of assuming an approximate distribution of forces and moments or an approximate distribution of stresses in a single element, which satisfies equilibrium but not necessarily compatibility. For a strengthened structure or element, all tensile forces/stresses arising from equilibrium must be resisted by the FRP system. However, these approximate distributions should be adopted with caution, as brittle local failures in the FRP–masonry system could lead to collapse, even under a statically admissible stress state.

All available types of FRP materials can be used to strengthen masonry structures. However, it is generally believed that glass fibre reinforced polymer (GFRP) is preferable to other types of FRP in masonry strengthening, not only because of its lower cost but also because of its lower modulus of elasticity, which makes it more compatible with masonry and makes premature debonding of the reinforcement after masonry cracking less critical (Li *et al.*, 2004).

FRPs can be used for masonry strengthening in different forms:

- *Externally bonded FRP laminates.* As for reinforced concrete (RC) beams, either FRP sheets impregnated using the wet lay-up technique, or pre-cured FRP plates or grids bonded in place with a suitable adhesive can be used (see Chapter 2).
- *NSM reinforcement* (see Chapter 1) (Fig. 9.1a).
- *Structural repointing* (see Section 9.4) (Fig. 9.1b).
- *Unbonded FRP systems* have been proposed, e.g. for strengthening out-of-plane loaded walls against overturning collapse, and strengthening buildings against global collapse mechanisms.

At present, the Italian guidelines on FRP strengthening of masonry structures only address the use of externally bonded FRP laminates, as the other techniques are more recent and have been less investigated.

In each of the above systems (except for the unbonded ones), the FRP composite is bonded to the masonry substrate with a suitable adhesive, which can be a polymer (typically epoxy) or a cement-based paste or mortar (note, however, that dry carbon or glass fibres embedded into a cementi-

Strengthening of masonry structures with FRP composites

(a) FRP bar in a saw-cut groove

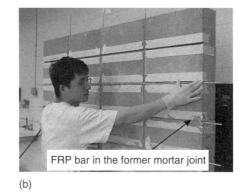

(b) FRP bar in the former mortar joint

9.1 Masonry walls strengthened with NSM FRP bars (a) or structural repointing (b) (from Tumialan and Nanni, 2001).

tious matrix cannot be classified as 'FRP' as the matrix is no longer a polymer; these systems are not further discussed in the following).

General rules valid for all *bonded* FRP strengthening systems can be summarised as follows:

- FRP laminates should not cover extended regions of a masonry surface, in order to ensure that moisture migration is not precluded.
- Any FRP system should be applied on a sound masonry substrate. If the masonry is damaged, non homogeneous or presents any defect that could compromise adequate stress transfer from the strengthening system, a consolidation of masonry with conventional techniques should be carried out prior to installation of the FRP.
- The surface on which the FRP system is applied should be reasonably regular, i.e. local curvatures and undulations should be small enough to be accommodated within the thickness of the adhesive. For substrates with global curvature (i.e. surfaces of arches and vaults) the only applicable FRP system are externally bonded wet lay-up laminates, whose small bending stiffness allows them to easily follow the geometry of the substrate.

Also, externally bonded FRP systems applied on external surfaces generally require UV protection by proper coatings or common plasters.

9.3 Bond of FRP systems to masonry

In bonded applications, the interfacial behaviour and strength between the FRP system and the masonry substrate plays a major role in the structural behaviour and capacity of a strengthened element. Debonding of the FRP is highly undesirable because of its brittle nature, and because it reduces the efficiency in exploitation of the FRP mechanical properties. In a structural design following the principles of the hierarchy of strength, the brittle debonding failure should not precede the development of inelastic stresses in the compressed masonry.

9.3.1 Bond of externally bonded laminates

In principle, the mechanics of bonding of FRP laminates to masonry is very similar to that of FRP–concrete bond. However, investigations into bonding of FRP to masonry are complicated by the large variety of materials used in masonry construction, and by the fact that the FRP system is bonded on a heterogeneous substrate made of blocks, mortar joints and sometimes even repair mortars used for consolidation. Different portions of the FRP along its length can hence be bonded to different materials having diverse interfacial properties. Nevertheless, the key aspects of the FRP–masonry bond, as indicated by research results, are similar to the FRP–concrete bond, as follows:

- As the tensile strength of the adhesives is normally larger than that of the substrate, debonding typically occurs as cohesive failure within the masonry.
- The maximum tensile force that can be applied to the FRP–masonry joint before debonding increases with the bond length. A limiting value of bond length exists, beyond which further increases in bond length do not produce any increase in the debonding load. This limiting bond length is termed effective bond length.
- In the case of FRP systems applied to flat surfaces, two main debonding mechanisms can be observed: one starting from the termination of the FRP (plate end shear debonding), and one starting from mortar joints or transverse cracks in the masonry. In both cases, debonding is initiated by high concentrations of shear stresses and normal (peeling) stresses due to the bond action, localised within a short region of the FRP.

The bond strength depends on the tensile strength of masonry, and on substrate surface characteristics such as roughness, soundness and porosity. Roko *et al.* (1999), for example, observed that less epoxy adhesive is absorbed into the surface of extruded brick units compared to moulded bricks, leading to a reduction of the bond strength at the FRP laminate–masonry interface. Substrate surface properties are also strongly influenced by surface preparation. Adequate preparation involves complete removal of all mortar residue, dust, dirt, plaster, paint and efflorescence

Strengthening of masonry structures with FRP composites

from the masonry surface. Smooth-faced epoxy-coated or glazed units must first be roughened by grinding or sand-blasting. For unspoiled new clay or concrete masonry surfaces, wire brushing proved to be adequate to remove any loose particles or dust. However, surface preparation of older clay or concrete masonry may require more intrusive techniques such as water-blasting or grinding. Concrete masonry units (CMU) may be lightly sand-blasted (Hamoush *et al.*, 2001), but this should be used with caution for clay units. For more details about surface preparation, see Chapter 3.

In the case of FRP laminates applied to a cracked masonry element, if the fibres are not orthogonal to the crack direction, stress concentrations may arise across the crack due to the relative displacement between the crack faces. Also, stresses acting in the direction normal to the bonded surface influence significantly the value of tensile force causing debonding of the FRP. These normal stresses are particularly large when the FRP has a significant bending stiffness, such as pre-cured plates, or is bonded to curved substrates, such as in strengthening of arches and vaults. In the latter case, the normal stresses are related from equilibrium to the bond shear stresses and the resulting stress is tensile for concave substrates (e.g. intrados of an arch) and compressive for convex substrates (e.g. extrados of an arch) (Fig. 9.2). Normal tensile and compressive stresses are, respectively, detrimental and beneficial to the bond strength of the FRP–masonry joint.

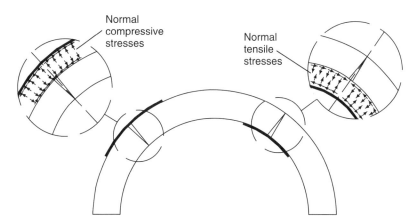

9.2 Shear bond stresses and normal peeling stresses between FRP and curved masonry substrates.

9.3.2 Bond of near-surface mounted reinforcement

Specific investigations into the bond of NSM bars to CMU have been conducted (De Lorenzis et al., 2000). In this case also results are qualitatively similar to those obtained for bond of NSM reinforcement in concrete. Splitting of the encapsulating epoxy and cracking of the masonry material adjacent to the groove control debonding of deformed bars, whereas bars with smooth or sand-blasted surface are susceptible to debonding at the bar–epoxy interface. The debonding load of a joint tested in pure shear normally increases with increasing bond length until, for a sufficiently long bond length, the tensile strength of the bar is reached. The splitting bond strength for ribbed bars increases with the tensile strength of the epoxy and of the masonry material. Bajpai and Duthinh (2003), by using epoxy reinforced with short glass fibres, were able to develop the full tensile strength of deformed bars which had limited embedment lengths. The splitting bond strength also increases with the groove size, but this is limited by the possibility of creating local damage in masonry walls during cutting of the grooves, especially for strengthening of hollow concrete masonry units. The bond behaviour of structural repointing systems is similar to that of NSM systems.

9.3.3 Design against debonding

The simplest approach to account for debonding in design is to compute the value of tensile force in the FRP required to cause its debonding from the substrate (maximum debonding force), and the corresponding minimum anchorage length (effective bond length). In design of FRP strengthening for the various possible applications, the maximum force that the FRP can resist is then taken as the minimum between the force required to rupture the FRP in tension, and the maximum debonding force (provided that the anchorage length is at least equal to the effective bond length). This approach is adopted by the Italian guidelines on masonry strengthening with FRP (CNR, 2004). These guidelines propose formulae, valid for externally bonded FRP laminates, to compute the effective bond length, the fracture energy of the FRP–masonry interface and the maximum stress sustainable by the joint, based on the model by Holzenkämpfer (1994), as modified by Neubauer and Rostásy (1997), previously proposed for FRP–concrete bond. The computation of the maximum debonding force in this model is based on a fracture mechanics approach.

Another approach consists in reducing the tensile strength of the FRP system with an empirical bond reduction factor (k_m), such as is done by current ACI guidelines on FRP strengthening of concrete structures and proposed by Tumialan and Nanni (2001) for strengthening of masonry structures. Galati *et al.* (2005) proposed values of k_m based on test results on URM walls strengthened with FRP laminates and NSM bars. However, a comprehensive calibration of the k_m factor is not yet feasible at this stage of research, and experimental testing is highly recommended to determine the k_m value for design purposes.

For substrates with modest curvature, the combined effect of shear bond stresses and normal peeling stresses can be approximately evaluated by assuming a linear interaction diagram (CNR, 2004). This can be done if reliable values for the bond (shear) strength and the normal (peeling) strength of the substrate are available; these can be found by experimental testing. Further research is still needed also in this area.

9.4 Strengthening of masonry panels under out-of-plane loads

9.4.1 General aspects

Masonry panels may be subjected to out-of-plane loads due to earth pressure, seismic actions, dynamic vibrations, verticality flaw, wind pressure and arch thrust (CNR, 2004). They can be strengthened with FRP in various forms:

- externally bonded laminates (pre-cured or wet lay-up systems);
- NSM reinforcement;
- structural repointing.

In all cases, the strengthening system is applied to the tension face of the panel with the fibre direction aligned with the principal tensile stress (with the exception of strengthening against over-turning, see next section). Externally bonded laminates are normally applied as discrete strips with a given width and spacing. Full bonding to the tension face of the wall is to be considered with caution as it could preclude moisture migration. In the NSM method, the bars are embedded into epoxy-filled grooves cut on the surface of the wall in the appropriate direction (e.g. perpendicular to the bed joints for vertical bending). Special care is needed during installation of NSM bars in walls made of hollow concrete masonry units, to avoid the creation of local damage due to the limited thickness of the concrete shell.

Structural repointing can be considered as a variant of the NSM reinforcement technique, applied to masonry (Tumialan and Nanni, 2001). Repointing is a traditional retrofitting technique for masonry structures, consisting of replacing the mortar missing from the joints. The term 'structural' is added to describe a strengthening system aimed at restoring the integrity and/or upgrading the shear and flexural capacity of masonry walls.

This is achieved by grooving the mortar joints, and bonding FRP rods into these grooves by means of a suitable adhesive. This technique offers several advantages over alternative methods, such as reduced surface preparation and minimal aesthetic impact and invasiveness. The difference with respect to the NSM system is that the grooves are cut in the mortar joints and not in the blocks. The most typical forms of reinforcement for structural repointing are round bars and strips. The latter offer better bond properties and have low thickness, aiding their embedment in thin mortar joints and reducing invasiveness. Epoxy adhesive is typically used for embedding the reinforcement. The use of less expensive pastes, such as latex-modified cementitious paste (Turco *et al.*, 2003), makes the FRP structural repointing technique more appealing, as the structural performance is not reduced and the appearance of the filled joints is similar to that of conventional mortar joints.

9.4.2 Collapse modes

Collapse of masonry panels under out-of-plane loads may be controlled by:

- loss of rigid-body equilibrium (over-turning);
- vertical bending;
- horizontal bending.

Loss of rigid-body equilibrium

This collapse mode consists in over-turning of a masonry panel about an ideal cylindrical hinge at its base. The hinge forms due to the limited tensile strength of masonry and can be assumed to be located on the external surface of the panel (Fig. 9.3a, b). The collapse by over-turning may happen for walls inadequately connected to the orthogonal walls and to the upper floor slab. The variables most relevant to this collapse mode are the restraint conditions and the slenderness of the panel.

For strengthening the wall against this collapse mode, one or more FRP horizontal laminate strips can be bonded to the upper region of the wall and anchored to the orthogonal walls. Particular care should be taken in rounding the corners in order to avoid stress concentrations that could lead to premature tensile rupture of the FRP. FRP sheets are more suited than pre-cured laminates to this application, as their limited thickness facilitates wrapping around corners and reduces possible debonding problems due to unevenness of the masonry surface and local curvatures. The highest effectiveness is reached if the building is entirely surrounded by FRP 'belts'.

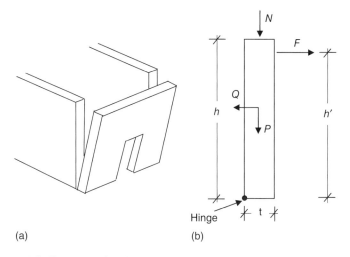

9.3 Collapse mode (a) and static scheme (b) of out-of-plane loaded unreinforced masonry walls for over-turning.

Triantafillou and Fardis (1997) proposed using FRP ties in the form of external unbonded tendons. These would be anchored to masonry only at their ends, applied to the external face of the structure and post-tensioned to provide horizontal confinement. An attractive feature of this method in the case of historic structures is its reversibility, as the bonded portion is limited to the anchorage zones.

In Fig. 9.3b, P is the panel self weight, N the normal force acting above the panel, Q the horizontal force due to the seismic action (in this example) and F is the restraining force exerted by the FRP (CNR, 2004). Other forces arising from wind or thrusts of arches or vaults could be present. Assuming that the restraining action of orthogonal floor slabs and walls is negligible, the static scheme is that in Fig. 9.3b, and the force in the FRP system can be easily computed from equilibrium of moments about the hinge. The FRP can then be designed against tensile rupture and debonding. The second verification is not necessary when the structure is entirely surrounded, provided that the extremities of the 'belt' are overlapped for an adequate length. The horizontal sections of the panel must then be verified against combined compression and bending and against shear. Possible further collapse modes (listed as follows) must also be checked.

Vertical bending

A URM wall adequately restrained at its upper and lower extremities and subjected to horizontal and vertical actions may collapse due to combined

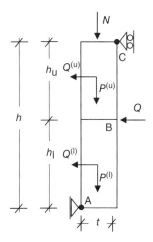

9.4 Static scheme of out-of-plane loaded unreinforced masonry walls under vertical bending.

compression and bending. In this case, collapse occurs by formation of three hinges, of which two are at the upper and lower extremities, and one at an intermediate height of the panel. A hinge forms when the pressure point at a certain cross-section falls outside the thickness of the wall.

This collapse mode is critical for high walls, connected to orthogonal walls at a large distance from each other, and for walls connected to floor slabs at different heights on the two sides, subjected to seismic actions. In these cases, the application of FRP systems with the fibres in the vertical direction realises an 'FRP-reinforced masonry' where compressive stresses are taken by the masonry material and tensile stresses by the FRP reinforcement.

Figure 9.4 shows an example of masonry panel subjected to: $P^{(u)}$ self-weight of the upper portion of the panel; $P^{(l)}$ self-weight of the lower portion of the panel; $Q^{(u)}$ seismic force acting on the upper portion of the panel; $Q^{(l)}$ seismic force acting on the lower portion of the panel; N normal force acting above the panel; Q additional horizontal force (e.g. from the lateral thrust of a barrel vault) (CNR, 2004). The restraint forces in A and C can be easily determined from equilibrium conditions. An FRP system with the fibres oriented vertically can be used to inhibit the formation of the hinge at the B cross-section. The FRP-reinforced member subjected to vertical bending may fail by combined compression and bending (by masonry crushing or FRP tensile rupture), by FRP debonding or by out-of-plane shear. These failure mechanisms and the methods to design against them are illustrated in the next sections.

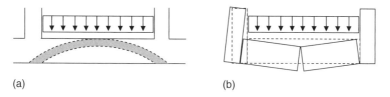

9.5 Static scheme (a) and collapse mode (b) of out-of-plane loaded unreinforced masonry walls under horizontal bending.

Horizontal bending

For a masonry wall, well restrained at its lower extremity and also well connected to transverse walls able to exert a buttressing action, the resistance to horizontal actions is provided by an arch-like functioning of the upper portion of the wall in the horizontal direction (Fig. 9.5a). This ideal arch may fail by masonry crushing, or by formation of a mechanism if the buttressing capacity of the transverse walls is exceeded (Fig. 9.5b). In contrast, FRP systems with the fibres in the horizontal direction may introduce tension resistance and transform the unit strip on the upper part of the masonry wall into a 'reinforced-masonry' beam. Again, the reinforced-masonry beam may fail in combined compression and bending (by masonry crushing or FRP tensile rupture), by FRP debonding or by out-of-plane shear, and it can be analysed and designed as illustrated in the next sections. One further verification to be conducted involves the resistance of the transverse walls subjected to tension close to the connection with the strengthened wall.

9.4.3 Failure mechanisms

As mentioned above, failure of an FRP-strengthened wall under vertical or horizontal bending may occur by three different mechanisms:

- failure under compression and bending by FRP rupture or masonry crushing;
- failure by FRP debonding;
- shear failure.

Failure under compression and bending

After developing flexural cracks primarily located at the mortar joints, the panel may fail by either rupture of the FRP reinforcement or masonry crushing. Typically, failure is controlled by crushing in walls strengthened with high area fractions of FRP and/or with high strength FRP materials such as carbon fibre reinforced polymer (CFRP), and by FRP rupture in walls strengthened with low area fractions and/or with low strength FRP materials such as GFRP. In intermediate situations FRP debonding often dominates. FRP rupture is less desirable than masonry crushing, the latter being more ductile (Triantafillou, 1998). However, both failure modes are acceptable in design.

FRP debonding

Loss of bond at the FRP–masonry interface may precede flexural failure. Debonding may start from the end of the strengthening system (in the absence of special anchorage) due to stress concentrations at the cut-off section, or from flexural cracks in the maximum bending moment region. Since the tensile strength of masonry is lower than that of the resin, failure typically occurs within the masonry substrate for walls strengthened with FRP laminates.

In the case of NSM FRP strengthening, after flexural cracking the tensile stresses at the mortar joints are taken by the FRP reinforcement and additional cracks due to the bond action may develop in the masonry units (oriented at approximately 45°) or in the head mortar joints. Some of these cracks follow the embedding paste–masonry interface causing debonding and subsequent wall failure. In the case of smooth NSM FRP bars, debonding failure can be due to the sliding of the bar inside the epoxy. Finally, if deep grooves are used, debonding can also be caused by splitting of the embedding material (De Lorenzis *et al.*, 2000; Galati *et al.*, 2005).

Shear failure

Cracking starts with the development of fine vertical cracks in the maximum bending region. Thereafter, two types of shear failure can be observed: flexural-shear or sliding shear. The first type is oriented at approximately 45°, and the second type occurs (for vertical bending) along a bed joint near the support, causing sliding of the wall at that location. The crack due to this flexural-shear mode causes a differential displacement in the shear plane, which often results in FRP debonding (Hamoush *et al.*, 2002; Tumialan *et al.*, 2003a). Shear failure should be prevented by appropriate design, due to its brittle nature.

9.4.4 Design of FRP strengthening under out-of-plane loads

For non-load bearing walls (subjected to bending with no axial force), the ultimate strength design criterion states that the design flexural capacity of

Strengthening of masonry structures with FRP composites   249

the member must exceed the flexural demand. For load-bearing walls, bending is coupled with axial force and the design criterion states that the point representative of the axial load and bending moment demand on the member must fall within the boundary of the axial load–bending moment interaction diagram.

In both cases, the 'reinforced masonry' section can be studied with the usual assumptions of reinforced concrete theory, and computations are based on force equilibrium and strain compatibility. The assumed tensile behaviour of the FRP strengthening is linearly elastic until failure, and the maximum usable strain is taken as the design FRP tensile rupture strain, multiplied by the bond dependent coefficient k_m. With this simplified approach, the verification against FRP debonding is built-in in the verification under bending or combined compression and bending. However, as mentioned in Section 9.3 on bond, the availability of the k_m factor is still limited to some particular cases.

The analytical expression of the capacity of FRP-strengthened masonry subject to out-of-plane loads is dependent on whether failure is governed by masonry crushing or FRP debonding or rupture. The failure mechanism can be determined *a priori* by comparing the FRP reinforcement ratio for a strip of masonry to the balanced reinforcement ratio, defined as the ratio for which masonry crushing and FRP debonding or rupture occur simultaneously.

An additional verification is needed against out-of-plane shear failure. In fact, if a large amount of FRP is applied, failure can be controlled by shear instead of flexure. The theoretical shear capacity of the FRP-strengthened masonry should be evaluated according to the design methods available for unreinforced masonry, neglecting the contribution of the FRP system. The shear strength capacity should exceed the shear demand associated to the maximum flexural capacity, in order to ensure that shear failure does not occur.

9.4.5 Strengthening limitations due to arching action

For walls with low slenderness ratio built between rigid supports, when the out-of-plane deflection increases, the wall is restrained from outward movement and rotation at its ends. This action induces an in-plane compressive force, accompanied by shear forces at the supports, which increase as the wall bends. Depending on the degree of support fixity, these in-plane axial forces can delay cracking and significantly increase the wall capacity. This mechanism is known as arching effect. Due to arching, the capacity of the unstrengthened wall can be much larger than that computed assuming simply-supported conditions and consequently the increase of capacity in walls strengthened with FRP reinforcement may be considerably less than

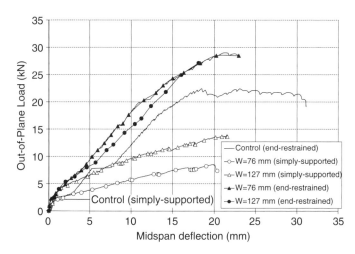

9.6 Effect of arching action on the load–displacement behaviour of an out-of-plane loaded masonry wall (from Galati, 2002).

expected. Experimental works (Tumialan *et al.*, 2003b), have shown that the resultant force between the out-of-plane load and the induced membrane force can cause the crushing of the masonry units at the boundary regions. The incidence of the arching effect in the load-resisting mechanism for FRP-strengthened URM walls depends on tensile and in-plane compressive strength of masonry, boundary conditions, wall slenderness ratio and material and bond properties of the FRP.

The arching mechanism must be considered in the quantification of the upgraded wall capacity to avoid over-estimating the contribution of the strengthening. Figure 9.6 illustrates a comparison between the load–deflection curves obtained in the case of simply-supported walls and walls with end axial restrains, tested under four-point bending (Galati *et al.*, 2002). If the wall behaves as a simply-supported element (i.e. large slenderness ratio or ends not restrained axially), the FRP reinforcement is very effective since the wall is in pure bending and the cracks are bridged by the reinforcement. The increase in the ultimate load for walls strengthened with 3 in. and 5 in. wide GFRP laminates is about 175 and 325%, respectively. If the wall is restrained axially (i.e. arching mechanism is observed) crushing of the masonry units at the boundary regions controls the strength of the wall. In this case, the capacity of the unstrengthened wall is far superior, and the increase in the out-of-plane capacity for strengthened specimens with 3 and 5 in. wide GFRP laminates is only about 25%.

When a masonry wall is built solidly between supports capable of resisting an arch thrust with no appreciable deformation or when walls are built continuously past vertical supports (horizontal spanning walls), the lateral load resistance of the wall can benefit from the arching action if the height-

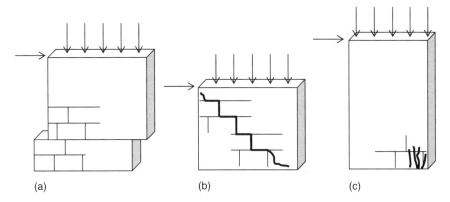

9.7 Failure modes of in-plane loaded unreinforced masonry walls: shear failure by sliding shear (a), shear failure by diagonal cracking (b) and flexural failure (c).

to-thickness ratio is less than 20. In such cases, a design procedure accounting for the influence of arching action has been proposed by Galati *et al.* (2005).

9.5 Strengthening of masonry panels under in-plane loads

9.5.1 General aspects

Collapse of masonry panels under in-plane loads is typically due to seismic actions or, less commonly, to the lateral thrust of arches and vaults. In-plane loaded panels can be strengthened with FRP in various forms:

- externally bonded laminates (pre-cured or wet lay-up systems);
- NSM reinforcement;
- structural repointing.

In-plane loading of a masonry panel induces two types of internal actions: in-plane shear and in-plane combined compression and bending. Correspondingly, UMR panels under in-plane loads may fail by in-plane shear (sliding shear, Fig. 9.7a), or diagonal cracking (Fig. 9.7b) or by combined compression and bending (masonry crushing, Fig. 9.7c).

For strengthening under in-plane combined compression and bending, FRP systems must be placed with the fibres oriented vertically and located symmetrically on the two external surfaces of the wall, in the tension region. These FRP systems should be properly anchored starting from the extreme sections of the panel.

For strengthening under in-plane shear, laminates and bars can be bonded to one or both sides of the panel, with the fibre orientation following different possible patterns: horizontal, vertical, horizontal and vertical (grid pattern), or diagonal (cross-pattern) (see Fig. 1.3 in Chapter 1). Complete bonding of laminates should be avoided as it could preclude moisture migration. The in-plane shear strength of a masonry panel strengthened by FRP systems applied symmetrically on the two external surfaces is due to two contributions: the shear resisted through friction in presence of a normal compression, and the shear resisted through the formation of an ideal resisting truss within the panel. This second contribution is made possible by the presence of the FRP introducing tension resistance.

When designing in-plane strengthening of a masonry panel, the FRP reinforcement should be adopted both to resist the tensile stresses generated by the bending moment associated with the shear, and to resist the shear force through the truss mechanism. If no FRP system is adopted for bending, shear strengthening can be achieved placing the FRP with the fibres along the diagonals of the panel.

Researchers comparing different strengthening patterns, including vertical and horizontal laminate strips, grid pattern and cross-pattern, found the cross-pattern layout to be the most effective configuration (Valluzzi *et al.*, 2002). In the case of walls strengthened only on one side with either FRP laminates or NSM bars, experimental results on panels under diagonal compression showed a limited increase in the in-plane capacity (Valluzzi *et al.*, 2002; Grando *et al.*, 2003). Such behaviour is due to the bending deformations induced during the loading phases along the diagonal on the unreinforced side, and caused by the noticeable difference of stiffness of the two sides of the panel as a result of the asymmetrical reinforcement. Therefore, strengthening on both sides is to be preferred.

Not only the FRP pattern but also the masonry topology has been observed to influence the in-plane wall behaviour. Grando *et al.* (2003) observed FRP strengthening to be more efficient with clay brick masonry walls than with concrete masonry. This can be attributed to characteristics of the parent material, such as height of masonry courses (smaller in the case of brick masonry) and improved mortar–masonry unit bond characteristics.

9.5.2 Failure mechanisms

For FRP-strengthened masonry panels under in-plane loads, the following failure mechanisms are possible:

- failure by in-plane compression and bending (by FRP rupture or masonry crushing);

- failure by in-plane shear (by sliding shear or diagonal cracking);
- failure by FRP debonding.

Failure under in-plane compression and bending

Failure of the strengthened wall due to this mechanism can occur by either rupture of the FRP in tension or masonry crushing. Typically, compressive crushing dominates in masonry walls strengthened with high reinforcement ratios. FRP rupture is less desirable than masonry crushing as the latter is a more ductile failure mode (Triantafillou, 1998).

Failure by in-plane shear

In-plane shear failure may occur either as sliding shear failure or as diagonal cracking. These failure modes occur for low amounts of FRP reinforcement and should be prevented with a proper design due to their brittle nature.

Failure by FRP debonding

Due to the bond transfer mechanisms at the interface, FRP debonding from the masonry substrate may occur before flexural or shear failure. Debonding typically starts from the location of shear cracks or from the horizontal flexural cracks due to the bond stress concentrations induced by cracking. Debonding is particularly critical in cases of insufficient anchorage of the reinforcement, such as with short anchorage lengths and in the absence of special anchorage devices at the laminate curtailment (see Section 9.10.2).

9.5.3 Design of FRP strengthening under in-plane loads

Based on the principles of capacity design, the FRP strengthening should be sized so that undesirable modes of failure in the masonry walls are avoided. In particular, the application of FRP can prevent the occurrence of the brittle shear failure modes, allowing the more ductile flexural modes to dominate. When considering flexural capacity, masonry crushing is preferable to FRP rupture, being more ductile. However, both failure modes are acceptable in design.

Computation of the capacity of FRP-strengthened walls under combined compression and bending can be conducted as for a standard reinforced concrete cross-section, accounting for the contribution of the FRP and introducing a bond-dependent coefficient k_m penalising the FRP tensile strength.

Less established is computation of the shear capacity of FRP-strengthened masonry walls. Design-oriented models of FRP strengthening of URM panels under in-plane actions have been proposed by Triantafillou (1998), Tumialan and Nanni (2001) and Galati *et al.* (2005). All these models compute the in-plane shear capacity of a strengthened panel adding the contribution of unstrengthened masonry and that of the FRP system. The first model is based on an extension of Eurocode 6 provision for steel reinforced masonry, where the term related to the reinforcement is modelled by the truss analogy. The only modification, tailored on externally bonded FRP laminates, is the introduction of an effective ultimate strain, to account for debonding of the laminate prior to its tensile failure. The second model, developed for structural repointing, is based on the assumption of formation of a diagonal splitting crack and on the ideal division of the strengthened panel into bond-controlled and rupture-controlled regions. In the third model, the FRP term is determined as the product of the total area of FRP reinforcement perpendicular to the shear crack and an effective stress in the FRP. This effective stress can be expressed as the FRP tensile strength, multiplied by a reduction factor accounting for the orientation of the fibres with respect to the direction of crack opening, and for the possibility of debonding mechanisms. In the absence of a comprehensive experimental campaign, conservative reduction factors have been proposed by the authors. All these models implicitly assume that splitting failure dominates and are valid in cases of symmetric strengthening.

The Italian guidelines on masonry strengthening (CNR, 2004) compute the shear capacity of the strengthened masonry panel as the sum of the masonry and FRP contributions. The upper limit of the shear strength is the value corresponding to crushing of the compressed struts of the ideal truss. The formulae given in these guidelines are taken from the analogous formulae valid for concrete members, and are valid if the fibres of the FRP system are parallel to the mortar bed joints. The minimum amount of FRP strengthening needed for the formation of the internal resisting truss is not specified. The large variety of masonry materials and typologies and the availability of various FRP strengthening systems are such that development of a general design-oriented model for the in-plane shear capacity of masonry walls is a very complex task and still requires a significant amount of research.

9.6 Strengthening of lintels and floor belts

The regions of connection between the different panels in a masonry wall are indicated as floor belts (Fig. 9.8). These belts carry the weight of the masonry located above the openings, and they force adjacent panels to deform compatibly under horizontal actions. The first resisting mechanism

Strengthening of masonry structures with FRP composites

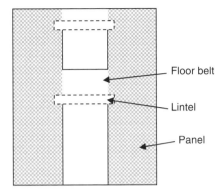

9.8 Panels and floor belts.

is achieved through the lintels, i.e. the portions of the floor belts situated above the openings, which are mainly subjected to vertical loads. The second resisting mechanism produces axial load, bending and shear in the belts and is activated primarily by seismic actions.

The portions of masonry located above the openings are not able to carry their own weight and must be carried by lintels, functioning as beams. Moreover, when the panels on the two sides of an opening are unable to bear the lateral thrust due to the presence of the opening, the lintel must also function as a tie-beam whose tensile stresses guarantee the global equilibrium of the wall. FRP systems can be applied to masonry lintels to transform them into reinforced masonry beams. It is recommended by CNR (2004) that the FRP be applied on the intrados of the lintel and not on the sides, and that it be adequately prolonged beyond the span of the opening, in order to ensure a proper anchorage into the adjacent panels.

The floor belts are subjected to combined axial force, bending and shear and can be strengthened with FRP applied on the two lateral surfaces, with the fibres parallel to the axis of the belt. In order to increase the shear resistance, the FRP system can also be applied along the diagonals of the panels above the openings, symmetrically on the two surfaces.

9.7 Strengthening of arches and vaults

9.7.1 Strengthening of arches

Analysis of masonry arches in the framework of limit analysis assumes that (i) masonry has no tensile strength and infinite compressive strength, and (ii) sliding failure does not occur (Heyman, 1982). The consequence of these assumptions is that failure of a masonry arch theoretically occurs by formation of a sufficient number of non-dissipative hinges transforming the arch into a mechanism, and stability under given loads depends essentially on the geometry of the structure. Due to the assumption of infinite compressive strength of masonry, the through-the-thickness location of the hinge at a given cross-section can be placed at the intrados or at the extrados of the element.

Externally bonded FRP laminates strengthen masonry arches by inhibiting the formation of the hinges. An FRP system prevents the opening of the crack faces and hence the relative rotation of the hinge sections. Hence, a hinge cannot form on the surface opposite to that on which the FRP is applied, i.e. the presence of FRP bonded on a portion of the intrados (extrados) inhibits the formation of hinges on the corresponding portion of extrados (intrados). Depending on the extension and location of the strengthened portions of the arch and on the loading pattern, the formation of hinges may be either altered (i.e. hinges form at locations different to those in the unstrengthened arch) or completely prevented. In any possible failure mechanism, the location of two consecutive hinges is always alternate between extrados and intrados, hence the application of FRP on the whole extrados or intrados completely prevents the formation of hinges. Less commonly, the FRP can be applied on both intrados and extrados. Bonding of FRP to limited portions of intrados and extrados, while not preventing completely the formation of hinges, can significantly increase the collapse load by modifying the controlling failure mechanism of the arch (Foraboschi, 2004).

If the formation of mechanisms is prevented, the capacity of the strengthened arch is controlled by material failures. The arch sections are subjected to combined compression and bending and may fail by masonry crushing or FRP debonding or rupture. Another possible failure mode is sliding shear at the mortar joints.

It is preferable that the FRP laminate be bonded to the extrados surface rather than to the intrados. At the extrados, the normal stresses accompanying the bond shear stresses are compressive and enhance the interfacial capacity. At the intrados, the same stresses are tensile and thus accelerate debonding failure. However, application of the FRP at the extrados requires prior removal of the floor finishes and spandrel fill and hence is less convenient from a practical standpoint. Early debonding of FRP laminates from the intrados can be prevented by using anchoring measures such as FRP anchor spikes (see Section 9.10.2). The experimental evidence shows that the application of FRP laminates on the side surfaces of the arch does not appreciably improve the behaviour of the element, due to premature FRP debonding triggered by local buckling of the laminate in the compressed regions.

If the applied load is substantially symmetrical, in the absence of tie-rods and with slender piers, collapse is usually due to the inability of the abutments to sustain the lateral thrust. The presence of the FRP reinforcement, introducing tension resistance, allows the line of thrust to fall outside the thickness of the arch. This fact has two important consequences: the capacity of the arch is increased (as discussed above), and the value of lateral thrust transmitted to the abutments is reduced. Also this second effect is very important for practical applications. For new structures, it implies that FRP-reinforced arches may not need tie-rods or massive piers. For strengthening of existing structures, it indicates that bonding FRP sheets can be an effective measure when the deficiency of the structure depends on the inability of the abutments to sustain the lateral thrust and/or on the removal of tie-rods.

De Lorenzis *et al.* (2007), studying a circular arch loaded symmetrically and strengthened with FRP, found that the application of FRP reinforcement allows a substantial reduction of the lateral thrust transmitted to the abutments. For this purpose, the FRP reinforcement should be placed either at the intrados spanning an angle centred at the crown, or at the extrados spanning two angles from the abutments towards the haunches (and anchored at the abutments). In the first case, if the FRP reinforcement ratio is such that the ultimate moment of the crown cross-section equals the maximum moment of the external loads, the minimum thrust becomes zero and the arch behaves like a simply-supported beam. In the second case, for an FRP reinforcement ratio such that the ultimate moment of the cross-section at the abutment equals the maximum moment of the external loads, the minimum thrust becomes zero and the arch behaves like two cantilever beams. In both cases, this theoretical limit condition may not be reached in practice because of prior failure by sliding of the mortar joints, which may then limit the possible reduction of the minimum thrust.

9.7.2 Strengthening of portal frames

In the portal frame, collapse mechanisms in addition to those in the arch are possible, associated with tilting or spreading of the abutments. In this case the FRP reinforcement must be properly anchored to the piers at the abutments or even prolonged on the surfaces of the piers. By studying a portal frame subjected to vertical (V) and horizontal (H) loads, strengthened with FRP at the whole intrados or at the whole extrados, Ianniruberto and Rinaldi (2004) found the failure domain to be substantially larger than that of the unstrengthened frame, and to be heavily influenced by the debonding strain assumed for the FRP (Fig. 9.9).

Another strengthening possibility is to place a tie-rod between the abutments to absorb the thrust. For this application, FRP bars present some advantages in place of the traditional steel bars (La Tegola *et al.*, 2000): they do not corrode, are lighter, and hence easier to handle on site, and have

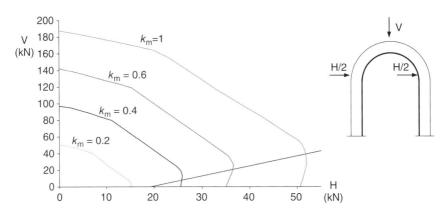

9.9 Failure domain of a portal frame strengthened with FRP at the whole intrados, for different assumed values of the bond reduction coefficient k_m (adapted from Ianniruberto and Rinaldi, 2004).

lower modulus of elasticity. In cases where the tie-rod is tensioned to close at least partially the existing cracks in a damaged structure, a lower modulus of elasticity implies lower tension losses during closure of the cracks and facilitates the indirect measurement of the applied load during tensioning.

9.7.3 Strengthening of vaults and domes

In the majority of cases, barrel vaults can be studied as arches and strengthened with FRP along the directrix, at the extrados or intrados (or both). The FRP system should be applied at regular intervals along the length of the vault, with an on-centre spacing not exceeding a limiting value. The Italian guidelines (CNR, 2004) indicate this limit as the width of the FRP strips plus three times the thickness of the vault. The same document suggests inserting FRP strips also along the generatrix of the vault, to enable a transverse collaboration between the adjacent arches forming the barrel vault. This function is particularly important for structures in seismic regions. It is suggested that the FRP amount per unit area in the transverse direction (along the generatrix) should be at least 10% (or 25% in seismic regions) of the amount placed along the directrix.

Vaults with double curvature include vaults on square or rectangular plans (such as cross and cloist vaults) and domes. All of these vaults, prior to reaching their ultimate limit states, develop cracking patterns which transform them into a series of adjacent arches. For this reason, FRP strengthening of these elements can to a certain extent be reduced to that of arches.

In domes, the typical cracking pattern consists of cracks along the meridians starting from the base of the dome, due to the circumferential tensile

Strengthening of masonry structures with FRP composites 259

(a)

(b)

9.10 Examples of strengthening of a masonry cross vault (a) and dome (b) with FRP laminates (from Foraboschi, 2005).

stresses arising in the lower parallels. FRP strips along the circumferential direction close to the base can be applied to absorb these tensile stresses and hence reduce the extension of the cracked region and the consequent increment of lateral thrust at the abutments. Figure 9.10 shows an example of a cross vault (a) and dome (b) strengthened with FRP laminates.

9.7.4 Design

Analysis and design procedures for strengthened arches have been proposed by different authors (Valluzzi *et al.*, 2001; Foraboschi, 2004; Ianniruberto and Rinaldi, 2004; De Lorenzis *et al.*, 2007) based on the principles of limit analysis implemented through the static and/or the kinematic theorem. These analyses can be partially extended to masonry vaults by reducing a vault to a system of arches, which is more or less realistic depending on the pattern of the pre-existing cracks. However, in the case of masonry vaults where the two- or three-dimensional state of stress has to be taken into account, limit analysis procedures are not yet of immediate design use, and the analyst must resort to numerical modelling accounting properly for the presence of the FRP reinforcement and for the non-linear behaviour of masonry. In design of the strengthening system, prediction of the failure mechanism by debonding of the FRP in presence of substrate curvature still needs considerable research.

9.8 Confinement of masonry columns

Confinement of masonry columns, such as that of reinforced concrete columns, aims at enhancing their strength and ductility. As most of the concepts applicable to confinement of concrete columns can also be extended to masonry columns, the reader is referred to Chapter 6 for further details. This section reports only some aspects specific to masonry columns.

260   Strengthening and rehabilitation of civil infrastructures

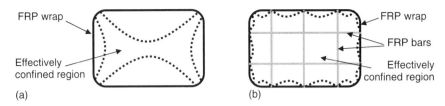

9.11 Effectively confined portion of a rectangular cross-section confined with FRP laminates (a) and FRP laminates and transverse bars (b).

Confinement for masonry columns is achieved by two different systems:

- by wrapping the column with one or more layers of FRP laminates, bonded to the column with the wet lay-up technique; the principal fibre direction is normally perpendicular to the axis of the member and can be applied either continuously over the surface (which, however, poses the problem of moisture migration) or as strips with certain width and spacing;
- by FRP bars inserted in holes drilled in the column perpendicular to its axis, in two orthogonal directions. The bars are then bonded in place with a suitable adhesive, commonly epoxy.

While the first system is widely used for confinement of concrete columns, FRP bars are only used for confinement of masonry columns. Unlike concrete columns, masonry columns are made of discrete blocks with different degrees of mutual connection depending on the quality of the mortar and on the masonry topology. The FRP bars have been shown to provide effective confinement (Micelli *et al.*, 2004; Aiello *et al.*, 2007), for different reasons:

- They connect together the blocks which constitute the masonry column and prevent their tensile rupture and expulsion under the transverse tensile stresses generated by the axial compression load.
- For confinement of rectangular columns, the use of FRP bars increases the effectively confined portion of the cross-section (Fig. 9.11), enhancing the effectiveness of external wrapping with FRP laminates.
- FRP bars can be used to confine columns of cross-sectional shapes for which external wrapping is less effective and sometimes completely inapplicable (e.g., columns with cross-shaped or star-shaped cross-sections). Columns of this type are frequent in historic masonry construction.

The Italian guidelines (CNR, 2004) provide design formulae to quantify the increase of compressive strength of a column confined with FRP laminates and/or bars. These formulae are based on formulae valid for concrete (see Chapter 6), with the exceptions that:

- The confined masonry compressive strength is taken as a linear function of the confinement pressure, while the most widely used models for confinement of concrete columns give the confined concrete strength as a non-linear function of the confinement pressure.
- The coefficient of the confinement pressure in the linear function mentioned above is taken proportional to the unit mass of masonry.
- The formulae account for the use of FRP bars in addition to or as an alternative to wrapping with FRP laminates. The contribution of the FRP bars is considered in computation of the reinforcement ratio, and in computation of the effectively confined portion of the cross-section.

It should be noted that research on confinement of masonry columns is still rather limited and much further work is needed on this topic from both the experimental and the theoretical standpoints.

9.9 Other applications

Other applications, less common than those illustrated above, are listed in Chapter 1. Among these applications, the use of unbonded FRP ties deserves a special mention.

One of the traditional retrofitting techniques for masonry structures is external or internal post-tensioning using steel ties, to tie structural elements together into an integrated three-dimensional system. The behaviour of the structure under static and dynamic loading is improved and global failure mechanisms are prevented. Triantafillou and Fardis (1997) first proposed using FRP ties in the form of external unbonded tendons for masonry domes and buildings (Fig. 9.12). These would be anchored to the masonry only at their ends, applied to the external face of the structure and post-tensioned to provide horizontal confinement. This method is particularly attractive in the case of historic structures due to its reversibility, as the bonded portion is limited to the anchorage zones. The same authors demonstrated that the transverse confining stresses applied by the FRP tendons are able to increase substantially the strength of masonry under both gravity and horizontal loads.

Bonded laminates can also be used as ties. FRP sheets are more suited than pre-cured laminates as the limited thickness facilitates wrapping around corners and irregular geometries and reduces possible delamination problems due to unevenness of the masonry surface and local curvatures.

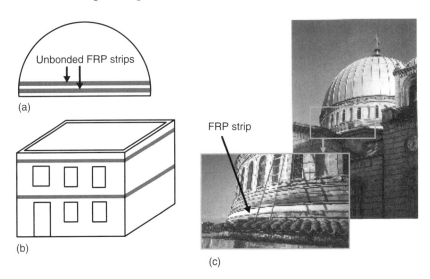

9.12 Circumferentially applied FRP tendons (a) at the base of domes, (b) in the form of belts around buildings; (c) application in the Church of Panaghia Faneromeni, Egion, Greece (from Triantafillou, 2001).

9.10 Detailing issues

This section outlines some detailing concepts and provisions available in the current literature and based in most cases on results of testing. It should be noted, however, that the quantitative provisions are susceptible to improvements as more research results become available, and should be tailored to the specific masonry structure and FRP system to be used for strengthening.

9.10.1 Minimum spacing and area of FRP reinforcement

The Italian guidelines on FRP strengthening of masonry (CNR, 2004) suggest that, when using FRP discrete vertical strips, their distance on centre should not exceed a limiting value, equal to the strip width plus three times the thickness of the wall. Galati *et al.* (2005) extended the validity of this provision to NSM FRP bars, suggesting that their spacing be not larger than three times the thickness of the wall.

The same authors provided guidelines on the minimum area of FRP reinforcement in strengthening of masonry panels under out-of-plane and in-plane loading. They suggested that, when FRP systems are used to enhance the in-plane strength of the member, the cross-sectional area of the main FRP system should not be less than 0.05% of the cross-sectional area of the member. In walls where the FRP system is used to enhance the resistance to lateral loads, the total FRP area should not be less than 0.03% of the cross-sectional area of the wall (i.e. 0.015% on each face).

9.10.2 Anchorage details

When the FRP system on a masonry structure is subjected to cyclic loading inducing tension and compression, e.g. as a result of seismic events or thermal variations, the bond between the FRP and the masonry substrate could significantly deteriorate during the life of the structure. It could be useful to adopt mechanical anchorages or to insert the FRP in grooves, to prevent local buckling during the phases of compression stress state. When the FRP system is always subjected to tensile stresses, it should preferably be anchored in the compressed regions of the masonry substrate. Also, proper detailing of the FRP reinforcement at wall boundaries is useful to delay or prevent premature failure by debonding. The FRP system should be correctly anchored at its ends using, for externally bonded laminates, an anchorage length at least equal to the effective bond length (see previous section of bond).

For externally bonded laminates, effective end anchorage may be attained using anchorage systems such as steel angles, steel bolts and FRP bars. Different systems offer their own advantages and disadvantages. Steel angles are easy to install but aesthetically problematic, and angles in direct contact with the masonry surface may locally fracture the wall due to displacement and rotation restraint. Steel bolts have shown high effectiveness, but require a significant installation effort (Schwegler, 1995).

Another technique consists of grooving a slot in the upper and lower boundary members. The ply is wrapped around an FRP bar and placed and bonded in the slot with a suitable epoxy-based paste (Carney and Myers, 2003). For NSM or structural repointing construction, bars can easily be anchored into adjacent concrete members by drilling and embedding their extended terminations (Tumialan *et al.*, 2003c).

As previously mentioned, application of FRP laminates at the intrados of arches and vaults makes debonding particularly critical. In this case, equilibrium dictates that the shear bond stresses between FRP and masonry are accompanied by normal tensile stresses, which accelerate debonding failure. In such cases, Eshwar *et al.* (2003) proposed the use of FRP spikes to enhance bond of FRP laminates to curved substrates and proved its effectiveness. The spikes are made of long continuous glass fibres, impregnated with epoxy resin for part of their total length. Diameter, length of the precured part and total length can be varied to fit the specific application. Their installation is carried out as follows. A hole with appropriate depth and diameter is drilled on the substrate surface at the location where the spike will be inserted, and is cleaned with air pressure. Then, the primer layer is applied on the substrate surface in the area where the laminate will be bonded. Once the primer has hardened and the first layer of epoxy resin is applied, the hole is partially filled with the same resin. The FRP laminate is then positioned and, at the location of the hole, its fibres are locally enlarged to insert the spike in the hole. Then, the impregnated part of the spike is inserted in the hole (if necessary, the hole is completely filled with resin) and the dry fibres are spread out on the FRP laminate. Finally, FRP laminate and dry fibres of the spike are impregnated with the second resin layer. The effectiveness of these anchoring spikes was recently proved by bond tests on specimens with curved substrates (Aiello *et al.*, 2004) and in strengthening of masonry vaults (De Lorenzis *et al.*, 2007).

9.11 References

AIELLO M A, DE LORENZIS L, GALATI N and LA TEGOLA A (2004) Bond between FRP laminates and curved concrete substrates with anchoring composite spikes, in La Tegola A and Nanni A (eds), *Proceedings Innovative Materials and Technologies for Construction and Restoration – IMTCR'04*, Lecce, Italy, Liguori Editore, Vol. 2, 45–56.

AIELLO M A, MICELLI F and VALENTE L (2007) Structural upgrading of masonry columns by using composite reinforcements, *Journal of Composites in Construction*, ASCE, **11**(6), 650–659.

BAJPAI K and DUTHINH D (2003) Flexural strengthening of masonry walls with external composite bars, *The Masonry Society Journal*, August, 9–20.

CARNEY P and MYERS J J (2003) *Static and Blast Resistance of FRP Strengthened Connections for Unreinforced Masonry Walls*, CIES Report No. 03-46, Rolla, MO, USA, University of Missouri–Rolla.

CNR (2004) CNR-DT 200/2004 *Guide for the design and construction of externally bonded FRP systems for strengthening existing structures – materials, RC and PC structures, masonry structures*, Rome, Italy, Consiglio Nazionale delle Ricerche (English version).

DE LORENZIS L and NANNI A (2004) *International Workshop on Preservation of Historical Structures with FRP Composites*, final report submitted to National Science Foundation, Arlington, VA, USA.

DE LORENZIS L, TINAZZI D and NANNI A (2000) NSM FRP rods for masonry strengthening: bond and flexural testing, Proc. Conf. on Mechanics of FRP-Strengthened Masonry Structures, Venezia, Italy.

DE LORENZIS L, DIMITRI R and LA TEGOLA A (2007) Reduction of the lateral thrust of masonry arches and vaults with FRP composites, *Construction and Building Materials*, **21**(7), 1415–1430.

ESHWAR N, IBELL T and NANNI A (2003) CFRP strengthening of concrete bridges with curved soffits, Int. Conf. Structural Faults + Repair 2003, 1–3 July, Commonwealth Institute, London, UK. M.C. Forde, Ed., CD-ROM version, 10 pp.

FORABOSCHI P (2004) Strengthening of masonry arches with fiber-reinforced polymer strips, *Journal of Composites in Construction*, ASCE, **8**(3), 191–202.

GALATI N (2002) *Arching Effect in Masonry Walls Reinforced with Fiber Reinforced Polymer (FRP) Materials*, MSc Thesis, Rolla, MO, USA, University of Missouri-Rolla.

GALATI N, TUMIALAN J G, LA TEGOLA A and NANNI A (2002) Influence of arching mechanism in masonry walls strengthened with FRP laminates, *Proceedings International Conference on Fibre Composites in Infrastructure ICCI-02*, San Francisco, CA, USA June 10–12 (CD-ROM).

GALATI N, GARBIN E and NANNI A (2005) *Design Guidelines for the Strengthening of Unreinforced Masonry Structures Using Fiber-reinforced Polymer (FRP) Systems*, CIES Report No. 53, Rolla, MO, USA, University of Missouri–Rolla.

GRANDO S, VALLUZZI M R, TUMIALAN J G and NANNI A (2003) Shear strengthening of URM clay walls with FRP systems in Tan K H (ed.), *Proceedings Fibre Reinforced Polymer Reinforcement for Concrete Structures – FRPRCS-6*, Singapore, World Scientific, Vol. II, 1229–1238.

HAMOUSH S A, MCGINLEY M W, MLAKAR P, SCOTT D and MURRAY K (2001) Out-of-plane strengthening of masonry walls with reinforced composites, *Journal of Composites in Construction*, ASCE, **5**(3), 139–145.

HAMOUSH S, MCGINLEY M, MLAKAR P and TERRO M J (2002) Out-of-plane behavior of surface-reinforced masonry walls, *Construction and Building Materials*, **16**(6), 341–351.

HEYMAN J (1982) *The Masonry Arch*, Chichester UK, Ellis Horwood-Wiley.

HOLZENKÄMPFER P (1994) *Ingenieurmodelle des verbundes geklebter bewehrung fürbetonbauteile*, Dissertation, TU Braunschweig (in German).

IANNIRUBERTO U and RINALDI Z (2004) Resistenza laterale di portali in muratura rinforzati con FRP all'intradosso., Atti del XI Convegno Nazionale L'Ingegneria Sismica in Italia, Genova, Italy, Gennaio 25–29 (CD ROM) (in Italian).

ICOMOS (2001) *Recommendations for the Analysis, Conservation and Structural Restoration of Architectural Heritage*, Paris, France, International Scientific Committee for Analysis and Restoration of Structures of Architectural Heritage.

LA TEGOLA A, LA TEGOLA AL, DE LORENZIS L and MICELLI F (2000) Applications of FRP materials for repair of masonry structures, *Workshop on Advanced FRP Materials For Civil Structures*, Bologna, Italy, Oct 19.

LI T, GALATI N and NANNI A (2004) In-plane resistance of UMR walls with opening strengthened by FRP composites, Advanced Composites Materials in Bridges and Structures, Calgary, Alberta, Canada July 20–23 (CD-ROM).

MICELLI F, DE LORENZIS L and LA TEGOLA A (2004) FRP-confined masonry columns under axial loads: analytical model and experimental results, *Masonry International*, **17**(3), 95–108.

NEUBAUER U and ROSTÁSY F S (1997) Design aspects of concrete structures strengthened with externally bonded CFRP-plates, in Forde M C (ed.), *Proceedings Seventh International Conference on Structural Faults and Repair*, Edinburgh, UK, Engineering Technics Press, Vol. II, 109–118.

ROKO K, BOOTHBY T E and BAKIS C E (1999) Failure modes of sheet bonded fibre reinforced polymer applied to brick masonry, in Dolan C W, Rizkalla S H and Nanni A (eds), *Proceedings Fibre Reinforced Polymer Reinforcement for Concrete Structures – FRPRCS-4*, Farmington Hills, MI, USA, American Concrete Institute, 305–311.

SCHWEGLER G (1995) Masonry construction strengthened with fibre composites in seismically endangered zones, in Duma G (ed.), *Proceedings Tenth European Conference on Earthquake Engineering*, Rotterdam, The Netherlands, Balkema, 2299–2303.

TRIANTAFILLOU T C (1998) Strengthening of masonry structures using epoxy-bonded FRP laminates, *Journal of Composites in Construction,* ASCE, **2**(2), 96–103.

TRIANTAFILLOU T C (2001) Strengthening of historic structures with composites, in Teng J G (ed.), *Proceedings International Conference on FRP Composites in Civil Engineering – CICE 2001*, Oxford, UK, Elsevier Science, 959–965.

TRIANTAFILLOU T C and FARDIS M N (1997) Strengthening of historic masonry structures with composite materials, *Materials and Structures*, **30**, 486–496.

TUMIALAN J G and NANNI A (2001) In-plane and out-of-plane behaviour of masonry walls strengthened with FRP systems, CIES Report No. 01-24, Rolla, MO, USA, University of Missouri–Rolla.

TUMIALAN J G, GALATI N and NANNI A (2003a) FRP strengthening of UMR walls subject to out-of-plane loads, *ACI Structural Journal*, **100**(3), 312–329.

TUMIALAN J G, GALATI N and NANNI A (2003b) Field assessement of unreinforced masonry walls strengthened with fiber reinforced polymer laminates, *ASCE Journal of Structural Engineeering*, **129**(8), 1047–1055.

TUMIALAN J G, SAN BARTOLOME A and NANNI A (2003c) Strengthening of URM infill walls by FRP structural repointing, *Proceedings Ninth North American Masonry Conference*, Clemson, SC, USA, June 1–3 (CD ROM).

TURCO V, GALATI N, TUMIALAN J G and NANNI A (2003) Flexural strengthening of URM walls with FRP systems, in Tan K H (ed.), *Proceedings Fibre Reinforced Polymer Reinforcement for Concrete Structures – FRPRCS-6*, Singapore, World Scientific, Vol. II, 1219–1228.

VALLUZZI M R, VALDEMARCA M and MODENA C (2001) Behaviour of brick masonry vaults strengthened by FRP laminates, *Journal of Composites in Construction*, ASCE, **5**(3), 163–169.

VALLUZZI M R, TINAZZI D and MODENA C (2002) Shear behaviour of masonry panels strengthened by FRP laminates, *Construction and Building Materials*, **16**(7), 409–416.

10
Flexural strengthening application of fibre-reinforced polymer (FRP) plates

L. CANNING and S. LUKE, Mouchel, UK

10.1 Introduction

The use of externally bonded fibre-reinforced polymer (FRP) composite systems for flexural strengthening of structures is the most common application of FRP composites in the civil infrastructure worldwide, and has received the greatest amount of research attention. The first uses of externally bonded FRP composite systems for flexural strengthening were on reinforced concrete; however, FRP composites are increasingly being used to strengthen cast iron, modern steel, timber and masonry in flexure and, in limited cases, wrought iron and riveted steel structures.

Most applications of FRP composite systems for flexural strengthening use unstressed FRP plates manufactured in the factory and adhesively bonded to the substrate, mainly due to the relative lack of research and more complicated installation procedure for prestressing technology. The application of unstressed FRP composite strengthening systems shall be discussed first, followed by current prestressing technology.

The successful use of externally bonded FRP composite systems for flexural strengthening is dependent upon a series of steps, each as important as the other:

- a detailed structural assessment of a structure including identifying the causes of any deterioration and the applicability of externally bonded FRP composites;
- design to accepted standards with a sufficiently detailed specification for materials and workmanship;
- choice of a suitably qualified and experienced contractor and operatives with communication between the designer and contractor to ensure an adequate method of installation ensuring a high level of workmanship;
- site supervision during the installation by experienced personnel;
- final inspection of the completed installation and identification of any necessary remedial measures.

This chapter focuses on those stages most closely associated with application of FRP strengthening; specification of materials and workmanship (and resolution of defective work or materials), contractor experience, application methods and sequence, site supervision and final inspection, with reference to current guidance and standards. In addition, case studies are used to highlight the practical issues involved in each step.

10.2 Unstressed FRP composite systems

The majority of FRP strengthening schemes on structures since the 1990s have used unstressed carbon fibre reinforced polymer (CFRP) composite systems. Unstressed CFRP composite systems typically fall into two categories:

- CFRP composite plates/sheets/rods manufactured in a factory by the pultrusion, resin infusion or pre-preg processes and adhesively bonded on site;
- CFRP composite fabrics manufactured and bonded *in situ* by wet lay-up (by manual or automatic methods), pre-preg or resin infusion.

The different application methods for unstressed FRP composite systems are described below.

10.2.1 Adhesively bonded FRP plates

This is the most common application method for flexural strengthening of structural elements, and has been used on reinforced concrete, prestressed concrete, modern steel, wrought iron and cast iron elements, in the forms of beams, slabs, columns and walls.

Pre-formed FRP plates manufactured from single or multiple bonded pultruded plates, factory resin infused plates or factory pre-preg vacuum cured plates are bonded to the tension face of the substrate to be strengthened. The bonding material is typically a two-part epoxy, vinylester or methacrylate adhesive, to provide adequate material properties, bond strength and working time during the bonding process. The bonding surface of the FRP plate is typically prepared by removal of a peel-ply layer immediately prior to application of the adhesive, or by light manual abrasion of the bonding surface. Preparation of the substrate surface, to achieve a suitable bonding surface free from contaminants, is generally carried out by grit-blasting, grinding or impact methods, depending on the substrate. For concrete substrates, needle-gunning, grinding and/or grit-blasting is usually carried out. For metallic substrates, grit-blasting is generally used, after manual methods if heavy contamination is present. In addition, a primer may be used to improve adhesion between the substrate and adhesive and to temporarily protect the prepared surface.

The adhesive is mixed, usually using a small powered tool, in batches of suitable size to avoid wastage or overly rapid curing of the adhesive, in accordance with the manufacturer's recommendations. The adhesive is then applied to the FRP plate in a controlled manner, preferably using a simple tool to create a domed profile, thicker at the centre than the edges – this minimises the risk of air voids occurring during the bonding process. In some cases, spacers of defined thickness are used to ensure a minimum thickness bond-line of typically 2 mm. The adhesive may also be applied to the substrate. Immediately prior to joining the bond, the surfaces of the adhesive should be checked to ensure there are no contaminants or moisture on the applied adhesive.

The FRP plate is then applied to the substrate surface, and the adhesive bond compressed with rollers in order to remove air voids. In addition, where the FRP plates are thick, the adhesive has a particularly low viscosity or significant live load will be present during adhesive cure, temporary propping of the FRP plates may be required. During the application of primers to the substrate, and adhesive bonding of the FRP plates, environmental controls such as space heaters may be used to ensure substrate and air temperatures, relative humidity and dew point are at acceptable levels.

In some cases, multiple FRP plates have been bonded to one another *in situ* as opposed to in the factory, which can increase the flexibility of the installation process. However, this increases the reliance on the quality of workmanship and also increases the installation time.

In general, the bonding surface of the substrate must be flat or have limited curvature, due to the stiffness of the pre-formed FRP plates. The presence of global curvature will increase the stresses within the adhesive bond (Porter *et al.* 2003) and requires careful consideration during the design phase – this is one reason why the flatness of the substrate should be checked prior to bonding, as it may differ from the assumptions at the design stage and as-built record information, and may have an effect on the efficacy of the design. Local deviations in flatness can be leveled by the use of high build cementitious mortars for concrete, or chemical metal adhesives for metals, or by use of the bonding adhesive if the local deviations from flatness are sufficiently small (typically less than 10 mm).

10.2.2 *In situ* laminated FRP fabric

In situ lamination of FRP fabrics has been undertaken on reinforced concrete, prestressed concrete and modern steel elements. Surface preparation of the substrate is undertaken in a similar manner to that for adhesively bonded FRP plates, but is typically required over a greater surface area. A primer is then applied, followed by lamination of the FRP fabric generally by one of two methods. The first method uses a two-part laminating resin, which is applied to the primer, followed by application of the dry fibres. Wetting out of the fibres and removal of air voids is undertaken by the use of rollers. This procedure is then repeated until the required thickness of FRP fabric is achieved. The second method uses FRP fabric pre-wetted by the laminating resin, which is applied in layers and compressed by rollers to remove air voids. For both methods, the number of layers of fabric that can be applied in a single operation is limited, to ensure voidage within the final cured laminate and misalignment of the fibres is minimised. Depending on the number of layers of laminated FRP fabric that are required and the general strengthening programme, the laminating operation may need to be carried out in a number of stages, with intervals between these stages. In these cases, some degree of curing of the previously laminated FRP fabric layers will have occurred; further laminations will typically require cleaning of the previously laminated FRP fabric layers to ensure a good bond between successive layers. Furthermore, it is generally recommended that if the laminated FRP fabric system is partially complete, simple measures to protect the surface are required to avoid the possible need for extensive cleaning and/or abrasion of the surface for further laminations to ensure a good bond.

This application method is particularly suitable for curved substrates, as the flexible FRP fabric can follow the profile of the substrate. It is also used where the shear stresses in the adhesive calculated during the design phase would be problematic for adhesive plate bonding, due to the generally smaller bond area – fully covering a surface with adhesively bonded plates will be less cost-effective or practical than using an *in situ* laminated FRP fabric system. Full coverage of an area which will prevent the drainage of water and other effluent, for example on the soffit of concrete bridges, requires careful consideration as this may exacerbate corrosion of steel reinforcement, and a loss in long-term durability of the strengthened structure. However, this can be overcome by careful detailing (e.g. installation of drainage channels, purposefully non-laminated areas to allow drainage, etc.)

10.2.3 Adhesively bonded FRP rods and bars (near-surface mounted reinforcement)

Near surface mounted reinforcement (NSMR) uses a similar application method to adhesively bonded FRP plates. It is described separately as it is a relatively new method for FRP strengthening which has only been used in the field in the past few years, following recent research Carolin (2003).

Pre-formed FRP bars or rods of circular, square or rectangular cross-section are bonded into pre-formed grooves within a concrete substrate. The grooves are typically formed by diamond saw cutting and the new surfaces prepared by grit-blasting in a similar manner to adhesively bonded FRP plates. A two-part resin adhesive is then applied into the groove, and the FRP bar or rod pushed into the adhesive, ensuring that the adhesive fully envelopes the FRP bar or rod. Research has also been undertaken into using a cementitious bonding agent as opposed to polymer resins, with some success (Carolin, 2003), although this has not been used in the field currently. The surface profile of the substrate can then be leveled by the use of a cementitious mortar or the bonding adhesive.

The main advantages of this strengthening method are that the FRP strengthening has increased protection from fire, impact or vandal attack, and the ratio of bond area to FRP cross-sectional area is generally increased with a corresponding reduction in adhesive stresses.

10.2.4 *In situ* vacuum cured FRP fabric

In situ vacuum curing of FRP fabric has been used to strengthen metallic elements with complex geometry. The method has not been used extensively in the civil infrastructure, mainly due to the lack of research and relative complexity in comparison with *in situ* lamination and adhesive plate bonding.

Surface preparation of the substrate is carried out in a similar manner to that for adhesively bonded FRP plates, and primer is used to promote adhesion and prevent rust blooming. Pre-preg FRP composite is applied to the prepared substrate in layers, and then cured under vacuum (using a sealed vacuum bag) under a raised temperature. The recent development of low temperature curing technology for pre-preg materials (curing at temperatures as low as $30\,°C$) has enabled this method to be used for strengthening of structural elements in the field. Forced curing at an elevated temperature can be achieved by a number of methods; tenting out of the structure and heating of the air and substrate with space heaters, or by the use of heating blankets placed directly against the vacuum bag.

10.2.5 *In situ* resin infusion

In situ resin infusion of FRP fabric has not been used within the civil infrastructure for strengthening purposes. However, there is no technical reason why this method cannot be used on metallic and concrete substrates. Many of the same requirements as for *in situ* vacuum cured FRP fabric exist. The dry fibre pack is placed on the substrate and the resin infused by a combination of vacuum and positive pressure through the placed fibre pack. The main limitation to its use in the field is the associated equipment required for the infusion operation and infusion/curing time of the resin. A general example of resin infusion, used in the civil infrastructure is described in Täljsten (2002).

10.2.6 Sprayed FRP composite

Sprayed FRP composite has been used to strengthen concrete structures, where the requirement for a high stiffness FRP composite is not as great as for metallic structures. Generally, chopped fibres are sprayed in combination with the resin matrix from a gun onto the prepared substrate. This method has not been extensively used due to the relatively low material properties achieved in comparison with other application methods. Furthermore, this application method requires a high level of protective equipment for operatives and environmental control.

10.2.7 Mechanically fixed FRP plates

This method, which is a relatively recent development, uses preformed FRP plates that are mechanically fixed to a concrete substrate by power-driven bolts (Bank, 2002). It is most suitable where very rapid strengthening of a structural element is required. Another advantage is that no surface preparation of the substrate is required (although the concrete in the anchorage region is required to be of sufficiently good quality to resist the bolt forces). Bolting of FRP plates is not usually a suitable method for anchorage; however, FRP plates used with this method have bidirectional or multidirectional fibres to prevent splitting of the plate.

10.3 Prestressed FRP composite systems

The use of prestressed FRP composites for flexural strengthening of concrete and metallic beams and slabs has developed over recent years, and a number of proprietary prestressed FRP systems are now available commercially. The design and application of prestressed FRP composite systems is a specialist field and suitably experienced designers and contractors should be used.

The main reasons for using prestressed FRP composite systems for strengthening structural elements are:

- to increase live load capacity;
- to reduce dead load deflections and mobilise otherwise locked-in dead load stresses;

- to reduce crack widths in concrete elements and to delay the onset of cracking in concrete and metallic elements (due to fatigue), and to arrest fatigue degradation of elements within a structure;
- to improve the behaviour of the structural element at the serviceability limit state (e.g. excessive deflection, high steel reinforcement stress in concrete elements);
- to improve fatigue strength by minimising the absolute tensile stress, and also the working stress range in metallic elements;
- to regain the original prestressed conditions in prestressed concrete elements – the original level of prestress may have been lost by damage to prestressing tendons (by impact, corrosion, etc.) or inadequate original design.

However, application of prestressed FRP composite systems is limited if the concrete or metallic elements have an existing high compressive stress at the serviceability limit state. In this case, prestressing with FRP composites will not provide a significant improvement in the behaviour of the structural element at serviceability loads and may reduce the existing ductility. Prestressed FRP composite systems also usually require a mechanical method of anchorage, although methods have been developed that rely solely on an adhesive bond. It should be noted that all prestressing methods currently in use in the field use proprietary prestressing systems.

10.3.1 General aspects of prestressed FRP composite systems

The design and application of prestressed FRP composite systems is a specialist activity due to the relative lack of design guidance and accepted application methods in comparison with those available for unstressed FRP composite systems.

The use of prestressed FRP composite strengthening systems may not be appropriate, or at least will require more detailed planning and risk assessment in a number of situations. Some examples are where there are limited available installation periods (for example within railway possessions or in structures where there is a continuous industrial process) and where there is significant risk of vandalism, or exposure to a highly aggressive environment that would affect the long-term properties of the FRP composite.

Post-tensioning of existing structures can be undertaken using a variety of prestressing methods, such as bonded or unbonded FRP composite plates or sheets, bonded FRP near-surface mounted reinforcement (NSMR) and external FRP composite tendons. These are described in greater detail below.

Prestressing of the FRP composite also allows a greater proportion of the FRP composite tensile strength to be utilised, and is therefore a more technically efficient use of the material, although the overall efficiency of a prestressing application depends on a number of factors (Table 10.2). CFRP composite is generally the most suitable type of FRP composite for prestressing applications due to its superior creep and stress rupture properties compared to those of glass fibre reinforced polymer (GFRP) and aramid fibre-reinforced polymers (AFRP) composite. AFRP composites may also be suitable in certain applications, as their lower tensile elastic modulus enables greater control of elongations in the prestressing process to be achieved. However, as with GFRP, AFRP can be susceptible to stress rupture, and therefore the prestress levels should be limited.

The basic principle of strengthening concrete structures by post-tensioning with FRP composites is similar to that for conventional post-tensioned concrete structures. A portion or all of the existing dead load in the structural member is transferred to the FRP composite by creating a tensile force in the FRP composite prior to application, bonding and/or anchoring the FRP composite to the concrete substrate, and then releasing the prestress load. The prestress force is then transferred into the structural member by either mechanical anchors or the use of an adhesively bonded joint, or a combination of the two. The prestressed FRP composite (and adhesive bond in some cases) carries both a portion of the dead load and also live load, in contrast to unstressed FRP strengthening where the FRP composite carries only a portion of live load.

Anchorage of the prestressed FRP composite is a critical aspect of the application. Large shear stresses are present within the adhesive bond when the prestress in the FRP composite is released and transferred into the structural member. Therefore, in most cases, a mechanical anchorage at the ends of the FRP composite is required, although methods using solely an adhesive bond have been developed, and are described below.

On transfer of the prestress to the concrete member, some losses occur in a similar manner to those for conventional prestressed post-tensioned structures. The losses in prestress in the short term are due to:

- elastic shortening of the structural member (which may be increased by the presence of cracks within the structural member prior to the prestressing operation);
- creep effects in the adhesive, which may increase shear lag across the adhesive bond, for prestress of systems with no mechanical anchorage;
- drawing in of the FRP plate within the prestressing anchorage.

In general, the losses in prestress force in the FRP plate in the long-term are similar to those for post-tensioning with low relaxation steel.

The level of prestress force applied to the structural element should be such that the following conditions are acceptable:

- tensile stress and cracking away from the prestressed FRP composite under dead and superimposed dead load, after transfer of the prestress – this will particularly be the case for continuous structural elements;
- compressive stresses in the structural member under dead, superimposed dead and live loads;
- tensile stress in the FRP composite under dead and superimposed dead load (for stress rupture) and live load (for fatigue);
- anchorage force (due to dead and live loads) in the adhesive or mechanical bond.

The acceptable prestress force, as a percentage of the ultimate tensile strength of the FRP composite, will vary depending on the material properties of FRP composite used. CFRP composite is not susceptible to stress rupture and therefore the greatest degree of prestress can be achieved, followed by AFRP composite and GFRP composite, which both exhibit stress rupture. As an indication, CFRP composite can be stressed up to 50% of the ultimate tensile strength, AFRP composite up to 30% of the ultimate tensile strength and GFRP composite up to only 15–20% of the ultimate tensile strength. For prestressing systems where the adhesive bond is relied upon to transfer prestress forces, the stress within the adhesive should be limited, as the prestressing force will increase both the shear and peel stresses within the adhesive bond. In particular, the permanent stress should be significantly limited as polymeric adhesive materials can exhibit a significant amount of creep depending on the permanent stress.

The main application methods for FRP prestressing systems, to concrete and metallic structural members, are described below. Prestressed FRP composites have also been applied to timber glulam beams (Triantafillou and Deskovic, 1992), although this application is relatively rare.

10.3.2 Mechanically anchored prestressed FRP plates

Mechanically anchored prestressed FRP plates are the most common application method for prestressed FRP composite systems, and can be used on both concrete and metallic substrates.

10.3.3 Application to concrete elements

Two basic methods of prestress application are used for strengthening concrete elements:

- Stressing and anchorage of the FRP plate in a steel anchorage, placed in a formed recess in the concrete substrate. The anchorage comprises a base plate for force transfer (typically bonded and bolted to the concrete substrate), a tensioning plate for the hydraulic jack at the live anchorage and levelling aids to ensure that out-of-plane prestress forces in the FRP plate are minimised. The FRP plate is located through a slot in one anchorage, bolted and fixed at the 'dead' anchorage, and the prestressing force is applied by reaction against a temporary anchorage, or against an existing, structurally adequate, part of the structure. Temporary tensile forces in the concrete element induced during the prestressing operation, local to the hydraulic jack, should be considered during the design phase to ensure structural adequacy. Prestressing of the FRP plate is undertaken in two stages using temporary and permanent anchorages.
- Stressing and anchorage of the FRP plate in a steel anchor block, placed in a recess in the concrete substrate. The FRP plate is located through a slot in one anchorage, bolted and fixed at the dead anchorage, and the FRP plate prestressed by means of a CFRP anchorhead and hydraulic jack. The prestressing operation is carried out in a single stage.

In general, the FRP plate is locally strengthened by steel or FRP end tabs where it is bolted through into the anchorage zone, to avoid local bearing failure or splitting of the FRP plate during the prestressing process, and also in the long-term once it is anchored.

10.3.4 Application to metallic elements

In general, the application of prestressed FRP plates to metallic elements is similar to that for concrete elements, by the use of steel anchorages bolted into predrilled holes in the flanges of a steel beam. However, drilling of certain types of metallic elements is not advised due to the risk of cracking, e.g. cast iron. In this case, the prestress in the FRP plate may be transferred to the metallic element by a combination of clamping using high strength friction grip bolts and adhesive bonding. Anchorages for prestressed FRP composite systems are also typically grouted on completion of the prestressing operation to ensure a high level of durability.

10.3.5 Jacking or dead load reduction

Jacking out dead load deflection of a concrete or metallic element, or physically reducing existing dead load temporarily (e.g. by removal of fill materials and/or replacement with lightweight fill), prior to the application of unstressed FRP composite is also essentially a prestressing solution as the

FRP composite carries a portion of the dead load in addition to live load. However, the additional permanent stresses within the adhesive require careful consideration during the design phase, and in some cases may require the use of mechanical anchorages.

10.3.6 Adhesively bonded prestressed FRP plates

Methods for transferring the force in prestressed FRP plates into a structural element have recently been developed at EMPA, Switzerland (Stoecklin and Meier, 2003). The FRP plate is stretched over a set distance between two large wheels, the entire mechanical system is lifted up to the soffit of the concrete substrate and the laminate is bonded to the structure. A gradually-anchored system was devised to reduce the longitudinal shear stresses and delay the onset of anchorage failure. Using this method, the plate was bonded to the concrete from the centre, moving outwards in stages. As each portion of the plate was bonded to the concrete at successive stages, the prestressing force was slightly reduced to a nominal value at the end of the plate. In order to speed up the curing process, so that the step-wise technique is economic and practical, heating devices within each portion of the plate were used to reduce the adhesive bond curing time. This method has not been used in the field, but may be suitable for strengthening both concrete and metallic elements.

10.3.7 Prestressed near-surface mounted reinforcement

Research has been undertaken recently (Nordin, 2003) to develop an application method using adhesively bonded prestressed NSMR to strengthen concrete elements. This may exhibit several advantages over adhesively bonded prestressed FRP plates, as the bond area is generally greater, with a corresponding reduction in adhesive stresses. However, further work is required on developing practical methods to introduce the prestress force into the FRP bars or rods; this method has not been used in the field at present.

10.3.8 Prestressed FRP tendons and cables

Prestressed FRP composites have also been used as an alternative to conventional external post-tensioning of a structure to improve the flexural behaviour. The main advantages over external post-tensioning using high strength steel tendons are higher strength, ease of application due to the lightweight nature of FRP composites and improved durability.

10.3.9 Summary

A number of unstressed and prestressed FRP composite application methods are available for flexurally strengthening concrete and metallic structural elements, and also timber (De Lorenzis *et al.*, 2001) and masonry (Borri *et al.*, 2000) in some cases. The design phase of a strengthening scheme should take cognisance of the most suitable FRP strengthening method to provide a cost-effective and durable strengthening solution. The advantages and disadvantages of unstressed and prestressed FRP composite application methods are summarised in Tables 10.1a, 10.1b and 10.2.

Examples of reinforced concrete and conventionally prestressed and post-tensioned concrete structures strengthened using prestressed FRP composites exist throughout Europe, particularly in Germany and Switzerland. The first full-scale application of a FRP composite prestressing system on a concrete structure in the field was on Lauterbridge, Gomadingen, Germany in October 1998. The first full-scale FRP composite prestressing application of a metallic bridge was on Hythe Bridge, Oxfordshire, UK, in March 1998. The case study for this bridge is given in Chapter 13.

10.4 Stages within the FRP strengthening installation process

Any FRP strengthening installation process should incorporate the following stages, to ensure adequate material properties, workmanship during the installation and durability of the completed scheme. The design of an FRP strengthening needs to take cognisance of these stages, to ensure that the chosen type of FRP strengthening is suitable for a particular structure.

10.4.1 Specification of materials and workmanship

The successful application of an FRP strengthening scheme is critically dependent on the level of workmanship, and also on achieving the required level of environmental control and material properties. This can be achieved, in part, by the use of a suitably detailed specification. Advice on specifying FRP strengthening schemes has been published in a number of guides (e.g., Concrete Society, 2004; CompClass, 2006). The following will be important items in a suitable specification.

10.4.2 Experience and track record of contractor

As FRP strengthening is so dependent on the level of workmanship, suitably experienced specialist contractors should be used. In particular, evidence of recent successful projects using similar FRP materials and

Table 10.1a Comparison of unstressed FRP composite systems

Adhesively bonded FRP plates	Near surface mounted reinforcement	In situ laminated FRP fabric	In situ vacuum cured FRP fabric
Fast installation	Fast installation	Relatively slow installation	Relatively slow installation
Strict limits on flatness of substrate profile	Flat grooves can be cut for application in uneven substrate profiles	Can follow substrate profile	Can follow substrate profile
Normal level of site quality control required	Normal level of site quality control required	High level of site quality control required	High level of site quality control required
Requires controlled environment during and immediately after installation	Requires controlled environment during and immediately after installation	Requires controlled environment during and immediately after installation	Requires controlled environment during and immediately after installation
Site test samples representative of installed system	Site test samples more difficult to produce to be representative of installed system	Site test samples reasonably representative of installed system	Site test samples reasonably representative of installed system
All FRP composite materials suitable	All FRP composite materials suitable	Use of fragile, ultra-high modulus carbon fibres on site relatively difficult	Use of fragile, ultra-high modulus carbon fibres on site difficult, but better than in situ laminated FRP fabric
Applicable to all types of substrate	Not applicable to metallic substrates	Applicable to all types of substrate	Applicable to all types of substrate
Small risk of contamination of bonding surface	Small risk of contamination of bonding surface	Greater risk of contamination of laminating surfaces	Greater risk of contamination of laminating surfaces
Void defects in bond can be rectified	Difficult to rectify void defects due to embedded nature	Difficult to rectify void defects in many laminations	Vacuum curing generally ensures minimal voidage
High level of site supervision required. Personal protective equipment required	High level of site supervision required. Personal protective equipment required	High level of site supervision required. Personal protective equipment required	High level of site supervision required. Personal protective equipment required
Application standards and procedures well developed	Application standards and procedures partially developed, but similar to adhesively bonded plates	Application standards and procedures well developed	Application standards and procedures generally undeveloped

Table 10.1b Comparison of unstressed FRP composite systems

In situ resin infused FRP fabric	Sprayed FRP composite	Mechanically fixed FRP plates
Fast installation	Fast installation	Very fast installation
Can follow substrate profile	Can follow substrate profile	Can follow substrate profile to some extent depending on FRP plate thickness
Normal level of site quality control required	High level of site quality control required	Low level of site quality control required
Requires controlled environment during and immediately after installation	Requires controlled environment during and immediately after installation	No requirements for controlled environment
Site test samples reasonably representative of installed system	Site test samples difficult to produce to be representative of installed system	No site test samples required (although pull-out tests may be required)
Use of fragile, ultra-high modulus carbon fibres on site difficult, but better than insitu laminated FRP fabric	All FRP composite materials suitable	All FRP composite materials suitable, but need to be locally strengthened at fixing locations or multiaxial fibres
Applicable to all types of substrate	Generally not applicable to metallic substrates where high stiffness FRP required	Not applicable to metallic substrates
Small risk of contamination of bonding surface	Greater risk of contamination of bonding surface	Risk of surface contamination during fixing not important
Vacuum resin infusion generally ensures minimal voidage	Difficult to rectify void defects	Lower likelihood of defects due to simple fixing method
High level of site supervision required. Personal protective equipment required	High level of site supervision required. High level of personal protective equipment also required	Normal level of site supervision required
Application standards and procedures not developed	Application standards and procedures partially developed	Application standards and procedures not developed

application methods should be submitted, together with proof of experience of the proposed key site operatives, in addition to the experience of the company.

10.4.3 Assessment of existing quality of the substrate

The design of an FRP strengthening scheme, even if using mechanical fixings, is highly dependent on the substrate properties for two reasons; firstly, the durability and strength of the bond or mechanical fixing will be dependent on the substrate quality (poor tensile strength, the presence of air voids and inclusions can all affect the bond strength) and secondly, the overall behaviour of the strengthened member is dependent to some degree on the existing material properties, regardless of the FRP material properties. Typically, for concrete substrates, limits are specified on compressive strength, chloride level, carbonation depth, crack widths, rusting of steel reinforcement, etc. For metallic substrates, limits are specified on surface contamination and the presence of cracks. If necessary, remedial work should be undertaken on the substrate to meet the specified limits, using an approved method. For example, cracks in concrete substrates can be filled with a cementitious repair mortar or injected grout; cracks in metallic substrates may be repaired by stitching. The underlying cause of the deterioration of the substrate should be identified and steps taken to improve the quality of the substrate, otherwise a durable strengthening solution will not be achieved.

In some cases, where there is significant concern over the achievable bond strength to the substrate, it is advisable to undertake preliminary pull-off tests using the bonding materials and surface preparation to be used in the strengthening installation. This will confirm or otherwise the suitability of the substrate (and surface preparation method) for FRP strengthening, and avoid the possibility of abortive work.

10.4.4 Specification of material properties and use

Specification of the material properties of the FRP composite and adhesive, or dry fibres and laminating resin, and the basis for determining these material properties (i.e. minimum number of test samples, test procedure, characteristic values) is critical in order to ensure that the correct type and quality of material is incorporated into a strengthening scheme. For preformed FRP plates, tensile strength, modulus, strain to failure and glass transition temperature of the composite material will generally need to be specified, together with the type, tensile and lap shear strength, tensile modulus, moisture uptake, glass transition temperature and maximum and minimum allowed thicknesses for the cured adhesive. For *in situ* laminated

Table 10.2 Comparison of prestressed FRP composite systems

Mechanically anchored prestressed FRP plates	Prestressed near surface mounted reinforcement	Gradually anchored prestressed FRP plates	Prestressed FRP tendons and cables
Relatively slow installation due to prestressing and anchorage process	Fast installation	Relatively slow installation due to gradual prestressing and anchorage process	Relatively slow installation due to gradual prestressing and anchorage process
Can follow substrate profile to some extent	Flat grooves can be cut for application in uneven substrate profiles	Can follow substrate profile to some extent	Not required to follow substrate profile
Normal level of site quality control required	High level of site quality control required	High level of site quality control required	Normal level of site quality control required
Low requirement for controlled environment as mechanical anchorage primary fixing method	Requires controlled environment during and immediately after installation	Requires controlled environment during and immediately after installation due to adhesive bonding	Requirement for controlled environment dependent on mechanical or adhesive bond fixing method
No requirement for site test samples, unless adhesive bonding also used	Site test samples difficult to produce to be representative of installed system	Site test samples reasonably representative of installed system	Site test samples reasonably representative of installed system if adhesive bonding used

CFRP composite materials generally most suitable, but GFRP and AFRP can also be used	CFRP composite materials generally most suitable, but GFRP and AFRP can also be used	CFRP composite materials generally most suitable, but GFRP and AFRP can also be used	CFRP composite materials generally most suitable, but GFRP and AFRP can also be used
Applicable to all types of substrate	Not applicable to metallic substrates	Applicable to all types of substrate	Applicable to all types of substrate
Risk of surface contamination during fixing not important	Small risk of contamination of bonding surface	Small risk of contamination of bonding surface	Risk of surface contamination dependent on fixing method
Lower likelihood of defects due to mechanical fixing method	Difficult to rectify void defects due to embedded nature	Difficult to rectify void defects in bond due to existing prestress	Difficult to rectify defects due to existing prestress
High level of site supervision required. Personal protective equipment required. Specialist application	High level of site supervision required. High level of personal protective equipment also required. Specialist application	High level of site supervision required. High level of personal protective equipment also required. Specialist application	High level of site supervision required. Personal protective equipment required. Specialist application
Application standards and procedures partially developed	Application standards and procedures not developed	Application standards and procedures not developed	Application standards and procedures partially developed

AFRP = aramid fibre reinforced polymer
CFRP = carbon fibre reinforced polymer
GFRP = glass fibre reinforced polymer

FRP fabric, the effective tensile strength, modulus and strain to failure of the fibres will need to be specified, and the laminating resin specified in a similar manner to the adhesive for preformed FRP plates. In addition, a minimum fibre volume fraction for the final laminated FRP fabric may be specified.

10.4.5 Specification of the environment controls during the bonding process and cure of the adhesive or laminating resin

This will generally require specification of the acceptable air and substrate temperatures (or required elevated temperatures), relative humidity and dew point, together with limitations on the time between mixing of the adhesive (or laminating resin) and jointing of the bond (or lamination of the FRP fabric). Correct specification of these parameters will minimise the risk of a poor quality adhesive bond or laminated FRP fabric.

10.4.6 Preparation of the substrate and assessment of the prepared surface

Acceptable methods for preparation of the substrate surface should be specified, such as needle-gunning, grinding or grit-blasting (which generally provides the best bonding surface). In addition, the use of primers may be specified to promote adhesion, or protect the prepared surface for a period. Assessment of the suitability of the surface preparation is usually undertaken by visual examination, surface moisture meters and ultimately pull-off tests using discs bonded to the prepared surface. At present, there is no simple and quick method for assessing the quality of the surface preparation (and likely adhesion strength); pull-off tests require the bonding adhesive to be adequately cured to provide useful results. Work by CompClass (2006) is currently developing simple field methods to assess surface preparation. Guidance on the surface preparation of concrete is provided in EN 1504 (BSI, 2003) and in Technical Report 55 of the Concrete Society (2004). For metallic substrates, prepared surface quality can be specified in accordance with ISO 8501, ISO 8502 and ISO 8503, which provide a variety of methods for the specification and assessment of different types of preparation method and quality.

10.4.7 Method and sequence of installation

The different application methods, depending on the type of FRP strengthening, are described above. Depending on the type and complexity of the

FRP strengthening scheme, the installation sequence and minimum curing period prior to live loading of the structure may need to be stated. In particular, where associated repair works are undertaken with FRP strengthening, a detailed installation sequence may be required. An example of this is where live load restrictions are in place on a structure, or where fill removal and replacement is undertaken and permanent stresses may inadvertently be induced into the FRP strengthening.

10.4.8 Quality assurance using test samples

The adhesion strength of the adhesive to the FRP and substrate surfaces and the material properties of the adhesive significantly affect the integrity of the strengthened structural element. Therefore, test samples manufactured on site during the installation process should be used to measure and provide confirmation of the quality of the bond. This is typically achieved by using single lap shear test samples made from samples of the FRP plate and the bonding adhesive, and bulk adhesive test samples to measure adhesion, shear strength, tensile strength and modulus and glass transition temperature of the adhesive. The samples should be cured in the same environment as the installed FRP strengthening for a minimum period, dependent on the materials used and ambient environmental conditions, and then tested in the laboratory to recognised standards. It is crucial that the importance of the test samples is realised, as they are meant to be representative of the installed materials and workmanship. Poor quality test samples may lead to unacceptable test results, possibly requiring further design verification, remedial work or replacement of the FRP strengthening installation. Furthermore, the testing house should be nationally accredited or experienced in testing FRP composites and adhesive materials, to ensure that experienced staff are used, test equipment is adequately calibrated and test results correctly reported.

10.4.9 Defect limits

It is not unusual for an installed FRP strengthening scheme to have minor defects. These may comprise air voids or inclusions within the adhesive bond, or laminated FRP fabric or straightness of the FRP plates or laminated fibres. Therefore, the level of acceptable defect should be specified, together with possible acceptable remediation methods. For voids within the adhesive bond or laminated FRP fabric, a maximum allowable area of a single void, and also maximum percentage area of voids, should be specified. Acceptable remedial methods for FRP strengthening schemes should also be stated, and usually include adhesive injection into air voids or the application of additional FRP plates. In severe cases of poor workmanship or inadequate material properties, installed FRP plates may need to be removed and the work carried out again. Quality control, maintenance and repair are discussed in Chapter 12. It should be noted that in some cases of poor workmanship or low material properties, the original design may be reviewed with knowledge of the as-built condition, and found to be acceptable in terms of structural adequacy. This course of action is generally worthwhile, although there is no certainty that the design based on the as-built condition will be found to be acceptable.

10.4.10 Bonding records

To enable deficient workmanship or materials to be traced on site, records for each bonding or lamination process, and also site test samples, should be taken, detailing the location, date and time, environmental conditions and any other pertinent information including pull-of testing. This will enable a record for further maintenance to be produced, and any failures of site test samples to be identified with a particular strengthened area on the structure, or period of time when the installation was undertaken.

10.5 FRP strengthening site activities

Once a fully detailed specification for workmanship and materials is in place, site activities can be undertaken in a controlled manner. The following site activities are typically undertaken during the installation of an FRP strengthening scheme.

10.5.1 Site set-up

The contractor will need to set up the site to enable suitable access to the part of the structure to be strengthened and to allow delivered materials to be adequately stored (particularly the case for adhesives and laminating resins, the application of which are temperature-sensitive). For highway bridges, and railway bridges in particular, access to the bridge may require extensive scaffolding or other method of access. However, FRP materials, due to their lightweight nature, are well suited to difficult access situations where other strengthening methods would be difficult or impossible to employ. In cases where the access period is limited, for example during highway lane rental or railway bridge possessions, the contractor should consider ways to ensure the most efficient use of the time available. Some examples of this would be:

- surface preparation of the substrate in previous access periods;
- maximising the forward preparation of any scaffolding or other access methods;
- back-up plant for powered tools, task lighting and heaters;
- pre-heating of areas where raised temperatures are required during the FRP strengthening installation.

10.5.2 Bonding, laminating or fixing of FRP strengthening

Prior to starting the FRP strengthening installation, the specialist contractor should produce a detailed method statement, in general accordance with the specification, to be approved by the designer and other relevant parties. A fully detailed method statement enables any installation risks or issues to be identified prior to the installation process, possibly saving time and additional costs. The method statement should include, as a minimum:

- mechanical and working properties of the FRP composite, adhesive, laminating resin and related materials (e.g. primers, repair mortars);
- method of surface preparation, how to achieve the required environmental conditions;
- FRP strengthening installation method and sequence with hold points for checking;
- requirements for personal protective equipment and site test sampling.

Once the method statement is approved, the installation works can proceed as planned. Depending on the ambient conditions, local forced heating and/or sheeting of the structure may be required to achieve the correct environmental conditions.

10.5.3 Site supervision

Site supervision, either continuous or occasional depending on the type of scheme, by engineers with experience of FRP strengthening is paramount to ensure the success of an FRP strengthening scheme. Site supervision ensures that the level of workmanship is in general accordance with the specification, and also that any technical problems encountered on site are efficiently and appropriately resolved. This is particularly the case where FRP strengthening is undertaken in restricted time periods, for example during railway possessions or temporary closure of highways. Lack of site supervision is probably the most significant reason for poor quality installation of an FRP strengthening scheme, regardless of the size or complexity of the strengthening scheme. Ongoing inspection of the installed FRP strengthening by site supervisors can be undertaken on larger jobs to avoid over-run of the project programme and extensive remediation at the end of a project, and to correct any inadequate installation methods or procedures at an early stage.

10.6 Inspection, maintenance and monitoring

An FRP strengthening scheme does not end on completion of site activities and strengthening of the structure. For the proper management of strengthened structures, the FRP strengthened structure should be inspected in accordance with a maintenance plan developed by the FRP strengthening designer, and applied by the structure owner or operator. FRP strengthening materials are recognised as potentially having a high level of durability; however, they still require an inspection programme to identify any defects. It is typical for visual inspections to be undertaken on bridges, tunnels and associated infrastructure on an annual basis, at least initially. Building structures, due to the difficulties of access, are usually inspected at less frequent intervals, with advantage being taken to inspect the FRP strengthening during other maintenance work, or on a change of use or occupancy.

Guidance has recently been published on defects that can be identified during close visual inspection (Concrete Society, 2003) in the FRP composite, adhesive, laminating resins and the strengthened structure as a whole, and include:

- evidence of crack initiation, or further cracking, in the substrate adjacent to the FRP strengthening;
- cracks within the adhesive bond or laminated FRP fabric;
- voids within the adhesive bond or laminated FRP fabric (detected by tap testing);
- discolouration of the adhesive or FRP composite material;
- damage to the FRP composite material (e.g. splitting, vandal attack, impact damage);
- ingress of moisture creating permanent wet or damp areas (and corrosion on metallic substrates).

In general, with an FRP strengthening scheme installed in accordance with best practice, very little maintenance will be required, and may comprise painting to restore an acceptable visual appearance, or minor remedial works to the adhesive bond line.

To monitor the long-term durability of the installed FRP strengthening scheme, sacrificial long-term test samples may have been specified, which can be tested as part of the structure maintenance plan. This is particularly important where access to and visual inspection of the FRP strengthening may be difficult; the test samples can be located in a similar, more accessible environment, enabling the condition of the FRP strengthening to be inferred. In a small number of cases, permanent monitoring measures such as strain gauges have been installed. This will only be worthy of consideration for particularly novel applications, or for structures where safety is a particularly high priority.

Maintenance and monitoring of FRP strengthened structures is discussed in greater detail in Chapter 12.

10.7 Relationship between application and design

During the installation of an FRP strengthening scheme, visual inspection and test results may show the original design assumptions to be incorrect, or the behaviour of the materials to be different to that assumed. This is generally the case for all construction activities, and can be overcome to some extent by the use of safety factors and flexible designs. Therefore, differences between the existing or as-built condition, and those assumed during the design process (made evident by the design drawings, specifications, risk assessments, etc. provided to the contractor) should be recorded and notified to the relevant party. This is one area where site supervision by experienced personnel can be very helpful. Some instances where the application conditions may impact on the design are described below:

- poor quality or contaminated substrate (which may become apparent during the surface preparation process);
- dimensional variations (which may hinder the location and application of the FRP strengthening);
- the presence of significant global curvature (which may increase the stresses within the bond);
- thickness of the final adhesive bond for adhesively bonded FRP plates;
- installed material properties (which may be inadvertently affected by the environmental conditions);
- Condition of the structural element (for example, corroded or non-existent steel reinforcement, or metallic elements with variations in section sizes and thickness).

Many of these possible problems can be avoided by a detailed inspection of the structure to be strengthened, preferably by the FRP designer. However, in some cases access can be difficult and previous inspection and record information will have to be relied upon.

10.8 Conclusions

The preceding discussion of application methods for flexural FRP strengthening of structural elements has shown the wide range of applications undertaken worldwide. In particular, the dependence of FRP strengthening applications on good workmanship and experienced contractors, and the requirement for a suitable specification has been highlighted. It should be noted that many of the aspects described above for flexural strengthening with FRP composite systems are also generally applicable for other types of strengthening, such as shear strengthening applications.

10.9 References

BANK L C (2002) *Report on Rapid Strengthening of RC Bridges*, Program SPR-0010 (36), Madison, WI, USA, Wisconsin Department of Transportation.

BORRI A AVORIO A and BOTTARDI M (2000) Theoretical analysis and a case study of Historic masonry vaults strengthened by using advanced FRP, in Humar JL and Razaqpur AG (eds), *Proceedings Advanced Composite Materials in Bridges and Structures ACMBS-III*, Montreal, Quebec, Canada, Canadian Society For Civil Engineering, 577–584.

BSI (2003) *BS EN 1504-10: 2003 Products and systems for the protection and repair of concrete structures. Definitions. Requirements. Quality control and evaluation of conformity. Site application of products and systems and quality control of the works*, London UK, British Standards Institution.

CAROLIN A (2003) *Carbon Fibre Reinforced Polymers For Strengthening of Structural Elements*, PhD Thesis, Luleå University of Technology, Luleå, Sweden.

COMPCLASS (2006) *Composite Materials Systems – Classification and Qualification*, www.compclass.org, Last updated November 14th, 2006.

CONCRETE SOCIETY (2004) *Design Guidance For Strengthening Concrete Structures Using Fibre Composite Materials*, TR 55, 2nd edn, Camberley, UK, The Concrete Society.

CONCRETE SOCIETY (2003) *Strengthening Concrete Structures Using Fibre Composite Materials: Acceptance, Inspection and Monitoring*, TR 57, Camberley, UK, The Concrete Society.

DE LORENZIS L, GALATI N and LA TEGOLA A, (2001) Strengthening of glulam beams with externally bonded FRP laminates, in Figueiras J, Ferreira A, Juvandes L, Faria R and Torres Marques A (eds), *Composites in Construction: Proceedings of the First International Conference*, Lisse, The Netherlands Balkema, 517–522.

ISO (2006) *ISO 8501-4: 2006 Preparation of steel substrates before application of paints and related products – visual assessment of surface cleanliness*, Geneva, Switzerland, International Organization for Standardization.

ISO (2003) *ISO 8502-12: 2003, Preparation of steel substrates before application of paints and related products – tests for the assessment of surface cleanliness*, Geneva, Switzerland, International Organization for Standardization.

ISO (1988) *ISO 8503-1: 1988, Preparation of steel substrates before application of paints and related products – surface roughness characteristics of blast-cleaned steel substrates*, Geneva, Switzerland, International Organization for Standardization.

NORDIN H (2003) *Flexural Strengthening of Concrete Structures With Prestressed FRP Rods*, PhD Thesis, Luleå University of Technology, Luleå, Sweden.

PORTER A D, DENTON S R, NANNI A and IBELL T (2003) Effectiveness of FRP plate strengthening on curved soffits, in Tan K H (ed.), *Proceedings Fibre Reinforced Polymer Reinforcement for Concrete Structures – FRPRCS-6*, Singapore, World Scientific, 1147–1156.

STOECKLIN I and MEIER U (2003) Strengthening of concrete structures with pre-stressed and gradually-anchored CFRP Strips, in Tan KH (ed.), *Proceedings of Sixth Fibre Reinforced Polymer Reinforcement For Concrete Structures – FRCPS-6*, Singapore, World Scientific Vol. 2, 1321–1330.

TÄLJSTEN B (2002) Construction of a full composite bridge – the ASSET West Mill Bridge, *Proceedings NGCC 2nd Annual Conference on Design, Specification and Construction of FRP Structures*, Solihull, UK, Oct 30.

TRIANTAFILLOU TC and DESKOVIC N (1992) Prestressed FRP sheets as external reinforcement of wood members, *Journal of Structural Engineering*, **5**(118), 1270–1284.

11
Durability of externally bonded fiber-reinforced polymer (FRP) composite systems

T. E. BOOTHBY and C. E. BAKIS, The Pennsylvania State University, USA

11.1 Introduction

The durability of externally bonded fiber reinforced polymer (FRP) composite systems refers to the ability of those materials to provide enhanced structural characteristics to existing structures over time frames of interest. The durability characteristics of a given FRP composite material applied to a given substrate material is influenced by numerous interrelated factors, such as:

- type and condition of substrate material;
- selection of the FRP material system, including individual constituents and fiber orientations;
- configuration details of the externally bonded system, including edges and ply drop-offs;
- quality of construction, including substrate preparation, fabrication and application of FRP material system, and maintenance;
- service environment of the rehabilitated structure, including applied load, temperature, moisture, chemicals, ultraviolet exposure, etc.

This review addresses durability issues for generic types of wet-laid and pre-cured thermosetting FRP material systems as well as the interaction between these material systems and four of the most commonly encountered substrates – concrete, wood, masonry and steel. FRP materials bonded into slots cut into a substrate (near-surface mounted reinforcement) are not specifically addressed in the review, although durability issues for these types of reinforcements are expected to overlap with those for internal FRP reinforcements and externally bonded FRP reinforcements. Also, the wide range of configuration details enabled by field-shaped FRP materials and the lack of published durability information related to these details prevents a meaningful general discussion of configuration-related issues, as well.

Externally bonded FRP reinforcements for civil infrastructure applications have been under extensive investigation only since the 1990s. Information on durability is therefore rather limited in terms of the duration and realism of testing conditions (Karbhari *et al.*, 2003). Scientific data from realistic testing conditions have been based primarily on observations made over several months or years, rather than more pertinent timeframes of the order of decades. In an attempt to shorten the time and expense of durability testing, researchers often adopt accelerated testing schemes wherein FRP degradation mechanisms are postulated to be accelerated by, for example, an increase of moisture, chemical concentration, temperature and/ or stress. This approach has been shown to succeed in cases where a single degradation mechanism is operative, such as the well-known example of glass reinforced concrete, where the loss of strength of glass fibers as a function of time and temperature has been modeled with a simple kinetic equation (Litherland *et al.*, 1981). However, in the typical substrate–FRP hybrid material system, there are multiple, simultaneous, interactive degradation mechanisms at play, each having different rate kinetics with respect to a given acceleration factor. In short, results from accelerated testing tend to be unique to the particular acceleration scheme followed and distinct from results one can expect in an actual service environment. Indeed, researchers have generally relegated the validation of accelerated testing schemes to 'future work.' To the extent that a given commercially available material system remains available well beyond the timespan of extended-duration testing of that system under realistic conditions, such future data will prove to be tremendously valuable to designers. If the suite of available material systems changes, however, the long-term data may be rendered less useful. Such uncertainty is one of the more important challenges in implementing and gaining confidence in the use of rapidly evolving, proprietary FRP materials in construction (Scalzi *et al.*, 1999).

Approaches for investigating the durability of externally bonded FRP materials on various substrates can be classified in two general types. In the first type of approach, experiments are carried out on scaled structural members where the overall failure tends to be dominated by one primary mechanism, but it is understood that other (less critical) degradation mechanisms are active as well. The advantage of this approach is that information useful to structural designers, such as ultimate load, is immediately apparent. The disadvantage is the potential lack of generality of the results due to a lack of accounting for shifting primary failure mechanisms and the interaction of secondary and primary failure mechanisms. Examples of the first approach are the measurement of failure loads in externally reinforced beams in flexure or wrapped columns in axial compression. The second type of approach contrives specimens and loading schemes to more strongly promote a single governing failure mechanism, which prevents shifting failure modes and reduces the impact of the secondary degradation mechanisms on the test results. The disadvantage of the second approach is that more information (e.g., effects of scaling and realistic secondary failure mechanisms) is needed to relate the test results to real structures. Examples of the second approach are lap shear tests for evaluating bond strength or interfacial fracture energy and circular ring or tensile coupon tests for evaluating fiber strength.

In summary, the prudent designer should place the most faith in durability data acquired under realistic environmental and loading conditions on full-scale structures rather than well-controlled, accelerated conditions on sub-scale specimens, inasmuch as such data exist. Carefully planned and executed accelerated tests on laboratory specimens can be useful for one-on-one comparisons of various FRP systems applied to a particular type of substrate under a fixed set of testing conditions, although such data may not accurately predict the field performance of the substrate–FRP hybrid system.

11.2 FRP composites

FRP composite material systems, like other structural materials, have durability issues worthy of consideration when selecting materials for a strengthening project. The mechanisms of degradation of FRP materials systems can be physical (e.g. cracking), chemical or, most often, both. Durability characteristics of a composite material system are most appropriately considered at the level of the composite rather than the individual constituents, as it is widely known that FRP system durability depends in complicated, non-linear ways on the constituents, such as the fibers, the type of sizing applied to the fibers and the resin used to impregnate the fibers (Liao *et al.*, 1999). Fiber sizings consist of proprietary chemical compounds attached to the surface of the fiber to facilitate handling of the fiber and to enhance the fiber–resin bond. While these 'linker' compounds greatly enhance the strength of FRP material systems under ideal conditions, they are often preferentially degraded by exposure to moisture, which could lead to blister-forming osmotic pressure, weakly bonded fibers, pathways for the increased ingress of aggressive chemicals into the composite material and, ultimately, reduced strength of the composite (Schutte, 1994; Kootsookos and Mouritz, 2004). It is advisable that fiber sizings used in FRP materials be treated with equal scrutiny to the fibers and resins themselves in material selection, particularly when the application includes moist, chemically aggressive environments (Gremel *et al.*, 2005).

The role of polymer resins in FRP material systems is to protect and support the fibers and to transmit loads between the substrate and the fibers by shear. Polymer resins suffer losses of stiffness and strength at elevated temperatures, embrittlement at low temperatures and losses of strength and/or stiffness by exposure to ultraviolet light or certain chemicals such as water, acidic solutions and alkaline solutions. The upper use temperature of a structural resin is closely linked to the primary glass transition temperature of the resin T_g which is a central measure of the range of temperatures over which the mechanical behavior of the polymer gradually transitions between the rigid (glassy) and rubbery ranges. The T_g depends primarily on the exact type of polymer and secondarily on the cure temperature of the material and environmental conditioning of the polymer. Higher cure temperatures (within reasonable limits) and longer cure times push the T_g asymptotically towards the maximum theoretical value dictated by the chemical makeup of the polymer, while absorbed fluids, including water, reduce the T_g. Time-dependent deformation of a resin is directly proportional to the magnitude of stress in a resin and the temperature, with dramatically increased strain rates occurring close to and above the T_g. Similarly, while higher temperatures increase the rate of absorption of potentially damaging fluids into polymers, the rate of absorption is particularly rapid above the T_g. Resins with relatively high T_g values tend to be relatively brittle at lower temperatures due to higher chemical cross-linking (less mobility) among the molecular chains comprising the polymer. In practice, resin selection is more likely to be governed by the upper use temperature rather than the lower use temperature.

Recognizing that the glassy–rubbery transition occurs over a range of temperatures near the T_g, that the T_g may change over time and with exposure to the environment and that the mechanical response of a resin to long-term loads and environments is complex and difficult to predict, it is recommended to select resin systems with T_g values at least 15 °C above the maximum expected service temperature. The T_g value used in this comparison should be the T_g in the expected service environment (e.g. wet or dry) and should reflect the expected degree of cure when the FRP material is expected to carry the intended service load. The T_g of a resin or an FRP material system may be measured using standardized methods such as dilatometry (E1545 – ASTM, 2005b), dynamic mechanical analysis, (E1640 – ASTM, 2004), or differential calorimetry (E1356 – ASTM, 2003d).

Contemporary FRP composite material systems for external strengthening generally use epoxy and vinylester resins as matrix materials on account of their advantageous durability characteristics in outdoor environments. Epoxies are used in practically all wet-laid FRP systems because of their desirable combination of field workability and service performance, while epoxies and vinylesters are both used in factory-produced, pre-cured FRP materials. Polyester matrix materials are generally found to be less durable in outdoor use (Chin *et al.*, 1999; Bank *et al.*, 2003), although the literature is not unanimous on this viewpoint (Dagher *et al.*, 2004).

Fillers are used in FRP composites and polymer top-coats for FRP composites to reduce the cost of the resin or to enhance particular durability attributes such as resistance to ultraviolet radiation, oxidation and fire (Katz and Milewski, 1987). Fillers have the disadvantages of increasing the viscosity and decreasing the strength of resins when added in sufficiently high quantities and may require that additional chemical compounds be used to enhance their bonding with resins. The durability of a filled resin system in a particular environment should be considered in the material selection process, although data of this sort for externally bonded FRP material systems are currently not readily available. Calcium carbonate and talc are examples of inert fillers that are widely used to reduce the cost of FRP materials, although the former also takes on the role of shrinkage reducer and the latter serves as a reinforcement on account of its plate-like shape. Carbon black and oxides of titanium, zinc and aluminum can be used for improving the resistance of polymers to photodegradation in sunlight (Wypych, 1999). Photodegradation of polymers is typically restricted to a skin depth of a few microns, which suggests that a thin protective coating should suffice for the protection of FRP materials against the direct effects of sunlight.

To increase the fire resistance and reduce the smoke generation rate of FRP materials, various intumescent coatings (Le Bras *et al*., 1998) and non-halogenated fillers (Hornsby, 2001) can be utilized. Intumescent coatings such as melamine phosphate and vermiculite function by releasing water vapor and expanding to a low conductivity, non-combustible foam barrier upon heating, which provides a measure of protection to the underlying material. Halogenated fillers have recently fallen out of favor on account of harmful emissions during exposure to fire (Zaikov and Lomakin, 2002). Non-halogenated fillers, such as aluminum trihydroxide and polymer-layered silicate-based nano-sized clay platelets, for instance, improve the fire resistance of polymers by several mechanisms, including the absorption of heat, reduction of polymer fuel, release of water and formation of char. Nanoclays are of increasing interest for improving fire resistance of resins at filler loadings believed to be low enough to avoid viscosity increases and strength reductions seen with conventional fillers at high loadings. Standard ASTM tests can be used to evaluate the fire response of FRP materials, either alone or bonded to structural elements (*ASTM Fire Standards*, 7th edn – ASTM, 2007a). Fillers may significantly improve the smoke and toxicity problems associated with burning FRP materials, but the low T_g values of currently used polymer matrix materials and adhesives prevent externally bonded FRP materials from contributing significantly to structural strength during a fire.

FRP material systems used for external bonding typically employ carbon or glass fibers as the principal tensile load-bearing elements, both of which are impervious to elevated temperatures normally encountered in construction applications. The polymeric matrix material effectively governs the upper use temperature of carbon and glass FRP materials. Bare carbon fibers are highly resistant to environmental degradation, excepting fire, and have exceedingly low sensitivity to cyclic and sustained stresses. Bare glass fibers are not combustible but suffer significant strength reductions in the presence of moisture and applied stress, particularly if the moisture is alkaline. Aramid fibers, consisting of oriented fibrils of a very strong, tough polymer, see less use than carbon or glass fibers. To date, the main market niche for aramid fibers appears to be in applications requiring resistance to extreme loading (e.g. blast, impact). Bare aramid fibers can lose strength in the presence of moisture, ultraviolet radiation and elevated temperature and, unlike bare carbon and glass fibers, can exhibit significant time-dependent deformation (creep and relaxation) in proportion to stress and temperature, particularly in wet environments (Wang et al., 1992).

As part of durability testing of FRP materials exposed to an exterior environment, outdoor exposure tests such as those described in ASTM D1435 can be carried out (ASTM, 2005a). Additionally, a realistic loading history should be imposed on the material along with the environmental exposure. For example, FRP materials bonded to structural elements carry a certain fraction of the live load applied to the structure, whereas FRP materials bonded to columns for passive confinement experience no load until they are activated by lateral expansion of the column during an extraordinary loading event. Stress limits for preventing the failure of all FRP materials under all service conditions are impossible to define within the confines of this review because of the confounding influences mentioned earlier. Some information on maximum recommended fiber stresses for representative FRP material systems subjected to representative service conditions has been developed in the context of concrete applications and can be found in documents such as those published by the Japan Society of Civil Engineers (JSCE, 2001), the ISIS network in Canada (ISIS Canada, 2001), the *fib* Task Group 9.3 (*fib*, 2001) and the American Concrete Institute (ACI, 2002). For example, to prevent creep and fatigue failure in glass, aramid, and carbon FRP materials, the ACI recommends that the maximum sustained plus cyclic stress levels be less than 20%, 30%, and 55% of the design ultimate strength of the respective materials. In turn, to account for exposure to various environments in the long term, the design ultimate strength is taken as 50–75% of the manufacturer's reported strength for glass FRP, 70–85% for aramid FRP, and 85–95% for carbon FRP. Such recommendations may not be directly applicable to multidirectionally reinforced FRP composites. Also, it must be emphasized that other modes of failure besides fiber fracture, such as debonding, can govern a design. Hence, the subsequent discussion covers durability issues that are particular to several types of bonded substrate–FRP hybrid systems currently in use.

11.3 Externally bonded concrete–FRP hybrid systems

11.3.1 Applications

Concrete has many advantages that motivate its widespread use in construction, such as low cost, long service life (when properly mixed, placed and cured), ease of construction and high compressive strength. However, concrete has a low tensile strength, requiring that it be reinforced with high tensile strength materials such as steel. Externally bonded FRP composite materials can function as supplementary tensile reinforcement provided there is sufficient load transfer between the concrete and FRP material. Externally bonded FRP reinforcements have been applied to concrete columns, beams, slabs, walls and connections for reasons such as strength degradation due to environmental exposure or extraordinary overload or insufficient strength due to defective construction, increased service load requirements, or updated codes (Rizkalla and Nanni, 2003; Triantafillou, 2007).

11.3.2 Durability issues for concrete

When expansive chemical reactions or phase changes occur among the materials present in concrete, internal tensile stresses are developed, which can lead to cracking, spalling and a general loss of integrity in structures (Neville, 1996). Adverse expansive actions in concrete, such as alkali–silica reaction and sulfite attack, are most often attributable to moisture penetration. In steel reinforced concrete and prestressed concrete, the presence of moisture and oxygen promotes the expansive corrosion of steel, particularly when the steel surface is depassivated by carbonation of the cover concrete or by the penetration of chlorides to the steel (Bentur, 1997). The freezing of water within concrete also causes expansive action. ASTM C666 (ASTM, 2003a) is a commonly used test method for comparing the cyclic freeze/thaw resistance of concrete under well-controlled testing conditions and has been applied to concrete–FRP bonded systems as well.

In the repair of environmentally degraded concrete, measures should be taken to reduce the ingress and accumulation of moisture using standard practices, such as (i) providing good drainage; (ii) properly locating and sealing expansion cracks; (iii) sealing the concrete on surfaces where corrosive agents can enter concrete and allowing trapped moisture to escape elsewhere; and (iv) providing a durable reinforcement cover consisting of low permeability, air-entrained concrete (Mailvaganam and Wiseman, 2003). Severely corroded steel should be repaired or replaced. Further corrosion protection measures may be needed, such as cathodic protection or electrochemical chloride extraction (Buenfeld et al., 1998).

11.3.3 Specific characteristics of bond between FRP and concrete

The two general types of applications of externally bonded FRP hybrid systems for concrete structures are known as contact-critical and bond-critical. In contact-critical applications, load is transferred between FRP and concrete by contact stress (pressure) across the interface, as in passive column confinement. In bond-critical applications, load is transferred by shear stress as well as peeling-stress, as in flexural and shear reinforcements for beams. High strength epoxies are generally required to transmit substantial shear and peeling stresses in bond-critical applications, whereas either epoxies or non-shrink cementitious grouts are suitable to ensure the intimate interfacial contact required for contact-critical applications. Proper initial bond quality is critical to the long-term durability of external reinforcements for concrete structures. Therefore, proper procedures for preparing the concrete and installing the FRP material should be followed. A report issued by the Transportation Research Board of the USA provides recommended construction specifications and a construction process control manual to ensure the intended performance of externally bonded FRP reinforcement systems applied to concrete structures (Mirmiran *et al.*, 2004).

The primary areas of concern regarding bond in externally bonded FRP reinforcement in bond-critical applications are at regions of bond stress concentration, such as near the cut-off (termination) points of the FRP reinforcement and near concrete cracks (Teng *et al.*, 2003). The low tensile strength of concrete combined with the high tensile stresses associated with bond stress concentrations can result in failure at the concrete–FRP interface or within the concrete – often manifested as cover separation from the internal reinforcement. Cracks should be grouted and uneven surfaces leveled before the FRP material is attached. Local areas of weak or no bonding resulting from installation problems or service-induced impact are also cause for concern, not only due to the immediate shear stress concentrations or loss of intimate contact but also due to the potential for degradation of concrete, bond and FRP material due to water accumulation in the poorly bonded area. An interfacial tension strength test, such as Method L1 in the ACI guide test methods for FRP materials (ACI, 2004), may be done in the field to characterize the strength of the concrete–FRP bond.

11.3.4 Durability considerations for externally bonded concrete–FRP hybrid systems

Currently available literature on the durability of FRP materials field-bonded to full-size reinforced concrete (RC) members is limited, necessitating that the following discussion of durability considerations be based significantly on investigations of reduced-scale specimens. The discussion emphasizes RC members, although results on plain concrete are included where no other information is available.

11.3.5 Degradation of reinforced concrete due to water

It has been demonstrated that externally bonded E-glass and carbon FRP (CFRP) materials can slow the expansion, cracking and spalling of RC members brought on by steel corrosion (Pantazopoulou *et al.*, 2001; Wootton *et al.*, 2003; Badawi and Soudki, 2005; Thaesler *et al.*, 2005). However, it is important to minimize the prospects of moisture accumulation inside concrete when selecting and configuring an external FRP reinforcement system. For example, for an RC pile submerged in seawater, complete coverage with a low-permeability FRP material can minimize entrapped moisture in the concrete by reducing the rate of intrusion from an abundant supply in the environment. The rate of corrosion in RC is also thought to be decreased by the reduction of oxygen transport rate through FRP material (Debaiky *et al.*, 2002). On the other hand, for an RC bridge girder normally not in contact with water, an open pattern of FRP strips minimizes moisture accumulation by facilitating the escape of internal moisture. Vapor-permeable epoxy resins have been developed for externally bonded FRP materials to enhance the transport of moisture, although the durability characteristics of such material systems are not well described at this time. Another potential source of degradation in RC is corrosion of steel reinforcement due to galvanic coupling with carbon fibers. Galvanic coupling can occur when there is a direct electrical connection between bare carbon fibers and uncoated steel when both materials are embedded in concrete containing moisture and chlorides (Torres-Acosta, 2002). In practice, it is extremely difficult to ensure that carbon or steel (or both) are electrically insulated from concrete. Therefore, efforts should be made to prevent the direct contact of externally bonded CFRP reinforcements and steel reinforcement. For example, a layer of electrically insulating E-glass FRP can be placed between CFRP and concrete.

11.3.6 Contact-critical concrete/FRP systems

A small subset of the abundant literature on contact-critical applications of externally bonded FRP materials (columns) is focused on durability. Toutanji and co-workers (Toutanji, 1999; Toutanji and Deng, 2002) showed evidence that a properly selected hoop-oriented, wet-laid FRP material system with E-glass, carbon, or aramid fibers wrapped on small-scale, plain concrete cylinders could provide significant increases in compressive strength and ductility (in comparison to the unwrapped case) with little or no sign of degradation after 300 cycles of wet/dry exposure in salt water. Several investigators have found that cyclic freezing and thawing degrades the compressive strength and/or ductility of small-scale, plain concrete cylinders confined with FRP materials, particularly when the specimens are thawed in water (Toutanji and Balaguru, 1998; Karbhari, 2002). Since concrete itself suffers degradation from moist freeze/thaw conditions, care must be exercised in interpreting test results on cylinders confined with external FRP reinforcements. Failure of FRP confinement due to creep rupture or fatigue is normally not an issue in columns due to the passive role of the reinforcement, although it is possible that prestress introduced by means such as expansive grout could lead to creep rupture. Among the limitations in the freeze/thaw results known to the authors are a lack of information on the resistance of the concrete to freeze/thaw conditions (e.g. use of air entrainment), a lack of internal steel reinforcement in many cases, the lack of data on size effects and variations in the method of experimentally dealing with the ends of specimens where environmental attack and stress concentrations during testing tend to be most severe. Additional data on the durability of externally bonded FRP wraps on RC columns, preferably developed with a future standard test method and specimen preparation procedure, should help to resolve these and other limitations.

11.3.7 Bond-critical concrete–FRP systems

Durability information is of better quality for bond-critical FRP applications than for contact-critical applications, particularly in terms of realism of the specimens and test conditions. Most of the literature focuses on RC beams with external flexural reinforcements, although external shear reinforcements are occasionally represented. It can be expected that bond durability issues for external flexural and shear reinforcements are fundamentally similar when no special efforts are made to anchor the reinforcements near bond stress concentrations (e.g. cut-offs and concrete cracks) and when no prestressing is applied to the reinforcements. When special anchoring or prestressing techniques are employed, the center of attention for failure and durability issues can shift from debonding at regions of bond stress concentration to fiber fracture at regions of fiber stress concentration (e.g. concrete cracks).

Fatigue effects

For cyclically-loaded RC beams fitted with non-prestressed CFRP reinforcements and tested at room and sub-ambient temperatures, the available data strongly suggest that the failure of internal steel reinforcements (flexural and shear), rather than the FRP or the concrete–FRP bond, controls fatigue lifetime unless the beam is extremely overloaded after application of the FRP (Inoue et al., 1994; Barnes and Mays, 1999; Lopez et al., 2003; Brena et al., 2005). These results also indicate a progressive softening of the bonded interface and a redistribution of stresses along the length of the external reinforcement during cyclic loading, which leads to a slight increase in beam deflection with continued load cycling. Since the current literature provides fatigue information only on non-prestressed FRP reinforcements, no inferences can made regarding the fatigue behavior of prestressed FRP reinforcements. Likewise, the literature is silent on the combined effects of cyclic temperature (including wet freeze/thaw) and cyclic loading. Additional research on these topics is needed.

Temperature effects

Two potential concerns regarding extreme temperature exposure of concrete–FRP hybrid systems are thermal stresses caused by mismatched thermal expansion coefficients and embrittlement of the materials at low temperatures. El-Hacha et al. (2004) investigated internally post-tensioned beams fitted with pre-cured, prestressed, carbon/epoxy plates and quasi-statically tested to failure at two temperatures (22 and $-28\,°C$). The failure load and mode (tensile failure of the prestressed plates) were not sensitive to test temperature. Lopez et al. (2003) observed no effect of low temperature ($-29\,°C$) on the quasi-static flexural bond strength or failure mode (debonding) of pre-cured, carbon/epoxy plates bonded to steel RC beams in comparison to control tests done at room temperature. Grace and Singh (2005) investigated the quasi-static behavior of RC beams flexurally reinforced with wet-laid and pre-cured carbon/epoxy reinforcements, subjected to warm/humid (100% RH, $38\,°C$) and dry heat ($60\,°C$) conditions for up to 10 000 h and found that flexural bond failure occurred at loads as much as 30% and 10%, respectively, below room temperature values.

Freeze/thaw

Several groups of investigators have evaluated the freeze/thaw behavior of carbon/epoxy and E-glass/epoxy FRP materials applied as flexural reinforcement to RC beams. Green and co-workers (Green et al., 2000; Bisby and Green, 2002; Green et al., 2003) found little or no deterioration of the flexural bond strength of carbon/epoxy and E-glass/epoxy materials after as many as 300 freeze/thaw cycles. Some of the E-glass FRP materials in these investigations failed by fiber tension when sufficient end anchorage was provided, but the failure loads in these cases also did not depend strongly on the number of cycles. Grace and Singh (2005) observed a

10–15% reduction in the flexural bond strength of two types of carbon/epoxy materials after 700 cycles. Researchers using specialized tests to evaluate concrete to E-glass FRP lap shear joints have noticed little change in the shear failure force yet noticeable changes in the shear stress distribution and fracture energy with freeze/thaw cycles (Mukhopadhyaya *et al.*, 1998; Jia *et al.*, 2005). The implications of such changes on design procedures remain to be determined.

11.4 Externally bonded wood–FRP hybrid systems

11.4.1 Applications

Although wood has nominally the same capacity in tension and compression parallel to the grain, bending members invariably fail in tension. FRP can thus be expected to contribute significantly to the flexural strength of wood beams, joists and other structural elements by increasing tensile capacity. Wood is also inherently very weak in tension perpendicular to the grain and, in applications giving rise to such perpendicular tension, judiciously applied FRP can be expected to increase the resistance of wood structural elements. FRP has had a number of applications as a material for the enhancement of the flexural strength of new engineered wood products, especially as reinforcement for horizontally laminated wood beams. The applications proposed to date for repair by externally bonded FRP have been more limited, although this limitation is not expected to remain for very long. To date, limited investigations of repaired general flexural members exist in the literature (Mosallam *et al.*, 2000), and the understanding of the durability of repaired members, in large part, will have to be based on available data on new manufactured glued-laminated flexural members. A further repair application that has been explored in significantly greater detail is the repair of existing marine piles by jacketing with FRP, and grouting the interstices (Lopez-Anido and Xu, 2002).

11.4.2 Durability issues for wood

Wood is a naturally-occurring, organic, relatively porous material. As such it is prone to various environmental effects to a greater degree than most other engineered materials. In addition to various forms of biological attack, wood also experiences significant creep under sustained compressive loading. The moisture content of wood varies up to weight proportions of well over 20%, and the change in volume between saturated and dry wood may be quite significant. The changes in volume due to alternate wetting and drying cycles have a deleterious effect on the strength of bonded connections between laminations of wood, or connections between wood and other materials. The impact resistance of wood is also quite low and, in most applications for new construction, it is recommended that an additional wood lamination, or 'bumper strip' be provided to shield the FRP and underlying wood substrate from impact.

11.4.3 Specific characteristics of wood–FRP bond

Unlike bonding to concrete-based or masonry materials, the wood–FRP bond has significant tensile resistance and remains intact during flexural loading. Interface shear stresses are high, however, during bending, and delamination can initiate at an air void remaining from the construction process or developed as a result of environmental conditioning. Davalos *et al.* (2000a) found mean shear strengths in the bond line ranging from 5.0–9.6 MPa using ASTM D905 (ASTM, 2003b). E-glass/epoxy-reinforced block shear specimens with a length of 30 mm along the loaded axis. In this study, the lowest strengths were obtained in wet specimens using a resorcinol formaldehyde coupling agent, while the highest were obtained in dry specimens using a hydroxymethylated resorcinol (HMR) coupling agent. For similar specimens using a phenolic resin adhesive the mean shear strengths obtained were 7.1 MPa dry, and 6.7 MPa wet. Lopez-Anido *et al.* (2004a) investigated wood piles repaired in an underwater environment by providing an FRP jacket bonded to a wood cylinder by combinations of grout and polymer coating. These researchers found mean ultimate shear strengths of 500–1000 kPa with failures recorded primarily at the grout–FRP interface. However, these shear strengths are averaged over a length of the order of 500 mm. Using a more realistic effective bond length for glass fiber reinforced polymer (GFRP) on cement-based grout of 50 mm results in a shear stress of 5.0–12.7 MPa, which is more consistent with other reported shear strengths, both for FRP–concrete and FRP–wood systems. Lyons and Ahmed (2005) reported on a study of bonding characteristics of GFRP sheets to pressure-treated Southern Yellow Pine and found that, for this application, one type of proprietary bonding agent is superior for dry conditions, and another for wet conditions. They also recommend the use of an HMR primer to improve the bond characteristics for most applications.

11.4.4 Durability considerations for externally bonded wood–FRP hybrid systems

On the basis of the very limited information in the literature, it may be reasonable to hypothesize that the initial bond strength of GFRP on a wood substrate or on a cement grout–wood substrate is similar to the bond strength of GFRP on concrete. The absence of flexural cracks as in FRP bonded to concrete or masonry implies fewer locations for initiation of delaminations; however, it has been noted that air bubbles provide similar initiators in FRP–wood systems. The effect of air bubbles can be exacerbated by shrinkage and swelling due to wet–dry cycles and by freezing under wet conditions. Freeze/thaw conditions were investigated for FRP jacketed piles by Lopez-Anido *et al.* (2004b). In the tests conducted by these authors, based on International Council of Building Officials acceptance procedures, single lap shear specimens were subjected to 20 cycles of freezing and thawing for investigation of retention of mean shear strength. An average strength retention value of 57% was found, and the authors attributed this rather poor freeze/thaw performance to entrapped air within the lap joint. Several studies have noted significant drops in interface shear strengths due to wetting and drying, especially the previously cited work of Davalos *et al.* (2000a). These authors also noted that bond strength increased significantly with increased clamping pressure and decreased assembly time. These authors also speak of the undesirable effect of air bubbles, and suggest that higher clamping pressures will most likely improve the durability performance of FRP–wood systems. In a second part of this study, Davalos *et al.* (2000b) found mode-I fracture energies for these systems. The values they found for the phenolic resin systems, which were applied in the Part I paper to flexural reinforcement, are 455 N/m dry and 721 N/m wet, which are surprisingly similar to values observed at FRP–concrete interfaces.

Lopez-Anido *et al.* (2000) tested transversely glued laminated wood reinforced with E-glass and a vinylester resin, with and without an HMR coupling agent, along with control specimens using a phenol-resorcinol-formaldehyde (PRF) resin-based adhesive. The authors observed comparable wet (approximately 6 MPa) and dry (approximately 4 MPa) shear strengths between the primed vinylester adhesive and the control specimens, but much lower strengths for the unprimed specimens. A further five cycle wetting and drying test found less than 1% delamination by ASTM D1101 (ASTM, 2006a) for the primed vinylester adhesive specimens, and complete delamination for the unprimed specimens. Battles *et al.* (2000) investigated code acceptance criteria of retained strength after 1000 and 3000 hours of exposure for an E-glass phenolic resin composite exposed to salt water, ASTM D1193 Type II reagent water (ASTM, 2006b), and calcium carbonate solution (pH 9.5). They found slightly deficient residual strengths in ultimate longitudinal tensile strength and interlaminar shear strength, especially in alkaline exposures, but overall strength retention values of over 80%. They hypothesize that the application of the primer and adhesive for bonding to wood will improve these strength retention values by closing voids in the FRP material.

The following recommendations can be made to maximize the long-term durability of externally reinforced wood structures subjected to flexural loading:

- FRP reinforcement applied to exposed wood structures is recommended to be provided with an additional wood external lamination or 'bumper strip' to minimize the effect of impact on the wood substrate.
- Environmental degradation is likely to promote delamination of the reinforcement due to shear at a much higher rate than other effects, such as loss of tensile strength of the FRP or interlaminar shear stress, due to the very different hygrothermal properties of the wood substrate and the FRP reinforcement.
- Shear stresses in the FRP–wood bond are recommended to be limited to relatively low values (1–2 MPa), and the shear resistance must be presumed to degrade with exposure to wet/dry cycles or to freezing and thawing, where moisture accumulation in the wood substrate is likely.
- Due to cost considerations, vinylester adhesive systems may be preferred over epoxy-based systems, but a primer is recommended to be used for the application of such systems to wood.
- A high clamping pressure is recommended, both to ensure proper curing of the adhesive and to inhibit the formation of air bubbles under the FRP system.
- As always with wood structures, proper detailing to resist moisture accumulation and moisture intrusion is necessary to ensure the durability of the structure, especially with the addition of FRP strengthening systems. This is critical where bonded repairs cover a large percentage of the exposed surface of the structure and inhibit proper drying of the wood substrate.

11.5 Externally bonded masonry–FRP hybrid systems

11.5.1 Applications

Masonry spanning horizontally in the form of arches or vaults is an important feature of a wide variety of ancient construction. However, in contemporary construction, masonry is only used for walls, piers or other vertical load-bearing element. The most common application of externally bonded FRP reinforcement for masonry structures is the strengthening of a wall (Almusallam and Al-Salloum, 2007). This strengthening may be either for in-plane forces due to the action of a wall as a shear wall in a building lateral force-resisting system, or for out-of-plane forces due to the bending induced by wind loading. As such, the FRP is not subject to constant tensile loading, although some transfer of compressive forces may occur over time due to creep.

FRP has also been proposed for the repair and strengthening of ancient arches, vaults and domes. Limited applications of this material have been documented (Valluzzi et al., 2001; Lourenço, 2004; Bati et al., 2007). In such instances, long-term loading effects may assume additional importance.

Types of masonry include:

- *Clay brick or tile masonry.* Clay material has historically been used in multi-wythe load-bearing applications, but is currently used exclusively as a veneer material backed by a wood, steel or concrete masonry supporting structure. Clay tile masonry has been used both for load-bearing structures and for walls infilling a concrete or steel frame. Its use is very rare in contemporary building practice.
- *Concrete masonry.* The use of this material is a relatively recent development, but reinforced or unreinforced concrete masonry accounts for the large majority of contemporary masonry construction.
- *Stone masonry.* This material appears in a wide variety of load-bearing applications, including horizontally spanning structures, such as arch bridges, and arches, domes and vaults in buildings. Although its use is very rare in contemporary construction, a large number of stone load-bearing masonry structures qualify for historic preservation treatment and may be in need of strengthening.

Through the first half of the twentieth century, large numbers of unreinforced masonry buildings have been built in regions susceptible to earthquakes. Whereas unreinforced masonry is very susceptible to failure in an earthquake, the strengthening of these structures is a major challenge to the engineering profession, especially in cases where the existing structure is required to remain intact due to historic preservation considerations.

11.5.2 Durability issues for masonry

Where FRP is used in exterior wall applications, the reinforcement material is subjected to direct environmental exposure, including driving rain, freeze/thaw action of moisture in the wall and exposure to direct solar radiation. The masonry substrate has a wide variability in its susceptibility to moisture and freezing effects, and the influence of moisture transport on both the masonry units and the mortar must be considered. Generally, the susceptibility of masonry materials depends on unit types.

11.5.3 Brick masonry units

The susceptibility of clay brick to moisture absorption and freeze/thaw action is determined on the basis of a combination of compressive strength and absorption characteristics. The total moisture absorbed during a 24 hour immersion in cold water is determined, followed by the moisture absorbed during a five hour immersion in boiling water. The ratio of cold water absorption to boiling water absorption, known as the saturation coefficient or C/B ratio, is determined. A low value is hypothesized to indicate the presence of small pores available for relief of expansion pressures, and is a partial indication of freeze/thaw resistance. The procedures for sampling and testing clay unit masonry, which need to be modified slightly for sampling of *in situ* brick masonry, are described in ASTM C67 (ASTM, 2007b).

11.5.4 Concrete masonry units

Concrete masonry is generally made of lightweight aggregate concrete. The susceptibility of concrete masonry units to freeze/thaw action is determined on the basis of a combination of unit compressive strength (higher compressive strength is roughly correlated to higher tensile strength) and moisture absorption. These values can be determined by testing on whole concrete masonry units in accordance with ASTM C140 (ASTM, 2007c). However, no specific correlation between moisture absorption, compressive strength and freeze/thaw susceptibility has been established for concrete unit masonry.

11.5.5 Stone masonry

Stone masonry appears in a very wide variety of stone types, ranging from very hard granite materials to very soft sandstones. The durability of these naturally-occurring materials depends on a wide variety of considerations, including the stratification, the orientation of the material in construction relative to natural bedding, the binding of material within the stone matrix, the porosity of the stone, etc. The absorption of stone material is determined by ASTM C97 (ASTM, 2002). Many types of stone, particularly sedimentary varieties such as sandstone or limestone, have limited resistance to freeze/thaw action, and may be severely degraded where exposed to the atmosphere. The application of bonded material to the exterior surface of degraded or weathered stone can only hasten the process of degradation, and should be avoided.

11.5.6 Masonry assemblies

In general, a masonry wall assembly must be considered as always having a pathway for moisture ingress, and the design of these assemblies takes this into account by including a way to remove moisture from within the wall assembly. The typical pathways are through the units themselves, if porous enough, as vapor transmitted from the interior of the building, which condenses within the units, through cracks in the mortar, through penetration into the top of the wall or through capillary action from the bottom of the wall.

Masonry walls are designed for and subjected to either in-plane or out-of-plane forces. Typically, exterior walls, are designed to resist bending and shear due to out-of-plane action of wind or earthquake motion. The design loading in such cases results from the action of either the wind or inertial forces on the particular wall element only. Other masonry walls are called on to function as shear walls in the lateral force-resisting system of an entire building, and may collect load from a number of levels of the building. These walls are subjected primarily to in-plane forces.

Use of column wrapping for confinement is a very large application area for concrete that has not yet been deeply explored for masonry. It can be expected that, following Krevaikas and Triantafillou (2005) and Corradi *et al.* (2007), this application will grow in the coming years.

11.5.7 Specific characteristics of bond between FRP and masonry

Masonry in general has a point of weakness in the bond between mortar and units. Although this joint is effective in transferring compression, its tensile capacity is, in most cases, well below that of either the units or the mortar. As a result of this effect, FRP reinforced masonry in out-of-plane bending cracks instantly at all of the mortar joints. This type of cracking has been reported by every author reviewed in this chapter. Mortar joints in shear walls are susceptible to cracking in a stair-step pattern, especially when the accompanying axial force is relatively low.

Mortar joints between units in masonry are generally concave. The joints require filling with either resin or other material prior to application of FRP strengthening material. The result is that either there is a variation in the stiffness of the substrate material of the FRP sheets along the length, or the sheets are rippled at each of the mortar joints.

11.5.8 Durability considerations for externally bonded masonry–FRP hybrid systems

At this point, the published literature provides a significant body of information on strength, failure modes and analytical models for unconditioned FRP composite systems for masonry rehabilitation. However, the literature appears to be silent on the topic of long-term durability studies of FRP–masonry systems. The authors' own thoughts on this issue are that the following are the most important considerations, and that the considerations can be grouped generally into issues for FRP applied to the exterior face of a masonry wall and those for FRP applied to the interior face of a masonry wall.

- Exterior application:
  - exposure of FRP material to the exterior environment;
  - failure or weakening of the bond between FRP and masonry due to freeze/thaw action;
  - weakening of the bond between FRP and masonry due to moisture penetration into wall;
  - weakening of the masonry substrate due to entrapment of moisture, or freeze/thaw action on entrapped moisture within the masonry substrate;
  - propagation of delaminations from flexural cracks at the mortar joints.
- Interior application:
  - weakening of the bond between FRP and masonry due to moisture penetration into wall;
  - effect of long-term compressive creep strains in the masonry;
  - propagation of delaminations from flexural cracks at the mortar joints.

In some cases, tentative inferences can be made as to the probable long-term durability effects of FRP–masonry systems based on the failure modes observed in FRP–masonry testing and information in the FRP–concrete literature. In the recent articles reviewed related to out-of-plane bending, the following failure modes have been observed:

- *Rupture of the FRP reinforcement.* This failure mode was observed in concrete masonry test specimens very lightly (reinforcement ratio 0.0015) reinforced with CFRP strap in tests conducted by Albert *et al.* (2001). This failure mode was also observed in two of six similar concrete masonry specimens reinforced with GFRP (reinforcement ratio 0.00040) tested by Hamilton and Dolan (2001).
- *Compression failure of the masonry.* This failure mode was observed in four laboratory specimens tested by Triantafillou (1998). A field application by Tumialan *et al.* (2003) resulted in this failure mode in infill walls influenced by confinement of the concrete frame around the wall perimeter.
- *Debonding failure of the reinforcement.* This is the most commonly observed failure mode. It was observed exclusively in tests by Roko *et al.* (2001) of CFRP reinforced brick prisms in bending. It was observed

in the tests of Kuzik *et al.* (2003). It was also the primary failure mode observed by Hamilton and Dolan (2001) and Tan and Patoary (2004). This failure mode has also been investigated for stone substrate materials by Accardi *et al.* (2004) to develop shear stress–slip curves for two types of calcarenite.

- *Shear failure of the masonry.* Various types of shear failure were also widely observed in the testing programs reported. Hamilton and Dolan (2001), Kuzik *et al.* (2003) and Hamoush *et al.* (2001) have noted the close connection between shear failure modes and debonding, providing various explanations of how a shear failure may initiate at a location where flexural cracks and debonding are present.

Although none of the above studies refers specifically to durability effects of FRP–masonry systems, information is available from other sources to allow the estimation of the influence of long-term environmental effects on the effectiveness of masonry systems reinforced to resist out-of-plane forces. The long-term effects of the environment on the FRP rupture failure mode can be estimated based on the material-specific and environment-specific estimates of long-term tensile strength capacity for the FRP material. In most masonry applications, the effect of long-term loading is negligible, but environmental exposure may result in a reduction in fiber strength of 15–50%, depending on the type of fibers, as explained in the earlier section on durability of FRP materials. By comparison, long-term changes in the crushing capacity of the masonry are smaller, except in cases where an incorrect weathering grade of brick or concrete masonry units has been applied. The compressive strength of saturated masonry may have some reduction in strength, compared to dry masonry, but this is not expected to be a significant factor in the durability of FRP–masonry systems.

On the other hand, the delamination failure mode and the shear failure mode, which typically follows delamination, are dependent on factors specific to the bond between masonry and FRP, which have been investigated in very few cases. From the available investigations, it can be observed that the bond strengths, effective bond lengths and slip at failure are similar to those observed in concrete systems – that is, a bond strength of approximately 5–10 MPa, an effective bond length of 5–10 cm and a slip at failure of 1–2 mm. No specific inferences concerning degradation of these properties have been made, although the absence of long-term loading would appear to be a mitigating factor. The bond is also influenced by the length of the masonry units in the fiber direction, which can be taken as an effective limit on the bonded length, due to the universal appearance of flexural cracks at each mortar joint in every test reported.

11.5.9 Failures of masonry walls reinforced in for in-plane shear and bending moments

The following recommendations can be made to ensure the long-term durability of externally bonded masonry structures subjected to out-of-plane or in-plane loading:

- Following ACI 440.2R-02, Section 9.2 (ACI, 2004) the strain in the reinforcement should be limited in order to mitigate the tensile strain in the substrate.
- A bonded length of no more than the length of a masonry unit should be used in order to evaluate the bond strength in flexure.
- The reinforcement material should be located at the interior surface of the wall where possible and exterior applications should be protected against direct sunlight.
- The presence of sustained loading should be carefully evaluated for each individual application; however, many masonry strengthening applications, particularly in-plane strengthening for seismic resistance, are not subject to sustained loading. In the absence of sustained loading, and given the weakness in tension of the masonry substrate, glass-based reinforcement materials are appropriate for reinforcement, provided the above guidelines can be observed.

11.6 Externally bonded steel–FRP hybrid systems

11.6.1 Applications

Much experience on the use of FRP materials to patch cracks and add strength in metallic structures (primarily aluminium) has been developed since the 1970s by the aircraft industry (Baker and Jones, 1988). For patching metallic substrates, carbon and boron fibers are generally favored over glass and aramid fibers because of the need to maximize the modulus of elasticity of the patch material. Recognizing the success of metal–FRP bonded systems in aerospace and the need for durable repair and upgrade methods for steel structures in infrastructure, researchers began to investigate externally bonded carbon/epoxy composites as supplementary tensile reinforcement for steel structures in the 1990s (Hollaway and Cadei, 2002; Zhao and Zhang, 2007). Guidance on design, implementation and maintenance is now available (Cadei *et al.*, 2004). In infrastructure applications, FRP reinforcements have been used to prevent or slow the growth of fatigue cracks in steel (Jones and Civjan, 2003) and to increase the post-yield stiffness and ultimate strength of steel beams rigidly connected at the compression flange to concrete slabs (Sen *et al.*, 2001; Tavakkolizadeh and Saadatmanesh, 2003). Thus far, experience in civil infrastructure has been primarily with standard modulus (230 GPa) carbon fibers, although high modulus (400–600 GPa) carbon fibers have recently attracted interest for augmenting the flexural strength of steel bridges (Schnerch *et al.*, 2007).

The attachment of external FRP reinforcements to metallic substrates is relatively simple in comparison to masonry and concrete substrates because of the substantial tensile strength of metals. Bolting and riveting can be viable methods of attachment of FRP materials if the FRP material contains multidirectional fiber reinforcement for added resistance to bearing and shear-out failures. To date, however, most interest has been in undirectionally reinforced composites that provide the highest modulus of elasticity in the loading direction with the least amount of material. Adhesive bonding is therefore the most relevant means of attachment of FRP material to metal substrates.

In keeping with the majority of civil infrastructure applications demonstrated to date, the following discussion focuses on carbon/epoxy composites reinforcements adhesively bonded to steel substrates. A current lack of published information on this particular hybrid material system, referred to as steel–CFRP in the remainder of this discussion, prevents a detailed exposition of durability issues, although inferences can be made based on analogies with bonded metal–FRP joints used in other industries.

11.6.2 Durability issues for metallic substrates

Corrosion and fatigue cracking are the archetypal types of degradation in steel structures under consideration for the use of external FRP reinforcement. Corrosion readily occurs when chloride-containing solutions such as de-icing agents or seawater come into contact with unprotected steel in the presence of oxygen (Bentur, 1997). Coatings provide some measure of environmental protection against corrosion, but must be inspected and repaired at sufficiently frequent intervals for full effectiveness. Fatigue cracking occurs near stress concentrations such as bolt holes and welds and can result in very sudden structural failures at surprisingly small lengths (Dowling, 1999). Timely inspections and repairs are recommended to prevent fatigue-related problems in steel structural members.

11.6.3 Specific characteristics of bond between FRP and metals

The amount of strengthening that can be accomplished by bonding an FRP composite to steel is limited by the tendency of the FRP material to debond at the termination (cut-off) points of the CFRP or at flaws in the bond line. Plausible measures to reduce the substantial shear and peeling stresses existing near the cut-offs include gradually reducing the thickness of the composite towards the cut-off (beveling, as was done in Schnerch and Rizkalla (2004), is one version of this idea) or applying compression through the thickness of the adhesive joint with a clamp-type anchorage placed near the cut-off (Nozaka *et al.*, 2005). Deviations in the flatness of the steel should be minimized with metal grinding and filling techniques. Typical transfer lengths (effective bond lengths) of CFRP reinforcements bonded to steel are in the range of 50–200 mm (Schnerch and Rizkalla, 2004; Nozaka *et al.*, 2005). All remnants of coatings and corrosion must be removed from steel by sand-blasting or wire-brushing prior to installation of FRP reinforcement.

11.6.4 Durability considerations for externally bonded steel–FRP hybrid systems

The main durability issues for externally bonded steel–CFRP hybrid systems concern fatigue, fire resistance, weather-induced degradation and galvanic corrosion between carbon and steel. As the durability issues pertaining to stand-alone FRP and steel have been covered earlier, only those issues pertaining to the hybrid system are discussed next.

Early research by Tucker and Brown (1989) revealed the potential for two specific durability problems when pre-cured CFRP materials are subject to a room-temperature seawater environment – namely, osmotic blistering and galvanic corrosion. Blistering occurred only in a poorly consolidated hybrid vinylester composite containing carbon and glass fibers, was associated more with the glass fibers than the carbon fibers and occurred only when the CFRP was in contact with steel. Blistering was not visibly evident in a well-consolidated carbon/epoxy composite tested under the same conditions. Galvanic corrosion of the steel occurred when placed in contact with either of the CFRP material systems in seawater. Separating the galvanically dissimilar steel and CFRP materials with a thin layer of electrically insulating polymer prevented corrosion. Tavakkolizadeh and Saadatmanesh (2001) observed that thicker layers of isolating resin between steel and carbon fibers reduced the rate of galvanic corrosion. The best means of ensuring that there is no electrical connection between CFRP and steel is to place a non-conductive layer of fiber, such as E-glass, between the carbon and steel.

The strength of steel–epoxy bonds is inversely related to the amount of moisture absorbed by the epoxy (Nguyen and Martin, 2004). A loss of bond strength attributed to moisture absorption is reversible if the moisture is removed from the epoxy, however. A layer of CFRP applied to steel greatly reduces the rate of diffusion of moisture and oxygen to the bonded surface of steel, although some amount of corrosion of the steel can be expected to occur over time. Voids at the steel/epoxy interface should be avoided as they promote cavity corrosion and blistering. The ASTM D3762 wedge test (ASTM, 2003c) is widely used to evaluate adhesion durability in metals and has been applied to the steel–FRP bond as well (Karbhari and Shulley, 1995).

For patching existing cracks in steel, prestressing the CFRP material and increasing the stiffness (i.e. modulus multiplied by thickness) of the CFRP patch can reduce the rate of growth of the steel crack and also reduce the likelihood of delamination of the patch (Colombi *et al.*, 2003). Moisture-induced reductions in the moduli of the resin and adhesive of the CFRP system reduce the effectiveness of the patch (Bouiadjra *et al.*, 2003). Field conditions permitting, applying CFRP patches to both surfaces (e.g. two-sided) of a steel substrate reduces induced out-of-plane plane bending and is therefore more effective than single-sided patching for reducing fatigue crack growth, particularly for thin-walled steel members (Duong and Wang, 2004).

11.7 Conclusion

There are indeed numerous reported instances of poor-performing FRP materials externally bonded to various substrates. However, it is the authors' opinion that, in many of these cases, where enough information is provided for a meaningful assessment of the experimental methodology, one could argue that the testing conditions were excessively harsh (e.g. imposed temperature close to or exceeding the T_g of the resin; continuous bath of highly acidic or alkaline solution), an improper material system was selected for the environment (e.g. low grade polyester in a moist environment; concrete with no entrained air used in freeze/thaw environments) or there was some type of deficiency in the preparation of the specimens (e.g. lack of air entrainment in concrete subjected to freeze/thaw; lack of full cure of the FRP material/adhesive system upon exposure to moisture). This opinion is not intended to cast aside the possibility that some of these very same deficiencies exist in current field applications of FRP materials, but it does prevent the authors from making blanket statements about how much degradation can be expected in a particular application and environment. As with conventional materials, the performance of FRP material systems depends on proper design, material selection and workmanship.

11.8 References

ACCARDI M, LA MENDOLA L and ZINGONE G (2004) CFRP sheets bonded to natural stone: interfacial phenomena, in Brebbia CA and Varvani-Farahani A (eds), *Damage and Fracture Mechanics VIII*, Southampton, UK, WIT Press, 73–184.

ACI (2002) *Guide for the Design and Construction of Externally Bonded FRP Systems for Strengthening Concrete Structures*, ACI 440.2R-02, Farmington Hills, MI, USA, American Concrete Institute.

ACI (2004) *Guide Test Methods FRPs for Reinforcing or Strengthening Concrete Structures*, ACI 440.3R-04, Farmington Hills, MI, USA, American Concrete Institute.

ALBERT M L, ELWI A E and CHENG J J R (2001) Strengthening of unreinforced masonry walls using FRPs, *Journal of Composites for Construction*, ASCE, **5**, 76–84.

ALMUSALLAM T H and AL-SALLOUM Y A (2007) Behavior of FRP strengthened infill walls under in-plane seismic loading, *Journal of Composites for Construction*, ASCE, **11**, 308–318.

ASTM (2002) *ASTM C97-02 Standard Test Methods for Absorption and Bulk Specific Gravity of Dimension Stone*, West Conshohocken, PA, USA, American Society for Testing and Materials.

ASTM (2003a) *ASTM C666/C666M-03 Standard Test Method for Resistance of Concrete to Rapid Freezing and Thawing*, West Conshohocken, PA, USA, American Society for Testing and Materials.

ASTM (2003b) *ASTM D905-03 Standard Test Method for Strength Properties of Adhesive Bonds in Shear by Compression Loading*, West Conshohocken, PA, USA, American Society for Testing and Materials.

ASTM (2003c) *ASTM D3762-03 Standard Test Method for Adhesive-Bonded Surface Durability of Aluminum (Wedge Test)*, West Conshohocken, PA, USA, American Society for Testing and Materials.

ASTM (2003d) *ASTM E1356-03 Standard Test Method for Assignment of the Glass Transition Temperatures by Differential Scanning Calorimetry*, West Conshohocken, PA, USA, American Society for Testing and Materials.

ASTM (2004) *ASTM E1640-04 Standard Test Method for Assignment of the Glass Transition Temperature By Dynamic Mechanical Analysis*, West Conshohocken, PA, USA, American Society for Testing and Materials.

ASTM (2005a) *ASTM D1435-05 Standard Practice for Outdoor Weathering of Plastics*, West Conshohocken, PA, USA, American Society for Testing and Materials.

ASTM (2005b) *ASTM E1545-05 Standard Test Method for Assignment of the Glass Transition Temperature by Thermomechanical Analysis*, West Conshohocken, PA, USA, American Society for Testing and Materials.

ASTM (2006a) *ASTM D1101-97a(2006) Standard Test Methods for Integrity of Adhesive Joints in Structural Laminated Wood Products for Exterior Use*, West Conshohocken, PA, USA, American Society for Testing and Materials.

ASTM (2006b) *ASTM D1193-06 Standard Specification for Reagent Water*, West Conshohocken, PA, USA, American Society for Testing and Materials.

ASTM (2007a) *ASTM Fire Standards*, 7th edn, West Conshohocken, PA, USA, American Society for Testing and Materials.

ASTM (2007b) *ASTM C67-07a Standard Test Methods for Sampling and Testing Brick and Structural Clay Tile*, West Conshohocken, PA, USA, American Society for Testing and Materials.

ASTM (2007c) *ASTM C140-07a Standard Test Methods for Sampling and Testing Concrete Masonry Units and Related Units*, West Conshohocken, PA, USA, American Society for Testing and Materials.

BADAWI M and SOUDKI K (2005) Control of corrosion-induced damage in reinforced concrete beams using carbon fiber-reinforced polymer laminates, *Journal of Composites for Construction*, ASCE, **9**, 195–201.

BAKER A A and JONES R (eds) (1998) *Bonded Repair of Aircraft Structures*, Dordrecht, The Netherlands, Martinus-Nijhoff Publishers.

BANK L C, RUSSELL G T, THOMPSON B P and RUSSELL J S (2003) A model specification for FRP composites for civil engineering structures, *Construction and Building Materials*, **17**, 405–437.

BARNES R A and MAYS G C (1999) Fatigue performance of concrete beams strengthened with CFRP plates, *Journal of Composites for Construction*, ASCE, **3**, 63–72.

BATI S B, ROVERO L and TONIETTI U (2007) Strengthening masonry arches with composite materials, *Journal of Composites for Construction*, ASCE, **11**, 33–41.

BATTLES E P, DAGHER H J and ABDEL-MAGID B (2000) Durability of composite reinforcement for timber bridges, *Journal of Transportation Research Board, Transportation Research Record 1696*, **2**, 131–135.

BENTUR A (1997) *Steel Corrosion in Concrete: Fundamentals and Civil Engineering Practice*, London, UK, E & FN Spon.

BISBY L A and GREEN M F (2002) Resistance to freezing and thawing of fiber-reinforced polymer-concrete bond, *ACI Structural Journal*, **99**, 215–223.

BOUIADJRA B B, MEGUENI A, TOUNSI A and SERIER B (2003) The effect of a bonded hygrothermal aged composite patch on the stress intensity factor for repairing cracked metallic structures, *Composite Structures*, **62**, 171–176.

BRENA S F, BENOUAICH M A, KREGER M E and WOOD S L (2005) Fatigue tests of reinforced concrete beams strengthened using carbon fiber-reinforced polymer composites, *ACI Structural Journal*, **102**, 305–313.

BUENFELD N R, GLASS G K, HASSANEIN A M and ZHANG J-Z (1998) Chloride transport in concrete subjected to electric field, *Journal of Materials in Civil Engineering*, **10**, 220–226.

CADEI J M C, STRATFORD T J, HOLLAWAY L C and DUCKETT W H (2004) *Strengthening metallic structures using externally bonded fibre-reinforced composites*, Report C595, London, UK, CIRIA.

CHIN J W, NGUYEN T and AOUADI K (1999) Sorption and diffusion of water, salt water, and concrete pore solution in composite matrices, *Journal of Applied Polymer Science*, **71**, 483–492.

COLOMBI P, BASSETTI A and NUSSBAUMER A (2003) Crack growth induced delamination on steel members reinforced by prestressed composite patch, *Fatigue and Fracture of Engineering Materials and Structures*, **26**, 429–437.

CORRADI M, GRAZINI A and BORRI A (2007) Confinement of brick masonry columns with CFRP materials, *Composites Science and Technology*, **67**, 1772–1783.

DAGHER H J, IQBAL A and BOGNER B (2004) Durability of isophthalic polyester composites used in civil engineering applications, *Polymers and Polymer Composites*, **12**, 169–182.

DAVALOS J, QIAO P and TRIMBLE B S (2000a) Fiber-reinforced composites and wood bonded interfaces: Part I. durability and shear strength, *Journal of Composites Technology and Research*, **22**, 224–231.

DAVALOS J F, QIAO P and TRIMBLE B S (2000b) Fiber-reinforced composite and wood bonded interfaces: Part 2. Fracture, *Journal of Composites Technology and Research*, **22**, 232–240.

DEBAIKY A, GREEN M F and HOPE B B (2002) CFRP wraps for corrosion control and rehabilitation of reinforced concrete columns, *ACI Materials Journal*, **99**, 129–137.

DOWLING N F (1999) *Mechanical Behavior of Materials: Engineering Methods for Deformation, Fracture, and Fatigue*, Upper Saddle River, NJ, USA, Prentice Hall.

DUONG C N and WANG C H (2004) On the characterization of fatigue crack growth in a plate with a single-sided repair, *Journal Engineering Materials and Technology, Transactions of the ASME*, **126**, 192–198.

EL-HACHA R, GREEN M F and WIGHT R G (2004) Flexural behaviour of concrete beams strengthened with prestressed carbon fibre reinforced polymer sheets subjected to sustained loading and low temperature, *Canadian Journal Civil Engineering*, **31**, 239–252.

FIB (2001) *Externally Bonded FRP Reinforcement for RC Structures*, Task Group 9.3, Bulletin No. 14, Lausanne, Switzerland, Federation Internationale du Beton.

GRACE N F and SINGH S B (2005) Durability evaluation of carbon fiber-reinforced polymer strengthened concrete beams: experimental study and design, *ACI Structural Journal*, **102**, 40–53.

GREEN M F, BISBY L A, BEAUDOIN Y and LABOSSIÈRE P (2000) Effect of freeze-thaw cycles on the bond durability between fibre reinforced polymer plate reinforcement and concrete, *Canadian Journal of Civil Engineering*, **27**, 949–959.

GREEN M F, DENT A J S and BISBY L A (2003) Effect of freeze-thaw cycling on the behaviour of reinforced concrete beams strengthened in flexure with fibre reinforced polymer sheets, *Canadian Journal of Civil Engineering*, **30**, 1081–1088.

GREMEL D, GALATI N and STULL J (2005) Method for Screening Durability and Constituent Materials in FRP Bars, in Shield C K, Busel J P, Walkup S L and Gremel D D (eds), *Proceedings Fibre Reinforced Polymer Reinforcement for Concrete Structures – FRPRCS-7*, Farmington Hills, MI, USA, American Concrete Institute, 153–164.

HAMILTON H R and DOLAN C W (2001) Flexural capacity of glass FRP strengthened concrete masonry walls, *Journal of Composites for Construction*, ASCE, **5**, 170–178.

HAMOUSH S A, MCGINLEY M W, MLAKAR P, SCOTT D and MURRAY K (2001) Out-of plane strengthening of masonry walls with reinforced composites, *Journal of Composites for Construction*, ASCE, **5**, 139–145.

HOLLAWAY L C and CADEI J (2002) Progress in the technique of upgrading metallic structures with advanced polymer composites, *Progress in Structural Engineering and Materials*, **4**, 131–148.

HORNSBY P R (2001) Fire retardant fillers for polymers, *International Materials Reviews*, **46**, 199–210.
INOUE S, NISHIBAYASHI S, YOSHINO A and OMATA F (1994) Strength and deformation characteristics of reinforced concrete beam strengthened with carbon fiber-reinforced plastics plate under static and fatigue loading, *Zairyo/Journal Society of Materials Science, Japan*, **43**, 1004–1009.
ISIS CANADA (2001) *Strengthening Reinforced Concrete Structures with Externally-Bonded Fibre Reinforced Polymers (FRPs)*, Manual No. 4, The Canadian Network of Centres of Excellence on Intelligent Sensing for Innovative Structures, Winnipeg, Manitoba, Canada, ISIS Canada Corporation.
JIA J, BOOTHBY T E, BAKIS C E and BROWN T L (2005) Durability evaluation of GFRP-concrete bonded interfaces, *Journal of Composites for Construction*, ASCE, **9**, 348–359.
JONES S C and CIVJAN S A (2003) Application of fiber reinforced polymer overlays to extend steel fatigue life, *Journal of Composites for Construction*, ASCE, **7**, 331–338.
JSCE (2001) *Recommendations for Upgrading of Concrete Structures with Use of Continuous Fiber Sheets*, Concrete Engineering Series 41, Tokyo, Japan, Japan Society of Civil Engineers.
KARBHARI V M (2002) Response of fiber reinforced polymer confined concrete exposed to freeze and freeze-thaw regimes, *Journal of Composites for Construction*, ASCE, **6**, 35–40.
KARBHARI V M and SHULLEY S B (1995) Use of composites for rehabilitation of steel structures – determination of bond durability, *Journal of Materials in Civil Engineering*, **7**, 239–245.
KARBHARI V M, CHIN J W, HUNSTON D, BENMOKRANE B, JUSKA T, MORGAN R, LESKO J J, SORATHIA U and REYNAUD D (2003) Durability gap analysis for fiber-reinforced polymer composites in civil infrastructure, *Journal of Composites for Construction*, ASCE, **7**, 238–247.
KATZ H S and MILEWSKI J V (1987) *Handbook of Fillers for Plastics*, New York, USA, Van Nostrand Reinhold Co.
KOOTSOOKOS A and MOURITZ A P (2004) Seawater durability of glass- and carbon-polymer composites, *Composites Science and Technology*, **64**, 1503–1511.
KREVAIKAS T D and TRIANTAFILLOU T C (2005) Masonry confinement with fiber-reinforced polymers, *Journal of Composites for Construction*, ASCE, **9**, 128–135.
KUZIK M D, ELWI A E and CHENG J J R (2003) Cyclic flexure tests of masonry walls reinforced with glass fiber reinforced polymer sheets, *Journal of Composites for Construction*, ASCE, **7**, 20–30.
LE BRAS M, CAMINO G, BOUBIGOT S and DELOBEL R (eds) (1998) *Fire Retardancy of Polymers: The Use of Intumescence*, Cambridge, UK, The Royal Society of Chemistry.
LIAO K, SCHULTHEISZ C R and HUNSTON D L (1999) Effects of environmental aging on the properties of pultruded GFRP, *Composites Part B: Engineering*, **30**, 485–493.
LITHERLAND K L, OAKLEY D R and PROCTOR B A (1981) The use of accelerated ageing procedures to predict the long term strength of GRC composites, *Cement and Concrete Research*, **11**, 455–466.

LOPEZ M, NAAMAN A E, TILL R D and PINKERTON L (2003) Behavior of RC beams strengthened with FRP laminates and ested under cyclic loading at low temperatures, *International Journal of Materials and Product Technology*, **19**, 108–117.

LOPEZ-ANIDO R and XU H (2002) Repair of wood piles using prefabricated fiber-reinforced polymer composite shells, *Journal of Composites for Construction*, ASCE, **6**, 194–203.

LOPEZ-ANIDO R, GARDNER D J and HENSLEY J L (2000) Adhesive bonding of eastern hemlock glulam panels with E-glass/vinylester reinforcement, *Forest Products Journal*, **50**, 43–47.

LOPEZ-ANIDO R, MICHAEL A P and SANDFORD T C (2004a) Fiber reinforced polymer composite-wood pile interface characterization by push-out tests, *Journal of Composites for Construction*, ASCE, **8**, 360–368.

LOPEZ-ANIDO R, MICHAEL A P and SANDFORD T C (2004b) Freeze-thaw resistance of fiber-reinforced polymer composites adhesive bonds with underwater curing epoxy, *Journal of Materials in Civil Engineering*, **16**, 283–286.

LOURENÇO P (2004) Analysis and restoration of ancient masonry structures. Guidelines and examples, in La Tegola A and Nanni A (eds), *Proceedings Innovative Materials and Technologies for Construction and Restoration – IMTCR'04*, Naples, Italy, Liguori Editore, 23–41.

LYONS J S and AHMED M R (2005) Factors affecting the bond between polymer composites and wood, *Journal of Reinforced Plastics and Composites*, **24**, 405–412.

MAILVAGANAM N P and WISEMAN A (2003) Axioms for building and repairing durable concrete structures, *Journal Performance Constructed Facilities*, **17**, 163–166.

MIRMIRAN A, SHAHAWY M, NANNI A and KARBHARI V (2004) *Bonded Repair and Retrofit of Concrete Structures Using FRP Composites. Recommended Construction Specifications and Process Control Manual*, NCHRP Report 514, Washington, DC, USA, Transportation Research Board.

MOSALLAM A, KREINER J and GILLETE K (2000) Structural upgrade and repair of wood members using cross-ply carbon/epoxy, in *Proceedings 45th International SAMPE Symposium and Exhibition*, Covina, CA, USA, Society for the Advancement of Material and Process Engineering, 861–869.

MUKHOPADHYAYA P, SWAMY R N and LYNSDALE C J (1998) Influence of aggressive exposure conditions on the behaviour of adhesive bonded concrete–GFRP joints, *Construction and Building Materials*, **12**, 427–446.

NEVILLE A M (1996) *Properties of Concrete*, 4th edn, New York, USA, John Wiley and Sons.

NGUYEN T and MARTIN J W (2004) Modes and mechanisms for the degradation of fusion-bonded epoxy-coated steel in a marine concrete environment, *Journal of Coatings Technology Research*, **1**, 81–92.

NOZAKA K, SHIELD C K and HAJJAR J F (2005) Effective bond length of carbon-fiber-reinforced polymer strips bonded to fatigued steel bridge I-girders, *Journal of Bridge Engineering*, **10**, 195–205.

PANTAZOPOULOU S J, BONACCI J F, SHEIKH S, THOMAS M D A and HEARN N (2001) Repair of corrosion-damaged columns with FRP wraps, *Journal of Composites for Construction*, ASCE, **5**, 3–11.

RIZKALLA S and NANNI A (eds) (2003) *Field Applications of FRP Reinforcement: Case Studies*, SP-215, Farmington Hills, MI, USA, American Concrete Institute.

ROKO K E, BOOTHBY T E and BAKIS C E (2001) Strain transfer analysis of masonry prisms reinforced with bonded carbon fiber reinforced polymer sheets, *The Masonry Society Journal*, **19**, 57–68.

SCALZI J B, PODOLNY W, MUNLEY E and TANG B (1999) Importance of documenting FRP composite data, *Journal of Composites for Construction*, ASCE, **3**, 107.

SCHNERCH D and RIZKALLA S (2004) Behavior of scaled steel-concrete composite girders and steel monopole towers strengthened with CFRP, in La Tegola A and Nanni A (eds), *Proceedings Innovative Materials and Technologies for Construction and Restoration – IMTCR'04*, Naples, Italy, Liguori Editore, 42–65.

SCHNERCH D, DAWOOD M, RIZKALLA S and SUMNER E (2007) Proposed design guidelines for strengthening of steel bridges with FRP materials, *Construction and Building Materials*, **21**, 1001–1010.

SCHUTTE C L (1994) Environmental durability of glass-fiber composites, *Materials Science and Engineering: R: Reports*, **13**, 265–324.

SEN R, LIBY L and MULLINS G (2001) Strengthening steel bridge sections using CFRP laminates, *Composites Part B: Engineering*, **32**, 309–322.

TAN K H and PATOARY M K H (2004) Strengthening of masonry walls against out-of-plane loads using fiber-reinforced polymer reinforcement, *Journal of Composites for Construction*, ASCE, **8**, 79–87.

TAVAKKOLIZADEH M and SAADATMANESH H (2001) Galvanic corrosion of carbon and steel in aggressive environments, *Journal of Composites for Construction*, ASCE, **5**, 200–210.

TAVAKKOLIZADEH M and SAADATMANESH H (2003) Strengthening of steel–concrete composite girders using carbon fiber reinforced polymers sheets, *Journal of Structural Engineering*, **129**, 30–40.

TENG J G, SMITH S T, YAO J and CHEN J F (2003) Intermediate crack-induced debonding in RC beams and slabs, *Construction and Building Materials*, **17**, 447–462.

THAESLER P, KAHN L, OBERLE R and DEMERS C E (2005) Durable repairs on marine bridge piles, *Journal of Performance of Constructed Facilities*, **19**, 88–92.

TORRES-ACOSTA A A (2002) Galvanic corrosion of steel in contact with carbon-polymer composites. I: Experiments in mortar, *Journal of Composites for Construction*, ASCE, **6**, 112–115.

TOUTANJI H A (1999) Durability characteristics of concrete columns confined with advanced composite materials, *Composite Structures*, **44**, 155–161.

TOUTANJI H and BALAGURU P (1998) Durability characteristics of concrete columns wrapped with FRP tow sheets, *Journal of Materials in Civil Engineering*, **10**, 52–57.

TOUTANJI H and DENG Y (2002) Strength and durability performance of concrete axially loaded members confined with AFRP composite sheets, *Composites Part B: Engineering*, **33**, 255–261.

TRIANTAFILLOU T C (1998) Strengthening of masonry structures using epoxy-bonded FRP laminates, *Journal of Composites for Construction*, ASCE, **5**, 96–104.

TRIANTAFILLOU T C (ed.) (2007) *Proceedings Fibre Reinforced Polymer Reinforcement for Concrete Structures–FRPRCS-8*, University of Patras, Patras, Greece, July 16–18 (CD-ROM).

TUCKER W C and BROWN R (1989) Blister formation on graphite/polymer composites galvanically coupled with steel in seawater, *Journal of Composite Materials*, **23**, 389–395.

TUMIALAN J G, GALATI N and NANNI A (2003) Field assessment of unreinforced masonry walls strengthened with fiber reinforced polymer laminates, *Journal of Structural Engineering*, **129**, 1047–1056.

VALLUZZI M R, VALDEMARCA M and MODENA C (2001) Behavior of brick masonry vaults strengthened by FRP laminates, *Journal of Composites for Construction*, ASCE, **5**, 163–169.

WANG J Z, DILLARD D A and WARD T C (1992) Temperature and stress effects in the creep of aramid fibers under transient moisture conditions and discussions on the mechanisms, *Journal of Polymer Science, Part B: Polymer Physics*, **30**, 1391–1400.

WOOTTON I A, SPAINHOUR L K and YAZDANI N (2003) Corrosion of steel reinforcement in carbon fiber-reinforced polymer wrapped concrete cylinders, *Journal of Composites for Construction*, ASCE, **7**, 339–347.

WYPYCH G (1999) *Handbook of Fillers*, 2nd edn, Toronto, Ontario, Canada, ChemTec.

ZAIKOV G E and LOMAKIN S M (2002) Ecological issue of polymer flame retardancy, *Journal of Applied Polymer Science*, **86**, 2449–2462.

ZHAO X-L and ZHANG L (2007) State-of-the-art review on FRP strengthened steel structures, *Engineering Structures*, **29**, 1808–1823.

12
Quality assurance/quality control, maintenance and repair

J. L. CLARKE, Concrete Society, UK

12.1 Introduction

In all but a few cases, a fibre reinforced polymer (FRP) strengthening system relies entirely on the ability of the adhesive to transfer loads between the fibre composite and the surface of the structure. The system must perform satisfactorily throughout the intended life of the structure. Thus all the component materials in the FRP system must be correctly specified, to give adequate strength and durability in the anticipated working environment. The quality of the installation of the FRP and, in particular, the preparation of the surface of the parent material (and of the FRP in many cases) will be crucial to the performance in both the short and long term. Thus quality assurance and quality control schemes must be in place, overseeing both the materials and the workmanship. In addition, because FRP strengthening is a relatively new technique, without the long-term experience associated with more traditional materials, routine maintenance regimes should be set up and procedures identified for repairing the FRP should damage or deterioration occur.

Much of this chapter is based on Technical Report 57 (2003) published by the UK Concrete Society. Clearly the Society's guidance was specific to concrete structures, but the principles involved will apply to bonding FRP materials to any substrate, although the details will obviously vary. There is generally less guidance available for materials other than concrete, but similar approaches were adopted in the CIRIA Report C595 (Cadei et al., 2004).

One key message is that it is essential to maintain detailed records of all stages of the FRP strengthening process and of subsequent inspections and any repairs. In many countries there is a growing legal requirement for such records; in the UK the information must be placed in the Health and Safety File, as required under the Construction (Design and Management) Regulations (available from http://www.hse.gov.uk/pubns/conindex.htm).

Strengthening with FRP materials is now being carried out in many different countries. Any materials testing should be carried out in accordance with the appropriate national or international standard. Clearly it is not practical to list all possible national standards within this chapter; reference is therefore made only to British, European and American standards as they are either used internationally or else form the basis for national standards.

12.2 Deterioration and damage

It is clearly important to be aware of the causes of deterioration and the damage that may occur to the FRP in service.

12.2.1 Environmental degradation

The durability of fibre composite systems is covered in Chapter 11 and hence only a brief summary is included here. The long-term integrity of bonded joints requires both chemical and mechanical durability in the presence of varying temperature, moisture and other environmental factors. Externally, these may include spray from de-icing salts or from the sea and, for buried structures, contaminants in groundwater. Adhesive bonded joints with equivalent bond strengths in short-term static tests may differ markedly in durability. All the materials in a strengthening system used in a civil engineering environment may be subjected to a variety of conditions in service. Those normally considered are:

- moisture (humidity, liquid water, salt spray);
- temperature (which may include freezing and thawing);
- chemical attack (alkalis, oil, fuel, chemical spills);
- exposure to ultraviolet light.

Correct specification of the materials used, and any additional protection system applied, should ensure that the performance is adequate. However, this will only be so if the environment does not change significantly during the design life of the structure. It is therefore important that any such changes are identified during routine inspections of the structure (see Section 12.6.2).

12.2.2 Accidental mechanical damage

Impact is the most likely cause of mechanical damage. Vulnerable locations include strengthened soffits of bridges, which are prone to impact from over-height vehicles, and strengthened columns in car parks. Another form of mechanical damage is that caused by subsequent work on the structure,

Quality assurance/quality control, maintenance and repair 325

12.1 Example of warning printed on carbon fibre plate [Courtesy of BASF (formerly Weber SBD)].

Contact Tel: 01XXX XXXXXX

12.2 Examples of proposed warning plates fixed to structure adjacent to strengthened area.

for example drilling through the composite or removal of a protective layer. (Because of stress concentrations around the hole, drilling through the composite causes a disproportionate loss of strength. It is reasonable to assume that drilling a single hole through a laminate reduces its strength by an amount equivalent to a reduction in section of two or three hole diameters.) To avoid such damage, suitable warning signs should be fixed on or alongside the composite, such as that shown in Fig. 12.1. When the composite is covered by a protective or decorative layer, warning notices should be fitted adjacent to the strengthened area. Figure 12.2 shows an example, which could be modified to suit particular applications.

Fire presents a serious risk to structures strengthened with fibre composite materials as the high temperatures rapidly soften the adhesive, and the composite action between the FRP and the structure will be lost. To date only limited experimental work has been carried out on the performance in fire of FRP-strengthened members (Bisby *et al.*, 2004). After a fire, the structure will have to be completely reappraised to determine the effect of the fire both on the main structure and on the FRP. It is likely that all the

FRP in the affected area will have to be removed and replaced with new material.

12.2.3 Anticipated mechanical damage

Sometimes, the FRP will be covered by a surfacing with a limited life, which will be removed at some stage, perhaps damaging the FRP. For example, FRP bonded to the top surface of a bridge deck may be covered by waterproofing and asphalt. After 15–20 years, the asphalt will be removed and there is obviously a high risk that the FRP will be damaged by the planing machine. Similarly, in buildings, the FRP on the top surface of a floor slab may be covered with a screed or by a decorative coating – materials that may be removed when the building is refurbished. In all cases, the owner must be made aware of the consequences of damage to the FRP and the remedial actions to take in the event of damage.

12.3 The need for an inspection regime

12.3.1 Introduction

As with all structural elements, the fibre composite strengthening system will need checking as part of the regular inspection and monitoring regime. Such inspections are already normally carried out for bridges. However, few buildings are regularly checked, inspections often being carried out only when there is a change of use or of ownership. It is recommended that *all* building owners instigate a regular inspection regime for strengthened elements. The structural engineer responsible for designing the strengthening should indicate what is to be done in the event of damage to the composite material. The action taken will be structure-specific: it will depend on the amount and location of damage and the extent to which the structure has been strengthened.

12.3.2 Types of inspection

Inspection should be carried out by a suitably qualified inspector at all stages of the strengthening process, as part of the agreed quality control procedure. As yet there are no universally recognised qualifications for those involved with adhesive bonding. Currently, the suppliers of the various materials provide training, but this will be material-specific. More general schemes, appropriate to all materials, are being developed.

Immediately after FRP installation, a full inspection should be undertaken to determine the 'as installed' condition. This will act as a benchmark for subsequent inspections. Depending on the acceptance criteria, there

Table 12.1 Aspects to be considered at various stages

Stage/type of inspection	Aspect to be considered
Installation	Conformity of strengthening system (fibres, adhesives/resins, etc.)
	Concrete condition
	Surface preparation
	Application of composite
	Appearance of composite
	Integrity of composite
Routine visual inspection	Appearance of composite, including local damage
	Change of use of structure or change in environment
Detailed inspection	Appearance of composite, including local damage
	Change of use of structure or change in environment
	Integrity of composite
	Overall performance of structure
Special inspection	Overall performance of structure

may be some initial imperfections, such as small areas of debonding. The extent to which these may be acceptable will very much depend on the type of strengthening and where the structure is. For example, an area of delamination in the wrapping of a column will probably have a limited effect on the performance, whereas debonding of a plate on the soffit of a beam, particularly at points of high interlaminar shear, will have a significant effect. In subsequent inspections, it will be important to identify new areas of delamination or increases in existing areas. Table 12.1 summarises the key aspects to be considered at each stage.

Any instrumentation installed before strengthening should be used to check that the change in response of the structure after strengthening is in accordance with predictions. Again, the 'as installed' readings from the instrumentation will form the basis for comparison with later readings. However, depending on the type of instrumentation, any change in the results will generally indicate a change in the overall response of the structure, which may not necessarily be due to a change in the FRP.

All strengthened structures should be routinely inspected visually, to check on the condition of the composite. This may be part of a general inspection of the structure, in which case it is likely to be carried out by an inspector who may not be familiar with composite materials. Key points to look for are given in subsequent sections. More detailed inspections, possibly including testing, should be carried out at agreed intervals.

12.3.3 Frequency of inspections

The intervals between inspections recommended below should be taken only as a guide. Structures in aggressive environments will require more frequent inspection. Special structures may require a special inspection regime, the frequency and extent being determined by a risk assessment.

Routine, visual inspection

The recommended intervals for routine visual inspection are as follows:

- bridges   every year
- buildings   every year
- other structures   depends on the use of the structure but ideally every year

With buildings, it will probably only be practicable to inspect a portion of the strengthened area. Industrial structures should be inspected when the plant is shut down. For example, the cooling towers of power stations are typically shut down annually for inspection and servicing.

Detailed inspection, with testing

The complete structure should be regularly tested, including specialist testing of the FRP. In the absence of other guidance, which most major bridge owners will already have, detailed inspections should be carried out at as follows:

- bridges   at least every six years
- buildings   at change of occupancy or change of use, when structural work or refurbishment is carried out on the building, but at intervals not exceeding ten years
- other structures   depends on the use of the structure but at least every ten years

It is recommended that detailed inspection of the FRP should be carried out more frequently in the first few years after installation, to give the owner of the structure confidence that the strengthening has been carried out satisfactorily. For example, for Barnes Bridge (Sadka, 2000) – one of the first major UK highway bridges to be strengthened – tests were scheduled every three months for the first year, every six months during years two and three, and annually during years four to ten. Thereafter, the frequency of testing would depend on the results obtained previously.

Special inspection

A variety of circumstances may require a special inspection, including:

- when a detailed inspection has indicated that there is a problem;
- when it is known that damage has occurred to the fibre composite, for example by a bridge strike or a fire;
- following repairs to the FRP;
- when the structure is to be, or has been, subjected to an abnormal load.

The extent of such an inspection will depend on the type of structure and the circumstances.

12.4 Inspection during strengthening

12.4.1 Introduction

Maintaining records

Full details of the materials used in the strengthening process must be recorded to ensure traceability. A referencing system should be developed so that individual parts of the FRP strengthening system can be identified – for example, a particular FRP plate, its properties and location and the adhesive. Table 12.2 shows a typical installation record for use with pultruded FRP plates or with fabric. Separate sheets should be used for each type of material. Reference should be made to certificates of conformity provided by the suppliers and to tests carried out on samples of the materials used on site. This form would need to be adapted for preformed shells.

Staff training

The operatives who will be carrying out the strengthening work, their supervisors and the inspectors should be suitably trained in the materials to be used. In the absence of a recognised independent body, training will generally be provided by the materials suppliers. Staff should provide evidence of appropriate training.

Site trials

Site trials should be carried out on a representative area of the structure or on a dummy structure that replicates, as far as possible, the actual structure. Such work must be carried out in the correct orientation, e.g. overhead if the material is to be applied to a soffit. The purpose of the trials is to show

Table 12.2 Typical installation record for FRP plates or fabric

Structure	
Location of strengthening within structure	
Date(s) work carried out	
Surface preparation of plates (if used)	Method
	Solvents (if used)
Surface preparation of concrete	Method
Filler material (if used)	Supplier
	Trade name
	Batch number(s)[2]
Primer (if used)	Supplier
	Trade name
	Batch number(s)[2]
FRP plates/fabric[1]	Supplier
	Trade name
	Fibre type
	Width
	Thickness/areal weight
	Strength
	Batch number(s)[2]
Adhesive	Supplier
	Trade name
	Batch number(s)[2]
Surface coating (if used)	Supplier
	Trade name
	Batch number(s)[2]
Weather during installation	Wet/dry Temperature Relative humidity
Additional information[3]	
Reference nos for inspection records	

Notes:
[1] Delete as appropriate
[2] Include details of any certificates of conformity supplied with the materials and any tests carried out on samples taken on site
[3] For example, this might include information on prestressing of the plates, load relief jacking, etc.

that the proposed materials and working method are satisfactory and that the operatives are sufficiently skilled. Work should not start on the main strengthening until tests have shown that the work is of the required standard. If operatives change, it may be necessary to repeat the trials.

Control samples

The making of control samples is recommended: they can be used as a benchmark for representative coupons made during the progress of the work. Control samples and representative coupons should be fabricated under laboratory conditions using adhesive mix ratios specified by manufacturers, good surface preparation techniques and appropriate curing. A comparison of data obtained from control samples and representative coupons can help identify why a joint fabricated on site has performed poorly. Benchmarking is therefore capable of identifying on site application problems such as poor surface preparation, insufficient mixing of adhesive, voiding within adhesive, incomplete spreading of adhesive, inadequate cure and durability issues.

12.4.2 Material conformity before strengthening

Introduction

The design process will involve certain assumptions about FRP properties. Discussion should be held with material suppliers so that appropriate and achievable values can be included in the job specification. It should be appreciated that, as in any manufacturing process, properties of notionally similar composite materials can vary. This section discusses property and performance requirements for FRP reinforcement materials, and gives details of appropriate test methods.

FRP plates

FRP plates are generally made by pultrusion, although other manufacturing processes, such as resin infusion, are used. Pultrusions may consist of unidirectional fibre reinforcement only, or a combination of unidirectional and woven rovings. Strengthening plates are generally unidirectional, but some include a limited amount of transverse fibres. Plates should carry permanent identification marks to indicate the type and grade of material, particularly when more than one type is used. The identification should be checked against the specification to ensure conformity.

Before installation, plates should be visually checked for signs of damage, such as cracks or delamination. If the plate is manufactured with a peel ply,

Table 12.3 Permissible tolerances in pultruded plate geometry

Plate length (m)	Geometrical tolerances (±)				
	Straightness of edge[1] (mm)	Width (mm)	Flatness[2] (mm)	Squareness[3] (mm)	Thickness (mm)
<2	3	1	3	3	0.1
2–4	4	1	3	4	0.1
4–6	6	1	3	6	0.1
>6	7	1	3	7	0.1

Notes:
[1] Measure the deviation from the intended line along the edge of the plate. There should be no sudden steps and any deviation should be uniformly spread along the length
[2] Deviation under a metre straight-edge laid flat along the length of the plate
[3] Measurement along the longer side of the plate. Measure the deviation from 90° with the shorter side taken as the base

it should be checked to ensure that the ply is uniform over the whole area, with no signs of kinking or other variations. Although pultrusion is a controlled manufacturing process, variations in dimensions occur and so plates should be checked against a specification for dimensional tolerances. Pultrusions are not always perfectly straight (in plan), which will result in misalignment of the fibres when they are installed. In addition, they may curl across the width of the plate. Further, it has been found that strips supplied in rolls do not always unroll to form a flat section, particularly if they were rolled up before the material had fully cured. Twisted or curved plates can be forced into an apparently correct alignment when fixed on to the adhesive, but this will almost certainly lead to built-in stresses, delamination before full curing of the adhesive or premature failure. If tolerances have not been specified, the values in Table 12.3 are recommended. Generally, thickness should be within 0.1 mm of specification. Standard tests, such as those in ASTM D3039 (ASTM, 2007) and BS EN 2561 (BSI, 1995), should be used to determine the tensile strength, tensile modulus and tensile strain. Typical specimen sizes are 250 × 15 mm, different recommendations being associated with each standard. Given that the width/thickness ratio should be at least ten, the specimen width should be determined from a consideration of thickness. (Owing to size effects, the thickness of the test specimen should be close to that of the laminate.) In practice, an appropriate width may be limited by the capacity of the testing house's test machines. It is recommended that lengths of pultruded plate be provided to testing houses, where they should be cut to appropriate width with either a diamond-tipped slitting disc or a water-jet knife to ensure defect-free edges.

Fabrics

Rolls of fabric should carry identification labels to indicate their type and grade. These should be checked against the specification. Before being installed, appropriate lengths of fabric should be checked visually for obvious signs of damage, such as significant faults in the weave or tears. Tensile test procedures can be undertaken on specimens using a representative number of fabric layers.

Shells

Shells are factory-made structures that are bonded around columns to provide confinement. Representative coupons may be cut from the shells and tensile tests carried out. Shells may be typically 4 or 5 mm thick, and made from bidirectional woven fabrics. The orientation of the principal fibre direction(s) with respect to the test direction therefore requires consideration.

Adhesives and resins

The conformity of the adhesive or laminating resin to particular physical and mechanical requirements is an essential part of the design process. Requirements are contained in Part 4 of BS EN 1504 (BSI, 2004a), along with a list of approved test methods and acceptance criteria. An Annex details the properties to be declared.

The quality of the materials of the adhesive system (adhesive and any primer) is important. They must be in accordance with the specification. Products should not be substituted without satisfactory evidence of their capability to fulfil all requirements, including compatibility with the remainder of the strengthening system. All products must be stored, handled and used in accordance with the manufacturer's recommendations.

12.4.3 Inspection during installation

Introduction

Inspection of fabrication processes and testing of representative samples are essential quality control procedures for all FRP reinforcement applications. Quality control methods can be divided into three categories, involving:

- site inspection of materials and fabrication procedures – this would typically include keeping records of procedures and materials, checking the suitability of surfaces to be bonded, and visual/manual inspections of the bond integrity during and after fabrication;
- on site manufacture of representative bonded joints for off site evaluation (by testing house);
- on site manufacture of bulk material samples for off site evaluation (by testing house).

Testing adhesives and representative adhesive systems requires good-quality specimens to produce reliable and reproducible results. How test specimens are made is therefore critical. Fabrication methods need to be carefully considered so that specimens can be manufactured accurately and quickly. Substantial defects in a specimen will result in data that may not reflect the quality of the actual bonding process. On site supervision of sample fabrication may be necessary if the contractor has limited experience.

Where possible, samples should be tested by laboratories having third-party accreditation for the tests, but this may not be possible because of their specialised nature. However, the laboratory must be able to demonstrate a suitable record of testing, with appropriate quality procedures, calibrated equipment, etc., and be able to interpret test data.

Surface condition (before application of FRP)

Pull-off tests are used to check that sufficient preparation has been carried out on the surface of the structure to ensure a good adhesive bond and that the adhesive is compatible with the substrate. A cylindrical steel 'dolly' is bonded to the surface of the concrete, a tensile load is applied and is increased to failure. Pull-off tests are carried out in accordance with BS EN 1542 (BSI, 1999), ACI 503R (ACI, 1998), ASTM D4541 (ASTM, 2002) or similar.

Surface preparation and application of composite

The surface of the concrete should be prepared in accordance with the specification. The process of installing the fibre composite should be closely monitored. The main points to be considered are summarised in Table 12.4, but this list is not exhaustive.

Inspectors should look for signs of contamination on the surfaces to be bonded, such as debris from grit-blasting, laitance on concrete structures

Table 12.4 Summary of main points to monitor during installation

Materials
Were all materials (plates, fabrics, adhesives, resins, etc.) in accordance with the specification?
Were certificates of conformity supplied with the materials?
Were all materials handled and stored according to manufacturers' guidelines?

Surface preparation
Was the concrete surface prepared in accordance with the specification?
Was the surface regularity in accordance with the specification?
Where applicable, were the corners of the concrete elements rounded?
Was the surface of FRP plates prepared in accordance with the specification?

Adhesive
Was the correct ratio of components used?
Was the colour uniform (within a batch and between batches), indicating consistent mixing?
Was application completed before the end of the open time?
Was the adhesive thickness controlled? If spacers were used, were they at the specified locations?
Was the structure subject to significant vibration while the adhesive was curing?

Plates
Where adhesive thickness varied (e.g. where plates cross) did the plates spring away from the surface?
Were the plates correctly oriented?
Were the plates moved after first coming into contact with the concrete?
For carbon-fibre plates, was there any contact with metallic parts?

Fabrics
Was the right amount of resin used?
Was the fabric correctly oriented?
Were there any wrinkles or irregularities in the fabric after compaction?
Was the fabric moved after coming into contact with the concrete or the previous layer?
Was the quality of each layer checked before the subsequent layer was installed?
For carbon-fibre fabrics, was there any contact with metallic parts?

Curing
Was the FRP system correctly cured?

Testing
Were test specimens prepared in accordance with the requirements?

Inspection
Was the completed strengthening inspected for voids by tapping or other means?

Records
Were the appropriate records maintained?

and rust or paint on metallic structures. In addition, care must be taken that moisture from the atmosphere does not condense onto the cold surface of the structure, by maintaining a suitable local environment. The adhesive should be mixed strictly in accordance with the manufacturer's recommendations and the thickness of the adhesive layer should be in accordance with the specification.

With wet lay-up systems, the inspector should check that fibre orientation complies with the specification. Misalignments of more than 10 mm from the true location should be reported to the engineer and recorded, as should any kinks, folds or waviness. FRP plates should inherently be easier than fabrics to install in the correct location and orientation. However, as indicated earlier, plates as delivered may not be perfectly straight in plan. Any deviation of more than 10 mm from the intended line should be recorded and reported.

Curing of the adhesive

Maintaining the correct temperature in the adhesive while it is curing is essential if it is to achieve its anticipated properties. The inspector should check that the temperature of the structure and the surrounding air are within the specified limits for the required period. There is some evidence that dynamic loads applied during the curing process can prevent full strength being achieved. The inspector should ensure that any restrictions on the loads applied to the structure are complied with.

Visual and sonic inspection methods for installed FRP

There should be no direct contact between any carbon–FRP system and any metallic parts, such as steel brackets, plates or bolts, because of the possible risk of bimetallic corrosion. Any such contact should be reported to the engineer and a layer of insulation inserted.

For all methods of fabrication, the FRP should be inspected after installation to check for evidence of debonding or other imperfections. Tapping the structure *gently* with a light hammer or coin (i.e. without causing damage) is a simple, established method and relies on a difference in sound between well-bonded and debonded areas. The technique depends on the skill of the operator but can be used to locate areas of large adhesive voiding or significant debonding. Other, more precise, inspection methods are being developed (see Section 12.6.3) but are not generally suitable for site use.

Small entrapped air pockets and voids (around 2 mm diameter) occur naturally in the mixed resin and do not require repair or treatment. For plates, there should be no significant voids in the adhesive, or areas of debonding, in critical locations such as anchorage regions. Elsewhere, limited voids or areas of debonding may be permitted; Technical Report 57 (Concrete Society, 2003) suggests that they should not exceed 5% of the width of the plate and should have a maximum aspect ratio of 2 (length/width). Where plates cross, the adhesive under the outer plate will be thicker than that under the inner plate. When the outer plate is rolled to compact the adhesive, care should be taken that it is not overcompacted, forcing it back towards the concrete surface. If this happens, there may be a tendency for the plate to spring back, leaving a void in the adhesive.

Larger defects may be acceptable for wet lay-up column-wrapping systems. In the USA the International Conference of Building Officials (ICBO, 2001) suggests that voids or areas of debonding of less than 1000 mm^2 (around 35 mm diameter) are permissible, except at edges or boundaries. However, not more than 10 debonds of this size should be allowed in any 1 m^2. Areas of debonding larger than 1000 mm^2 should be repaired by an approved method. The position and size of any significant voids that are identified should be recorded on 'as strengthened' drawings. These recommendations are largely intended for columns strengthened against seismic loading.

Testing quality control samples

Various performance tests may be carried out on adhesive and joint samples that are representative of materials and joints on a given structure. All are carried out on representative coupons and not on any part of the reinforced structure. Test samples must be carefully made, stored and transported to the testing house in accordance with an agreed procedure. The testing house should be asked to report on the quality of the sample as tested, not simply report the results. Most of the test methods are well-established in various industries, including construction. They offer an excellent and sometimes essential means of applying a quality control check to a bonding procedure that can be very difficult to carry out effectively.

The results obtained from tests on representative samples are applicable to a structure only if the materials and procedures are similar. The same adhesive should be applied to both the structural reinforcement and the quality control samples, and in the same manner, e.g. spread with a spatula to the same adhesive thickness. Test joints should therefore be fabricated on site at the same time that the reinforcement is bonded to the structure. After fabrication, quality control samples should be allowed to cure in the same environment (temperature, humidity, etc.) as the reinforced structure. Once sufficient cure has been achieved on site, the quality control samples can be sent for test.

For wet lay-up systems, samples no smaller than 300 × 300 mm and consisting of at least two layers of FRP should be fabricated each day the system is being applied. These should be cured on site under the same conditions as the FRP applied to the structure.

For each property, various standard test methods are available, e.g. British Standards and ASTM standards. Depending on the country in which the work is being carried out, one standard may be more commonly used than another, but all test methods may be considered to be equally valid. However, similar tests will not necessarily give the same numerical result. Specifications should therefore state the test method required and all confirmatory tests must be carried out in accordance with the same standard.

The principal tests are as follows:

- *Single-lap shear tests* are used for plates to check adhesion between the adhesive and the FRP, and to give an indication of the mechanical properties of the cured adhesive (expressed as average shear strength). The test is therefore a check both on surface preparation of the FRP and on correct mixing and curing. Several standard test methods can be used as a guide for testing single-lap shear joints. These include BS 5350-C5 (BSI, 2002), ASTM D1002 (ASTM, 2005) and ASTM D5868 (ASTM, 2001). The single-lap joint is recommended as it is easy and inexpensive to fabricate. Individual specimens may be fabricated (possibly from small samples of plate provided by the supplier) using a jig. Alternatively, lap joints may be made from large pieces of plate (e.g. the width of a pultrusion) that are then cut to the appropriate widths at the testing house, using a diamond-tipped slitting disc or water-jet knife to minimise damage. Either approach is acceptable, provided that resulting specimens are of adequate quality.
- *Tensile dumb-bell tests* are used to check the material properties of the adhesive (tensile modulus, tensile strength and tensile strain to failure), to confirm that the adhesive has been correctly mixed and cured. They cannot be used to determine adhesion characteristics. Specimen dimensions are defined in BS 2782-3 (BSI, 1976). The specimens are cast in moulds and are cured in the mould for a predetermined time under conditions similar to those of the structure. As with all tensile material tests, the results are sensitive to the quality of the specimens, which should be free from voids and surface defects.
- *Flexural modulus tests* are carried out on prism specimens, which are much larger than dumb-bell specimens but less susceptible to breakage on site and during transport to the laboratory. They are used to confirm that the adhesive has been correctly mixed and cured.
- The *glass transition temperature* T_g is an important material characteristic of adhesives. Its measurement provides assurance that the adhesive

has been correctly mixed and cured. BS EN 12614 (BSI, 2004b) covers two standard test methods, differential scanning calorimetry (DSC) and differential thermal analysis (DTA). Guidelines for both methods are given in BS EN 12614 (BSI, 2004b), covering apparatus, sample preparation, thermal cycling and interpretation of results. DSC is also covered by other standards including ASTM E1356 (ASTM, 2003b), ASTM D3418 (ASTM, 2003a) and BS EN ISO 11357 (BSI, 1997). An alternative test, dynamic mechanical analysis (DMA), also known as dynamic mechanical thermal analysis (DMTA), is covered by ASTM E1640 (ASTM, 2004). As the values obtained from the different tests vary, comparisons should only be made with values obtained using the same tests.

12.5 Instrumentation and load testing

12.5.1 Instrumentation

Depending on the relative costs and benefits, it may sometimes be appropriate to install instrumentation to assess the effectiveness of the strengthening and to monitor the performance of the strengthened structure in service. It will probably be most appropriate for highly critical structures or, as a monitoring tool, for those not readily accessible for other forms of inspection. It will obviously also be appropriate when the strengthening is innovative. It should be noted that vandalism may be a problem with long-term monitoring. In vulnerable locations, it will be necessary to protect instruments, cables, etc. to prevent them being damaged.

The instrumentation may be fitted to the main structure, in which case it will provide information on overall performance. Significant changes in the readings may indicate a change in the performance of the FRP, but probably will not give any direct information. Alternatively, instrumentation may be mounted on the surface of the FRP, and techniques are being developed so that the instrumentation can be embedded within the FRP. The instrumentation will thus give a direct indication of changes in the performance of the FRP.

Consideration should also be given to environmental monitoring; climatic data can be used to determine the influence of a service environment on the long-term performance and durability of a fibre composite. Principal environmental data such as temperature can be obtained from local meteorological stations or by using previously published daily maximum, minimum and mean temperatures. Alternatively, the local environment can be monitored on or near the structure. This may be particularly important when the FRP is to be used in inaccessible locations.

Allowances must be made for changes in temperature between successive readings. These changes will lead to movements within the structure and may affect the instrumentation. Where possible, instrumentation should be used to monitor parts of the structure not significantly affected by live loading as well as the critical areas. Readings from the former may be used to compensate for thermally induced changes in the readings from the latter areas.

The choice of instrumentation will depend on the nature of the structure and the type of information required. Examples include the following:

- *Electrical resistance strain gauges* may be used to measure the strain caused by bending or other structural loadings. The gauges tend to 'drift' with time and so are more suitable for short-term tests than for long-term monitoring. One limitation is that they are strongly influenced by local cracks. Hence, the readings obtained may not be representative of the overall level of strain in the structure.
- *Fibre-optic sensors* may be attached to structural components to measure strains. The measurement is extremely accurate, and additional software incorporated into the analytical device can be used to compensate for the effect of temperature change on the measurements. Unlike traditional sensing technologies, fibre-optic sensors are not vulnerable to electromagnetic interference from lightning, overhead electrical distribution cables, etc. Structural materials or elements containing fibre-optic sensing systems are now commonly referred to as 'smart' materials or structures. Further information may be found in ISIS Canada, Manual No. 1 (ISIS Canada, 2001a). One example of their use was on the Sainte Émélie-de-L'Énergie Bridge in Québec, which was fitted with fibre-optic sensors and electrical resistance strain gauges when it was strengthened with carbon FRP plates and glass–fibre shear reinforcement (Labossière *et al.*, 2000).
- The *demountable mechanical* or 'demec' gauge is a simple instrument for measuring strains or movements across a crack. The gauge measures the relative displacements between pairs of 'studs' or 'pips' bonded to the surface of the concrete. Obviously the operator must have access to the surface of the structure.
- Remote monitoring of strains is possible with *vibrating wire strain gauges*. These can be surface-mounted, are more stable than electrical resistance strain gauges and have been used extensively for monitoring bridges.
- *Displacement transducers* may be appropriate for structures in which deflections are critical, provided they can be mounted on a fixed reference frame. These generally require contact between the gauge and the

structure, though some proximity gauges do not. Deflection may also be determined optically, by means of targets mounted on the structure.

12.5.2 Structural testing

Large-scale load tests may be appropriate for some elements or even complete structures. However, they are costly and require considerable planning. In addition, they provide limited information. In general, they should be considered as proof load tests only, i.e. they demonstrate the ability to carry a particular load but give no indication of the ultimate capacity of the structure.

Bridges are probably the easiest type of structure to load rapidly and effectively. A standard road or railway vehicle can be moved slowly across the bridge to apply a known 'static' load at the required locations. This approach should lead to a minimal requirement for possession times and lane closures. ISIS Canada, Manual No. 2 (ISIS Canada, 2001b) reports on a short-span concrete T-beam bridge strengthened with carbon FRP plates. Load tests before and after strengthening confirmed that live load deflections were marginally reduced. However, the authors note that the tests could not provide realistic information about the enhancement of load-carrying capacity.

Loading buildings (apart from car parks where vehicles can again be used) will generally be difficult. Standard methods include water tanks or bags and sand bags, but these are costly and time-consuming. Other structures, such as cooling towers or offshore structures, may be impossible to load. Any measurements will therefore have to be in response to the natural loading imposed by the wind or waves.

An impulsive load, for example a falling weight, can be applied to a structure and the resulting accelerations measured to determine the response frequency. Changes in the response may be used to determine the effect of strengthening on structural stiffness, as in Sikorski *et al.* (2001). If carried out both before and after strengthening, vibration testing will give a good indication of the effectiveness of the FRP in improving the stiffness. Subsequent testing will indicate changes in the response of the structure but will give little information on local differences. It should be emphasised that, for complex structures such as buildings or bridges, vibration testing will give no information on ultimate strength or on potential modes of failure.

12.5.3 Interpreting test results

Measuring strains and/or deflections will generally monitor the overall response of the structure. They will give an indication of the change in the stiffness of the structure. However, strengthening will generally have little influence on the stiffness, unless the FRP was prestressed before application. Thus measurements of the change in the overall response will not indicate the changed load capacity of the structure. If significant changes are identified, either in the FRP or the structure, it will be necessary to determine the cause and, where possible, rectify the problem. Guidance on the principles involved in appraisal may be found in documents such as the Institution of Structural Engineers (ISE, 1996).

12.6 Inspection of strengthened structures

12.6.1 Introduction

This chapter is primarily concerned with the performance of FRP strengthening. However, any inspection must also consider the condition of structure itself. Details of each inspection should be recorded, noting which parts of the FRP have *not* been inspected as well as those that have. The reasons for not inspecting certain parts should be noted. It is particularly important to record any significant changes since the previous inspection. A competent structural engineer who can determine what remedial action, if any, should be taken should check the information.

12.6.2 Routine visual inspection

Routine visual inspections should be carried out at the intervals indicated earlier. Table 12.5 summarises the main points to look for during a routine visual inspection. Much of the time, the inspector will be looking for changes that have occurred since the previous inspection. Taking and maintaining detailed records is therefore essential.

The nature of FRP materials means that they should need little or no maintenance in service provided the composite and adhesive have been correctly specified and installed. However, the surface of the fibre composite should be inspected for signs of crazing, cracking or delamination, which would indicate some level of deterioration. In addition, the composite should be inspected for local damage, such as impact or abrasion. Particular attention should be paid to anchorage regions and also locally at cracks in the concrete. In addition, it is important to check locations where adhesive thickness changes, for example because of steps in the concrete or where pultruded laminates overlap. If several plates have been bonded together to form a thicker plate, all edges should be inspected to check they are not separating.

The nature of many buildings makes visual inspection difficult, with structural elements hidden behind false ceilings and cladding. Ceiling voids may

Quality assurance/quality control, maintenance and repair 343

Table 12.5 Summary of what to look for during a visual inspection

General condition

Is the surface of the composite and/or the structure wet? Are gutters, drains, gullies, etc. clear?

Has there been any change of use or change in the environmental conditions?

What is the condition of any paint or other coating applied to the composite?

Has any paint or other coating been applied to the composite since the last inspection?

Are the appropriate warning plates in place?

What is the general condition of the structure? Has there been any significant change?

Composite

Are there any signs of damage to the composite, such as impact or holes cut through the material?

If the composite is carbon-based, is there any contact with metallic parts (such as bolts or brackets)?

Is there any cracking or crazing on the surface of the composite?

Are there any cracks in the adhesive?

Is there any sign of the composite peeling away from the concrete, such as at the ends of plates?

Where multiple layers of plate have been used, is there any sign of the layers separating?

---

be filled with air-conditioning ducts, wiring conduits, etc. Inspection will be made easier by the provision of readily removable sections of fire protection and cladding. It is important that representative parts of the strengthening are visually inspected, not just those that are easily accessible.

Where the composite has been given a protective or decorative coating, it will not be possible to directly inspect the composite. Significant damage to the protective layer will obviously suggest the possibility of damage to the composite. However, in other areas it will not usually be practical to remove the protective layer as this may damage the fibre composite. FRP on the top surface of a highway bridge will be covered with an asphalt layer over most of its length. In this situation, inspection may be restricted to the material under removable panels in non-trafficked areas. If this is not possible, inspection will have to be limited to uncoated control samples.

The fibre composite may have been covered by paint or other layer (e.g. for protection from ultraviolet light), which will have a limited life. This layer must be checked and, if necessary, replaced, in accordance with the supplier's recommendations, by a material compatible with the fibre composite.

Any identification/warning labels (such as those in Fig. 12.2) should be checked, and missing ones replaced. This is particularly important where future work is likely that could damage the fibre composite material, such as the installation of fixings for services.

In addition, of course, the inspection should look for signs of deterioration of the structure itself, such as additional deflection, cracking in concrete or corrosion of steel. It will be particularly important to take notice of any deterioration close to the strengthening that may lead to debonding of the FRP from the parent concrete.

As moisture and temperature changes are the major causes of FRP deterioration, the inspector should look for signs of water running onto the FRP, dampness of the parent structure or, in extreme cases, mould growing on the FRP itself. It is obviously important that drains, gutters, etc. are checked regularly, as part of the maintenance of the structure, to keep the FRP as dry as possible. The inspector should record the temperature of the FRP and note any changes in the ambient temperature, for example through a change of use. In addition, the inspector should note any new coating over the composite; in sunlight, a dark paint will absorb more radiation, leading to higher temperatures.

12.6.3 Detailed inspection and testing

Detailed inspections should be carried out at the intervals indicated earlier. A professional engineer with appropriate experience of fibre composites should interpret the findings, ideally carrying out the inspection as well.

Detailed inspections will consist of a visual inspection, as above, that should identify areas for a more thorough inspection. In addition, some physical testing of the FRP or of previously prepared samples will generally be needed, at least for the first few inspections following installation.

Tests on FRP

The standard test for the integrity of FRP strengthening systems is *tapping* the surface to identify voids and areas of debonding. This technique is sensitive to the skill of the operative and so alternatives are being developed. Trials of a hand-held acoustic meter, basically an electronic 'ear' for use with the standard coin or hammer tapping technique, have shown promise.

The principle of *thermography* is that when the surface of a bonded structure is heated or cooled, heat transfer occurs more slowly in areas that have debonded than in areas that remain well-bonded. Thus debonded areas are at a slightly different temperature from the surrounding 'perfectly' bonded FRP. If the surface is scanned by an infrared camera, the different temperatures can be identified. A major advantage of the technique is that it can be carried out remotely. When applied to externally bonded composites, it will pick up voids and debonding; however, it will not give any indication of the quality of the adhesion. Trials have been carried out on FRP-strengthened concrete beams, deliberately incorporating voids in the adhesive. It has been found that 15 mm diameter voids can be identified from a distance of 20 m in the laboratory (Delpak *et al.*, 2001). This technique therefore shows promise as a way of determining gross problems in an FRP-strengthened structure for subsequent detailed investigation. Trials on an actual structure have been carried out with some success.

Ultrasonic technology can be used to detect areas of debonding of FRP plates from the parent material in plated concrete structures. The technique works well in the laboratory but has yet to be used on a structure. One sensor acts as a generator of stress pulses through the plate. The travelling pulse may be detected either by the generating sensor (reflection method) or by other sensors attached elsewhere along the plate (transmission method). By comparing the original and detected pulses, the sizes and locations of any defects in the bond may be deduced. The technology thus has strong potential for establishing the integrity of the bond in a finished plated operation before handing the plated structure over to the client, or for routine inspection of the bond in plated structures in service.

Acoustic emission techniques may be used to monitor debonding under service live loads. Acoustic emissions are packets of high-energy, high-frequency stress waves emitted by brittle materials during fracture. In FRP-strengthened structures, debonding or fracture of individual fibres in the matrix releases acoustic emissions. Small piezo-electric sensors are attached at various locations and are activated by any acoustic emissions travelling through the FRP. By connecting the sensors to suitable equipment, the location of an active debonding front may be determined. Acoustic emission technology is currently being developed from a laboratory tool to one suitable for site use.

CIRIA Report C595 (Cadei *et al.*, 2004) suggests that *radiographic techniques* (using X-ray or gamma-ray transmission) can be used to detect voids, inclusions and other defects, but such techniques are only likely to be effective for relatively thin members.

Tests of sacrificial test zones of strengthened structures

It is strongly recommended that additional samples of the fibre composite material should be bonded to the structure away from the region to be strengthened. This approach has been adopted on several structures including Barnes Bridge in Manchester, UK (Sadka, 2000) and the John Hart Bridge in British Columbia, Canada (Hutchinson, 2000). Samples at the Manchester bridge included specimens of fibre composite material bonded to concrete elements to demonstrate the response to hammer testing of a 'perfect' bond, a void in the adhesive or complete debonding. These various samples were used in the process of training the inspectors.

Additionally, or alternatively, FRP can be bonded to separate concrete, steel or other elements, which can be stored on or alongside the structure so that they are in the same environment. Ideally, the material properties in the samples should be representative of those in the main structure; for concrete the same grade and aggregate type should be used. The samples can be inspected and tested as part of the inspection regime. To aid inspection, some or all of the samples should be free from any protective layer. They should thus indicate a lower bound to the performance of the composites bonded to the main structure.

As indicated earlier, the area most likely to degrade will be the interface between the adhesive and the structure. Partially cored pull-off tests may be carried out on the additional samples. Failure should still be at the interface or, for concrete structures, in the surface layer, demonstrating that this still remains the weakest element and hence that there has been no significant change in the properties of the adhesive. Coupons may be cut from removed samples of FRP to be tested in tension to determine whether the properties of the composite material have changed.

12.6.4 Special inspection

As indicated earlier, a special inspection will be required when it is known that damage has occurred to the fibre composite or when a detailed inspection has revealed a problem. The scope of the work to be undertaken will depend on the nature of the damage and may require a detailed inspection of the complete structure, for example following a major fire, and a reappraisal of the performance of the strengthened structure.

12.7 Routine maintenance and repair

12.7.1 Routine maintenance

All gutters, drains, etc. must be kept clear of debris so that rainwater is carried off the structure and away from the FRP. If any cleaning is carried out near the FRP, it must be checked that any solvents used will not cause damage. (Some may not cause obvious problems but may soak into the fibre composite and lead to long-term deterioration.) For some exposed structures, there may be a requirement to clean the surface to remove graffiti or other contaminants. Techniques such as water-jetting or grit-blasting, while appropriate for concrete, masonry and other surfaces, are not appropriate for FRP as they are likely to cause damage. Similarly, steam cleaning should not be used as it is likely to soften the adhesive. The supplier of the FRP materials should provide guidance on appropriate cleaning methods. In addition, warning notices (similar to that in Fig. 12.2) should be installed alongside the FRP, instructing personnel to seek guidance before any work is carried out.

12.7.2 Repair

When major damage to the concrete structure is identified, a full structural assessment may be necessary before any remedial work. If a large void behind FRP wrapping has to be filled, this should be done by vacuum filling with a suitable resin. Pressure injection should be avoided as it will tend to cause additional debonding of the FRP on the periphery of the void.

Where a large area of wrapping has debonded, it may be necessary to remove the defective material over a large enough area such that material on the periphery is fully bonded. The surface should then be prepared again and further FRP installed. In the absence of more detailed information, the new fabric should overlap the old by at least 200 mm. If multiple layers are required, each should overlap the previous by 200 mm at the ends.

Where this technique is used, it is crucial to check that the repair material is compatible with the materials already in place. In addition, it must have similar characteristics to the material in place, including fibre orientation, volume fraction, strength, stiffness and total thickness. There must be an adequate overlap between the new and old material at the periphery of the repaired area. This may be a problem when the FRP adjacent to the reinstated area has previously been painted. Special testing will be needed to determine the bond capacity between the original FRP and the repair material.

If the original FRP was in the form of plates or strips of fabric in discrete bands, it is preferable to bond the additional FRP directly to the surface of the structure rather than to the existing material. Obviously, this will be possible only when there is enough space between the plates or bands of fabric and the surface can be prepared adequately. If this approach is not possible, an additional plate may be bonded over the damaged plate or fabric. Again, it must be ensured that where the new material bonds to the original the latter is fully bonded to the structure. In the absence of more detailed information, the overlap should be at least 300 mm.

Because of stress concentrations around the hole, drilling through a composite causes a strength loss equivalent to a loss in section of two or three hole diameters. A layer of fabric should be placed transversely on either side of the hole, to distribute stresses and minimise stress concentrations around the hole. Additional longitudinal fabric should then be bonded over the transverse material, equal in cross-sectional area to twice that removed by the hole. Again, in the absence of more detailed information, an anchorage of at least 200 mm should be provided on either side of the hole.

12.7.3 Training

Although not strictly part of inspection and monitoring, a very significant development would be the introduction of a compulsory certification scheme for operatives. It is essential that the concrete surface is prepared and the FRP applied by trained operatives working under experienced supervision. The major materials suppliers will provide some training when required. However, it would be preferable for training to be approved by some independent authority that could provide certification, probably covering a range of materials and application techniques.

12.8 Summary and conclusions

If correctly designed, specified and installed, FRP strengthening systems should be durable, with an adequate service life. However, the technique is highly susceptible to the quality of the workmanship and so a high level of supervision is required at all times. Testing should be carried out during installation. The frequency of testing will be a function of the type of structure, the nature of the strengthening work and the duration of the contract. The engineer should be satisfied that the testing is carried out on representative areas of the structure. At all times, a number of 'identical' samples should be tested to check repeatability.

As strengthening with FRP is still a relatively new technique, a plan should be developed for inspecting and monitoring the structure in service. The type and number of samples tested as part of detailed inspections will depend on the nature of the structure but should be based on Table 12.6. Recommended inspection intervals are given in Section 12.3.3. More frequent inspections should be carried out in the first few years after installation. Often, it will be appropriate to fabricate dummy specimens, with FRP bonded to non-critical parts of the structure. These can be tested at the required intervals. Detailed records must be kept of all inspection and testing, both at the time of installation and in service.

12.9 Acknowledgements

The material in this chapter is largely drawn from Technical Report 57 (Concrete Society, 2003). The author would like to thank all those who contributed to the development of the report, and in particular the following companies who provided financial support: BASF, Degussa Construction

Table 12.6 Guidance on tests to be carried out during installation

Property	Test type	Frequency
Surface preparation	Pull-off	Five per representative area
Adhesive		
Glass transition temperature	DSC (or DMA if specified)	Three per 50 kg batch
Tensile strength and modulus	Dumb-bell tension test or flexural modulus test	Three per 50 kg batch
Bond to FRP	Single-lap shear test	Three per 50 kg batch
FRP plates	Tensile strength and modulus	Five samples per batch
Fabric	Tensile strength and modulus (determined from laminate samples prepared adjacent to the work)	One sample per 100 m^2 of laminate, five tests per sample
Shells	Tensile strength and modulus (determined from laminate samples prepared adjacent to the work)	One sample per shell, five tests per sample
Bonding of FRP plates to concrete	Tapping test	Total area
Wet lay-up	Tapping test	Total area

DMA = dynamic thermal analysis
DSC = differential scanning calorimetry

Chemicals, Fyfe Co. LLC, The Highways Agency, London Underground Ltd, Network Rail.

12.10 References

ACI (1998) *Use of Epoxy Compounds with Concrete*, ACI 503R, Farmington Hills, MI USA, American Concrete Institute.

ASTM (2001) *ASTM D5868-01 Standard Test Method for Lap Shear Adhesion for Fiber Reinforced Plastic (FRP) Bonding*, West Conshohocken, PA, USA, American Society for Testing and Materials.

ASTM (2002) *ASTM D4541-02 Standard Test Method for Pull-Off Strength of Coatings Using Portable Adhesion Testers*, West Conshohocken, PA, USA, American Society for Testing and Materials.

ASTM (2003a) *ASTM D3418-03 Standard Test Method for Transition Temperatures and Enthalpies of Fusion and Crystallization of Polymers by Differential Scanning*

Calorimetry, West Conshohocken, PA, USA, American Society for Testing and Materials.
ASTM (2003b) *ASTM E1356-03 Standard Test Method for Assignment of the Glass Transition Temperatures by Differential Scanning Calorimetry*, West Conshohocken, PA, USA, American Society for Testing and Materials.
ASTM (2004) *ASTM E1640-04 Standard Test Method for Assignment of the Glass Transition Temperature By Dynamic Mechanical Analysis*, West Conshohocken, PA, USA, American Society for Testing and Materials.
ASTM (2005) *ASTM D1002-05 Standard Test Method for Apparent Shear Strength of Single-Lap-Joint Adhesively Bonded Metal Specimens by Tension Loading (Metal-to-Metal)*, West Conshohocken, PA, USA, American Society for Testing and Materials.
ASTM (2007) *ASTM D3039/D3039M-07 Standard Test Method for Tensile Properties of Polymer Matrix Composite Materials*, West Conshohocken, PA, USA, American Society for Testing and Materials.
BISBY L A, GREEN M F and KODUR V K R (2004). Fire performance of reinforced concrete columns confined with fibre-reinforced polymers, in Hollaway L C, Chryssanthopoulos M K and Moy S S J (eds), *Advanced Polymer Composites for Structural Applications in Construction: ACIC 4*, Cambridge, UK, Woodhead, 465–472.
BSI (1976) BS 2782-3: Methods 320A to 320F: 1976 *Methods of Testing Plastics. Mechanical properties. Tensile strength, elongation and elastic modulus*, London, UK, British Standards Institution.
BSI (1995) BS EN 2561: 1995 *Carbon fibre reinforced plastics. Unidirectional laminates. Tensile test parallel to the fibre direction*, London, UK, British Standards Institution.
BSI (1997) BS EN ISO 11357-1: 1997 *Plastics. Differential scanning calorimetry (DSC). General principles*, London, UK, British Standards Institution.
BSI (1999) BS EN 1542 *Products and systems for the protection and repair of concrete structures. Test methods. Measurement of bond strength by pull-off*, London, UK, British Standards Institution.
BSI (2002) BS 5350-C5: 2002 *Methods of tests for adhesives. Determination of bond strength in longitudinal shear for rigid adherences*, London, UK, British Standards Institution.
BSI (2004a) BS EN 1504-4: 2004 *Products and systems for the protection and repair of concrete structures. Definitions, requirements, quality control and evaluation of conformity. Structural bonding*, London, UK, British Standards Institution.
BSI (2004b) BS EN 12614: 2004 *Products and systems for the protection and repair of concrete structures. Test methods. Determination of glass transition temperature of polymers*, London, UK, British Standards Institution.
CADEI J M C, STRATFORD T J, HOLLAWAY L C and DUCKETT W G (2004) *Strengthening Metallic Structures using Externally Bonded Fibre-Reinforced Polymers*, Report C595, London, UK, CIRIA.
CONCRETE SOCIETY (2003) *Strengthening Concrete Structures Using Fiber Composite Materials: Acceptance, Inspection and Monitoring*, TR 57, Camberley, UK, The Concrete Society.
DELPAK R, SHIH J K C, ANDREOU, E, HU C W and TANN D B (2001) Thermographic blister detection in FRP strengthened RC elements and degradation effects on section performance, in Teng J G (ed.), *Proceedings International Conference on FRP*

Composites in Civil Engineering – CICE 2001, Oxford, UK, Elsevier Science, 1135–1142.

HUTCHINSON R (2000) Fibre for health; strengthening the Maryland Street Bridge, *Innovator – Newsletter of ISIS Canada*, February.

ICBO (2001) *Acceptance Criteria for Inspection and Verification of Concrete and Reinforced and Unreinforced Masonry Strengthened using Fiber-reinforced Polymer (FRP) Composite Systems*, Whittier, CA, USA, International Conference of Building Officials.

ISE (1996) *Appraisal of Existing Structures*, 2nd edn, London, UK, Institution of Structural Engineers.

ISIS CANADA (2001a). *Installation, Use and Repair of Fibre Optic Sensors*, Design Manual No. 1, Report ISIS-MO2-00, The Canadian Network of Centres of Excellence on Intelligent Sensing for Innovative Structures, Winnipeg, Manitoba, Canada, ISIS Canada Corporation.

ISIS CANADA (2001b) *Guidelines for Structural Health Monitoring*, Manual No. 2, Report ISIS-MO3-00, The Canadian Network of Centres of Excellence on Intelligent Sensing for Innovative Structures, Winnipeg, Manitoba, Canada, ISIS Canada Corporation.

LABOSSIÈRE P, NEALE K W, ROCHETTE P, DEMERS M, LAMOTHE P and DESGAGNÉ G (2000) FRP strengthening of the Ste-Émélie-de-L'Énergie Bridge: Design instrumentation and field testing, *Canadian Journal of Civil Engineering*, **27**/5, 916–927.

SADKA B (2000). Strengthening bridges with fibre-reinforced polymers, *Concrete*, **34**/2, 42–43.

SIKORSKY C, STUBBS N, KARBHARI V and SEIBLE F (2001). Capacity assessment of a bridge rehabilitated using FRP composites, in Burgoyne C J (ed.), *Proceedings Fibre Reinforced Polymer Reinforcement for Concrete Structures – FRPRCS-5*, London, UK, Thomas Telford, 137–146.

13
Case studies

L.C. HOLLAWAY, University of Surrey, UK

13.1 Introduction

The preceding chapters have discussed the utilisation, analysis and design of fibre-reinforced polymer (FRP) composites in the civil infrastructure. It has been shown that the composites have advantages over steel upgrading where high specific strengths and resistance to attack from road salts and chemicals are critical or where rapid installation times are paramount. However, FRP composites are still perceived as a relatively untested material by the practising civil engineer and there are concerns related to the overall durability of these materials and to the capacity for maintaining sustained performance under load when exposed to severe and changing environmental conditions.

A considerable amount of research has been undertaken on laboratory and large-scale models under short-term static and dynamic loading, but there is a lack of data relating to the durability and long-term loading characteristics of composites used separately or in conjunction with the more conventional civil engineering materials. However, in order to obtain durability properties of the polymer composite system in a reasonable timeframe, sometimes accelerated tests are undertaken on the material or structural unit whereby the test specimens are exposed to an accelerated environmental test. This would generally involve the specimen or structural unit being subjected to an environment many times more severe than that which would be experienced in practice. In addition, sometimes the test specimens are exposed to elevated temperatures; the results from any accelerated tests should be interpreted with caution. Temperatures above 60 °C which have sometimes been applied in order to accelerate the degradation of FRP composites are not relevant in the context of predicting the service life of civil engineering structures which operate normally at lower temperatures. As the temperature rises towards the glass transition temperature the polymer will lose some of its stiffness and strength with the result that the accelerated investigation will not be analysing the true material.

From a durability point of view the matrix is the most important component of the composite, as it is exposed to the environment and any breakdown in the polymer or ingress of moisture or salt solutions through it could affect the fibres and/or the fibre–matrix interface region and hence the loading characteristics of the composite over a period of time.

It must be concluded that, at the present time, the information regarding the durability and long-term loading characteristics of composite materials used in the construction industry must be obtained from case studies of actual structural systems exposed to their particular environmental conditions. To understand the results of case studies, there must be detailed information on the type of materials used, their manufacturing methods, their strength and stiffness characteristics, the type of environment to which they are exposed and the length of exposure.

The following case studies involve composites made from carbon fibre, glass fibre or aramid fibre in a matrix of epoxy or vinylester polymer. Structural members manufactured more than 10 years ago were fabricated using isophthalic polyester resins. Currently, epoxy or vinylester polymers and carbon, aramid or glass fibre materials are invariably used to rehabilitate and to retrofit reinforced, prestressed concrete, metallic systems and timber structures. The next section will briefly introduce the various manufacturing systems which have been used in the case studies.

13.2 The manufacturing systems used in the case studies considered

(1) *FRP plates.* Pre-cast plates are used for the majority of examples illustrated in this section. These are manufactured by either the pultrusion or the pre-impregnated techniques. High modulus carbon fibre and either epoxy or vinylester resin composites are utilised in the manufacture of the plates for RC or PC upgrading. In rehabilitating cast iron and steel members, ultra-high modulus (UHM) carbon fibres in epoxy matrix materials are generally utilised with fibre weight fractions of about 57%. The plates are bonded to the structural member using a cold setting adhesive polymer. (See Chapter 2 for further explanation.)

(2) *The hot melt factory-made pre-impregnated fibre (prepreg).* This system has been described in Chapter 2, Section 2.5.4. The example given in Section 13.7.1, which discusses the upgrading of a curved steel beam, used this type of composite manufacture. The factory-made FRP prepreg composite and the compatible film adhesive were cured simultaneously on site. The cure procedure consisted of wrapping a halar film and breather blanket around the FRP composite and film adhesive on the beam, and the whole composite system was

surrounded with a vacuum sheet properly sealed at its extremities. A vacuum-assisted pressure of 1 bar was applied and a heater blanket covered the whole composite. The curing temperature was 65 °C for 16 hours. The type of fibre and matrix used was the high modulus (HM) carbon fibre and an epoxy polymer. Canning (2002), Hulatt *et al.* (2004) and Hollaway *et al.* (2006) and Chapter 2 provide more information on the technique.

(3) *Wet lay-up method.* The example given in Section 13.7.3 used the wet lay-up manufacturing method as the cross-sections of the structural members to be upgraded were non-symmetric. The matrix polymer of the composite was also the adhesive polymer, consequently, the system was fabricated using cold cure resins. The composites were manufactured using high-modulus carbon fibres and epoxy polymer as the matrix and adhesive materials.

13.2.1 Adhesives

A two part cold cured epoxy adhesive would generally be used with the preformed plates; this material will have the following typical pristine properties.

- Two part solvent free:
  - density approximately in the range 1800 gm/m^3;
  - glass transition temperature about 65 °C;
  - flexural modulus approximately 13 GPa;
  - compressive strength approximately 90 MPa;
  - tensile strength approximately 30 MPa;
  - moisture uptake <0.5%.

In rehabilitating flexural and shear members using a hot melt prepreg site cure procedure, a film adhesive, is used this has been discussed in item (3) above. The advantage of this upgrading system is that both the composite and adhesive film cure at the same time under pressure, thus minimising the number of voids present and allowing an intermixing of the polymer of the composite and the adhesive film polymer.

13.2.2 Surface preparations for the adherents

- Substrate surfaces would be prepared by removing all contaminants:
  - *concrete*   coarse and fine aggregate exposed, achieved by grit-blasting; and surface cleaned to remove any lose particles;

- *metals*   grit-blasting using angular chilled iron grit and solvent degreasing, immediately before applying the adhesive to the surface of the steel;
- *timber*   surface planed and grit-blasted;
- *masonry*   fine gripping surface texture achieved by blast cleaning.
• Film adhesive is used with the factory-made prepreg; the surface of the adherends requires no preparation.
• Rigid FRP composite plates are given a light roughening of the surface, ensuring that the fibres of the composite are not exposed. An alternative method is to use a peel-ply with the pre-cast plates. The peel-ply is installed at the time of manufacture of the composite and removed immediately before the bonding operation.

When a film adhesive is used with the site-cured prepreg composite, the surface of the adherends on to which the composite and adhesive are bonded would be prepared as above. More information on preparing the surfaces of substrates can be obtained from Chapter 3.

There are numerous examples of the rehabilitation of beams, slabs and columns using FRP composites, which could be cited throughout the world (Meier, 2000): examples include bridges, precast prestressed (Rizkalla, 1999), curved concrete roof structures (Barboni *et al.*, 2000). However, due to limitations of space only a few specific examples will be discussed here.

13.3   Case study 1: Reinforced concrete beams strengthening with unstressed FRP composites

13.3.1   King's College Hospital – London, UK (1996)

King's College Hospital was the first project in the UK to use a carbon fibre reinforced polymer (CFRP) strengthening system for external strengthening. An extra floor was required to be constructed which involved converting the original roof slab to a floor slab thereby increasing the live load capacity to 3 kN/m^2. The construction of the original building consisted of a reinforced concrete frame with cast *in situ* trough floors. The design of the upper storey extension was a lightweight strut frame structure and, after calculations were undertaken, it was shown that the slab could not sustain the additional live load. Three solutions were put forward to increase the capacity of the slab and the one chosen was a plate strengthening system (CarboDur®–Sika AG, Switzerland) composed of 50 mm wide, 1.2 mm thick composite plates with an elastic modulus of 155 GPa. The CarboDur® laminates were delivered to site in 250 mm long rods pre-boxed for protection and cut on site using a guillotine, cleaned to remove surface contamination and coated with epoxy adhesive. The concrete bond surface was also coated with adhesive and the laminate plates were offered up to the beam and rolled into position to remove air voids and to obtain a good adhesive contact. After 24 hours, chemical anchors at each end of the strips were inserted by drilling through the CarboDur® plates.

The cost in 1996 was £60 000 and the total length of Carbodur® laminates used was 1100 m.

13.3.2 Strengthening of Route 378, reinforced concrete Bridge over Wynantskill Creek – New York, USA (2000)

The bridge carrying Route 378 over Wynantskill Creek New York State carries approximately 30 000 vehicles per day without weight restriction in five lanes of traffic; it is a vital link into the city of South Troy. The construction of the bridge consists of 26 parallel beams spaced 1.37 m centres and it is approximately 12 m long.

During routine inspection, deterioration from salt solution and moisture attack on the rebar reinforcement, concrete spalling and deterioration due to freeze/thaw cycles was identified. Furthermore, difficulties were compounded by the absence of the original design documentation; this prevented the bridge from being load rated in its current structural state. The State Department of Transport, New York, elected to rehabilitate this important bridge structure as opposed to reconstructing a new bridge.

Bonded FRP laminates were chosen as opposed to the traditional steel plates due to their versatility, ease of installation and more favourable durability. The FRP laminates were required to contain freeze/thaw cracking and to improve flexural and shear strength of the reinforced concrete T-beam bridge structure. Load tests were conducted before and after installation of the laminates to evaluate the effectiveness of the strengthening system and to investigate its influence on the structural behaviour of the bridge. The test results were analysed and compared with those obtained using classical analysis.

The analysis at live load serviceability showed that, as a result of the use of FRP plates, the stresses in the steel rebars were reduced in value, the concrete stresses were moderately increased and the transverse live load distribution to the beams was slightly increased. In addition the neutral axis of the section moved slightly towards the bottom of the beam. This application demonstrated that FRP laminates could be installed on to the soffit of bridge beams with minimal disruption to traffic. It should be said that the benefits of the FRP laminate system were not fully realised within the loading range used in the testing programme. The maximum load applied during the testing was about 2.75 MS–18 loading and was not sufficient to induce non-linear behaviour.

The cost in the year 2000 was US$300000 and this compared to an estimated cost of bridge replacement of US$1.2 million.

Further information on this project can be obtained from – Special Report 135 (Report FHWA/NY/SR-01/135) Transportation Research and Development Bureau, New York State Department of Transportation (Authors: Hag-Elsafi O, Kunin J, Alampalli S and Conway T).

13.3.3 The flexural repair and strengthening of the Louisa–Fort Gay reinforced concrete Bridge – Kentucky, USA (2003)

The Louisa–Fort Gay Bridge is located in Lawrence County, Kentucky. The multi-span bridge crosses the Big-Sandy River and is composed of steel plate girders at both spans at the ends of the bridge. Within the middle four spans, the bridge is constructed with cast-in-place reinforced concrete girders acting compositely with cast-in-place concrete bridge decks. During a routine inspection severe flexural cracks, particularly in the entire middle reinforced concrete section of the bridge, were observed. These cracks were most likely caused by heavily loaded lorries; five axle trucks up to 1000 kN have been recorded passing over the bridge. The load limit on the bridge is approximately 712 kN. A subsequent analysis confirmed that the girders had been overloaded to an extent that the service limits allowed in American Association of State Highway and Transportation Officials (AASHTO) provisions were close to or had been exceeded. It was clear that the bridge had to be rehabilitated.

Several repair technologies were investigated, but the eventual agreed solution (after studies at the University of Kentucky and encouragement from the state and federal agencies with programs such as the innovative Bridge Research and Construction–IBRC), was to use high strength CFRP laminates, manufactured by Sika Corporation. The strength and stiffness values were 2800 MPa and 150 GPa, respectively. The advantages of this particular technique are: (i) short construction time – the work was completed in weeks; (ii) minimal man-power was required – only two skilled workers were involved in the rehabilitation process – and in addition only lightweight tools were required; (iii) there was no traffic disruption during the rehabilitation with the bridge remaining open to traffic. The preparation of the concrete and composite surfaces for the bonding operation was undertaken in the accepted way as discussed in Chapter 3. Figure 13.1a

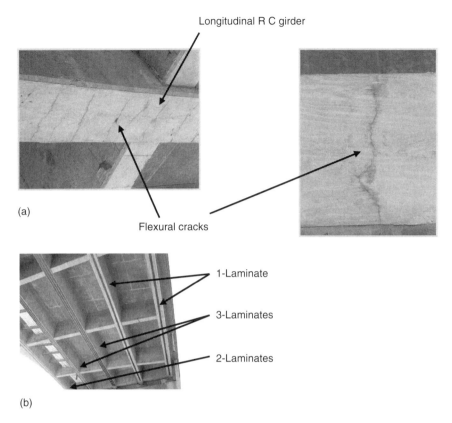

13.1 (a) Visual evidence of damage; (b) completed rehabilitation work in central span (kind permission of Professor I Harik, University of Kentucky, USA).

shows the visual evidence of damage in the reinforced concrete girder, Fig. 13.1b the completed rehabilitation work and Figs 13.2a and 13.2b crack and spalling pedestal and the rehabilitation of the reinforced concrete pedestal using CFRP composites.

The cost of the project (in 2003) was US $195 000; this is less than any other retrofit measures considered. The bridge replacement was not an option on this bridge because of its location and its importance to the economy of the local community.

The bridge was monitored on a regular basis for a period of three years from December 2003. Avnogard crack gauges were installed at different locations and these monitored any movement in the cracks. Two inspections have been carried out since the installation of the CFRP composites, none of the gauges has indicated any permanent crack movement.

(a) (b)

13.2 (a) Cracked and spalling pedestals; (b) retrofitted concrete pedestal (kind permission of Professor I Harik, University of Kentucky, USA).

Further information on this project can be obtained from Professor I Harik, University of Kentucky, Lexington, KY, USA.

13.3.4 Strengthening reinforced concrete slabs

Luke and Canning (2004) have described the application of multiple factory-bonded pultruded FRP plates to strengthen the slab of a reinforced concrete multiple span viaduct and the application of pultruded FRP plates and *in situ* laminated FRP fabrics to strengthen the cantilever slabs on a reinforced concrete motorway slip road (Figs 13.3a and 13.3b). The multiple span viaducts, at Theydon Bois, UK, required flexural strengthening to improve the hogging capacity of the slab over the supports. As new road surfacing would eventually cover the pultruded CFRP plates, the FRP composite materials and adhesives were specified to be compatible with the temporary high temperatures present during laying of the road surface. To enable the installation of the FRP strengthening, the road surface was planed off in one lane, leaving the other lane open to traffic, to reveal the concrete slab. Surface preparation was undertaken by grit-blasting and levelling with the epoxy bonding adhesive, followed by pull-off tests which found the surface preparation to be acceptable. To achieve an efficient installation process, a tent was erected over the installation area, to enable a warm, dry environment to be created in accordance with the specification. On completion of FRP strengthening in one lane, the road surface was reinstated, and work carried out on the other lane.

(a) (b)

13.3 (a) Adhesive FRP plate bonding on cantilever slab; (b) *in situ* FRP fabric lamination on cantilever slab (by kind permission of Mouchel Parkman, West Byfleet, UK).

The motorway slip road on the M1 required strengthening of the cantilever slabs in flexure due to the installation of new parapets, increasing the dead load and also the live load due to vehicle impact. The cantilever on one side of the deck had a generally flat profile, whereas the cantilever on the other side of the deck had a curved step, as shown in Fig. 13.3b. Due to the curvature at this location, an *in situ* laminated CFRP fabric was specified at this location, with pultruded CFRP plates on the other cantilever. Surface preparation was undertaken by planing of the road surfacing and verge concrete on the cantilever, followed by grit-blasting of the revealed concrete slab. Polytunnels were erected to enable strengthening work to be undertaken in inclement weather, and to aid in achieving the required environmental conditions. Installation of the *in situ* laminated CFRP fabric, in particular, required protection during the installation process. Permanent long-term single-lap shear test samples of the pultruded CFRP plates, laminated CFRP fabric and bulk adhesive samples were located within an enclosed concrete chamber in close proximity to the installed FRP strengthening. Long-term test samples were specified due to the difficulty in inspecting the covered pultruded CFRP plates and fabric and the importance of the structure.

Farmer (2004) describes two applications of near-surface mounted reinforcement (NSMR) to flexurally strengthen cantilever slabs in a car park. Near surface mounted reinforcement comprising CFRP bars was chosen as opposed to pultruded CFRP plates for practical reasons. Surfacing was required to be applied over the FRP strengthening, and therefore the use of NSMR reduced the amount of surface preparation work, which was undertaken using diamond saw cutting and limited grit-blasting. The embedded nature of the NSMR also enabled grit-blasting of the final surface for application of an elastomeric membrane to improve the durability of the strengthened structure.

13.3.5 Ring strengthening of cooling tower C1, West Burton Power Station – Retford, UK

West Burton Power Station, at Retford, is a 2000 MW coal fired power station constructed in the 1960s, it has eight reinforced concrete cooling towers. In the year 2000 the cooling tower C1 was strengthened, using state-of-the-art technology and materials, following major shape distortion and associated vertical cracking. The original C1 tower shell was 125 mm thick and it was 118 m high with a 46.3 m internal throat diameter and a hyperbolic profile. In the 1970s, the structure was strengthened by the application of gunite and, within the gunite, a reinforcing mesh was placed to form a mantle 50 mm thick. This added additional tensile capacity in the vertical direction but increased the dead weight of the tower by 40%; strengthening of the foundation with a ring beam was thus required. In 1997, after inspections of the tower by the Babtie Group, Consulting Engineers (now part of Jacobs and called Jacobs Babtie), it was revealed that the tower C1 was progressively deforming with an increase in concrete cracking. Analysis revealed that the tower was potentially unstable and could collapse in a moderate wind; consequently the tower required urgent remedial and strengthening work to be undertaken.

The majority of the concrete cracks were eventually treated with low viscosity epoxy resin, but trixotropic epoxy resins were required to be used in a few regions where the cracks were greater than 4 mm wide. Two strengthening techniques were considered; (i) concrete mantling similar to that adopted in 1980s and (ii) constructing concrete ring beams onto the outer tower, a construction which would increase the circumferential as well as the vertical strength. The latter technique was adopted as it offered the more cost-effective solution; in addition it was a lighter modification than that of item (i). The cost in the year 2000 was less than £1.7 million compared with building a new tower at a cost of £6 million. A soil nailing technique was used to increase the shear capacity of the soil.

There existed one further problem with the ring beams. The relative stiffness of the ring beams meant that they would attract load from the shell causing high peak stresses under a wind load. The beams were reinforced to limit the potential cracking in their outside but, as the beam was integral with the shell, there was insufficient reinforcement on the inside of the shell and this situation could have caused cracking. An aramid fibre fabric reinforcement was chosen to reinforce the inside of the tower, this was encapsulated into an epoxy resin and was light enough to be applied from a

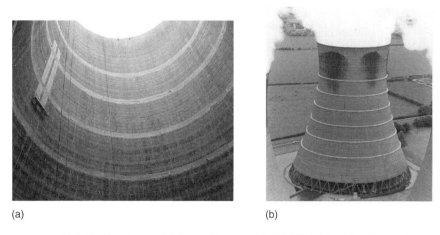

13.4 Cooling tower Cl, from Doyle *et al.* (2004): (a) bands of aramid fibre strengthening inside of tower; (b) completed tower C1 (by kind permission of Thompson Publishing Services and AA Balkema Publishers).

suspended cradle. The use of the aramid fibre reinforced polymer composite was the first time that FRP was used to reinforce a cooling tower; Fig. 13.4a shows the bands of aramid fibre strengthening inside the tower and Fig. 13.4b shows the completed tower C1. This work has demonstrated the importance of seeking the right solution in terms of materials, techniques used and economics.

Further information on this project can be obtained from J Hurst, Jacobs Babtie, UK, and from Doyle *et al.* (2004).

13.4 Case study 2: Repair and strengthening the infrastructure utilising FRP composites in cold climates

Canada and certain northern regions of the USA experience severe winter conditions and, combined with the extensive use of de-icing salts, this causes serious deterioration in a large number of structures. These conditions have imposed a limit, until recently, on the use of FRP composites for structural repair and strengthening. Researchers at the University of Sherbrooke have assessed the reality of composite retrofitting technologies in severe environments, and field applications have been undertaken on various structural elements such as buildings, highway columns, partially submerged bridge piers, beams and beam/column elements of a major carpark structure. This section will discuss two structures that have been strengthened in severe winter conditions.

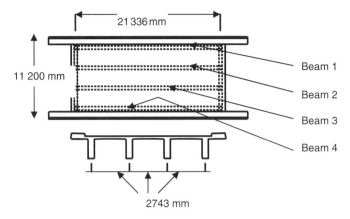

13.5 Section of bridge over Rivière Noire.

13.4.1 The bridge over the Rivière Noire on Route 131 near the village of Sainte-Émélie-de-l'Énergie in the Laurentides-Lanaudière Region – Québec, Canada

The first major bridge for experimental application of FRP composites rehabilitation was selected by the Ministère des Transports du Québec (MTQ) as the bridge over the Rivière Noire on route 131 near the village of Sainte-Émélie-de-l'Énergie in the Laurentides-Lanaudière region, about 130 km northeast of Montréal. The bridge elevation and cross-section is shown in Fig. 13.5. Although this one span river crossing is on a secondary road and therefore the traffic is not excessive, the bridge is required to carry heavy loads. Bridges in Canada are characterised by the live load rating factor (LLRF) defined in the Standard CAN/CSA-S6-88, 'Design of Highway Bridges' (CSA, 1998); the MTQ prefer an LLR > 0.85. This bridge was calculated to have a capacity factor LLRF of 0.94 and therefore its replacement was not necessary. However, upgrading the load-carrying capacity to the current standards was deemed necessary.

The bridge was built in 1951 to the standard 20 t truckload applied through two axles. Its span was 21.3 m and it was designed with four parallel reinforced concrete beams and integrated slab forming T-sections. To increase the load capacity of the bridge it was found that the flexural and shear strengths would need to be increased by 35% and 20%, respectively. The general condition of the bridge indicated neither significant spalling of the concrete nor degradation of the steel rebars through corrosion, these favourable conditions made the bridge ideal for experimental full-scale examination.

Experimental work was sponsored by the MTQ at the University of Sherbrooke to evaluate the potential for FRP strengthening; the tests were undertaken on a series of 1:3 scale models. In general the work undertaken confirmed the potential for reinforcing T-beams with carbon fibre composite materials.

Two design options for the rehabilitation of Sainte-Émélie bridge were proposed but differed only in the type of CFRP composites to be used; the options were in the shape of sheets or strips of approximately the same amount of reinforcement. The one chosen by the contractor had three longitudinal strips of Sika CarboDur® SI214 of width 120 mm and thickness 1.4 mm; no anchorage was required at the free end of the strips. U-shaped stirrups of glass fibre reinforced polymer (GFRP) and SikaWrap® Hex 100 G (Sika AG, Switzerland) were selected for the shear reinforcement; these were wrapped around the section with the fibres in the vertical direction. The width of the stirrups in the vicinity of the supports was 686 mm with spacings of 102 mm, and the width and spacings of these stirrups were decreased and increased, respectively, towards the centre of the beam.

The repair work took place during the autumn of 1998 and the bridge remained open to traffic during the repair in the autumn period, but the lanes under which the FRP was being installed were closed in rotation. The sand-blasting of the concrete surface took place as discussed in Chapter 3, but took longer than anticipated due to the more severe degradation of the concrete than was at first realised. The longitudinal installation was undertaken without incident, but the U-shaped shear reinforcement proved more difficult due to the large size of sheets and their weight once impregnated with epoxy resin, leading to potential slippage on the concrete surface. There were practical difficulties in eliminating the formation of voids between the concrete and FRP. Eventually, epoxy polymer was injected into the voids to expel the air. A layer of protection paint was finally applied to the upgraded beam. Due to the cold weather, the repair took place in an enclosure and the beam was heated in order to post-cure the epoxy polymer.

The bridge was tested after the completion of the rehabilitation procedure. The instrumentation incorporated into the bridge consisted of electrical resistance strain gauges and innovative fibre-optic sensors. The loading tests on the bridge were undertaken under the direction of the MTQ. Three four-axle trucks of 33 t weight were used to apply simulated traffic loading to the bridge before and after rehabilitation. These tests have confirmed that the behaviour of the bridge is consistent with the design hypothesis for the FRP reinforcement. The total cost of the rehabilitation of the structure was Canadian $108 500 in 2000; the MTQ estimated that this sum was 50% of the total cost for the replacement of the infrastructure. They suggested that this type of repair would be economically viable if it could be reduced to about 33% of the replacement cost.

The description of the upgrade of the bridge and tests undertaken before and after its rehabilitation are given in Labossière *et al.* (2000), and further

(a) (b)

13.6 (a) Illustrations of the corroded columns; (b) FRP composite rehabilitation of columns (by kind permission of Concrete International – Neale and Labossière 1998).

information may be obtained from Professor K Neale at the Université de Sherbrooke, Canada.

13.4.2 Overpass Bridge located on Highway 10 – Québec, Canada

In 1996 12 of the 18 circular reinforced concrete columns on the Highway 10 overpass in Québec, Canada, had become severely damaged by the splashing of salt-contaminated snow during the winter months. The columns measured 760 mm in diameter and approximately 6 m in height. The corrosion of the steel reinforcing the concrete columns was the main cause of cracking and spalling of the concrete; Fig. 13.6a shows the corroded columns and Fig. 13.6b the rehabilitation of the column. The columns had to be repaired, and it was decided to gain experience in (i) upgrading the structure using FRP composites in severe climatic conditions, (ii) familiarising the operatives with the installation of composite wraps in cold weather and (iii) utilising innovative measuring techniques by encapsulating fibre-optic techniques into the composite material and gaining experience in the long-term monitoring methods.

Nine of the 12 columns were repaired using FRP composite materials, and conventional materials were used to upgrade the remaining three columns. Five of the nine columns were upgraded using GFRP and the remaining four used carbon fibre reinforced polymers. For comparison purposes three different companies supplied the composite. All 12 columns are being monitored on a regular basis.

Although FRP composites are more expensive than the more conventional civil engineering materials, these increased costs are compensated for during the erection processes by a significant reduction in the labour costs (reducing the number of workers and the hours that they work), the absence of formwork and the non-closures of the highway. Traffic flow was not interrupted during the execution of this project.

Further information may be obtained from Neale and Labossière (1998) and from Professor K Neale or Dr P Labossière at the Université de Sherbrooke, Canada.

13.5 Case study 3: Prestressed concrete bridges strengthening with prestressed FRP composites

13.5.1 Carbon fibre reinforced polymer strengthening and monitoring of the Gröndals Bridge – Stockholm, Sweden (2002)

The Gröndals Bridge is a large freivorbau bridge (prestressed concrete box girder), approximately 400 m in length with a free span of 120 m. It was opened to tram traffic in 2000. During routine inspection cracks were observed in the webs, which have continued to increase in size; the largest exceeded 0.5 mm. The bridge was designed according to the Swedish Code BRO94 and BBK 94, which allowed the designers to erect the bridge with extremely slender webs. Consequently, relatively high shear stresses and principal stresses were generated by the small web widths. The permanent loads on the structure are the dominant ones.

Preliminary investigations revealed that the cause of the cracks was inadequate shear reinforcement in the webs, the dimensions of which had a thickness and a total height to the main span support of 350 mm and 7.5 m, respectively. The main reasons for the cracking were (i) slender webs, (ii) the cross-sections were under-reinforced.

The bridge was strengthened at several sections in both the ultimate limit state and in the service limit state but in different sections of the bridge. Strengthening in the former limit state was undertaken by prestressed DYWIDAG stays and in the latter by utilising CFRP laminates. The reason for strengthening the service limit state was to inhibit existing cracks and prevent new ones from developing.

Hejll and Norling (2002) have shown that the bridge strengthening can be undertaken without bridge closure; work proceeded on this structure with the bridge open to traffic. The laminates were bonded at 70° to the horizontal to allow the laminates to be perpendicular to the direction of the cracks.

A monitoring system has been installed on the bridge, consisting of the utilisation of (i) linear variable displacement transducers (LVDT) for continuous monitoring of the long-term effects of the crack development and (b) fibre-optic sensors (FOS) for the purpose of (a) periodic monitoring of the crack development and strain changes due to temperature and live load traffic and (b) gaining experience in the use of FOS in the field condition. In addition to these two monitoring techniques, two thermocouples have been installed, one on the inside and one on the outside of the east web of the bridge.

The results of the first year of monitoring showed that the strengthening of the webs had apparently been successful in that the cracks in the webs of the box girder did not show any signs of progressive behaviour. The data revealed a clear daily behaviour pattern of the cracks opening and closing; this was attributed to the daily temperature variations. The traffic traversing the bridge did not seem to influence the crack movement, which were of the order of one tenth of the daily thermal movement. During the installation of the sensors, further cracks were discovered between the junction of the top flange and the webs; movements across these cracks were also monitored, but these were of small value. However, the existence of these cracks indicated that the top flange and the webs were not acting compositely. Hejll and Norling (2002), Täljsten and Hejll (2004) and James (2004) have discussed the results of monitoring of the bridge, up to March 2004.

13.6 Case study 4: Rehabilitation of aluminium structural system

13.6.1 Reinforcing aluminium structures – aluminium sign support structures

At the beginning of the 1960s the New York State Department of Transport began to utilise welded aluminium sign support structures on many of their interstate and intrastate highways in both rural and metropolitan areas; Fig. 13.7 illustrates a tri-truss aluminium sign support structure. These structures are essentially a truss design utilising horizontal chord elements with aluminium cross members welded to the chord members; 6061-T6 aluminium is generally used for these systems. Aluminium upright members then hold the trusses at the supports. A fitted butt joint, with no hole bored into the chord, joins the cross members to the chords. The welded joints tend to

13.7 Tri-truss aluminium overhead sign support structure (by kind permission of Professor C Pantelides, University of Utah, USA).

crack at the connection between the aluminium cross members and the horizontal chord members. Typically, the crack follows the weld and remains completely within the weld pool or, due to poor penetration of the weld into the base metal, propagating along the interface between the weld and the base metal. Pantelides *et al.* (2003) and, Nadauld and Pantelides (2007) have reported on the static and fatigue series of experimental tests undertaken and have discussed the various causes for the cracks forming. They conclude that the greatest contributor to these cracks was a lack of code requirements for fatigue design during the original truss analysis undertaken in the 1960s. Other factors include a lack of shop inspection during fabrication, insufficient construction supervision and thermal strains in the welding process, all of which may cause internal stresses into the overhead sign structure before the sign is attached. Cracks in the welds of aluminium overhead sign structures can propagate to complete failure of members, which can cause signs to fall and cause injuries. These structures are exposed, throughout their lives, to freeze/thaw cycling, rain, wind and long-term oxidation.

In 1994 the American Association of State Highway and Transportation Officials (AASHTO) included fatigue as a design factor for these overhead signs. This development, combined with the observation of cracking and some weld failures in a number of these structures, alerted engineers to the possibility of a potentially life-threatening problem in states throughout the country. Several conventional methods have been used to prevent collapse (but not to restore the original strength) of these structures by using steel cable sling or splice plates. GFRP composites have been investigated to ascertain whether this material would be suitable for repair procedures.

New York State Department of Transport (NYSDOT), Utah Department of Transportation and the University of Utah have developed a FRP repair system to restore the original design strength of a cracked connection, and Air Logistics Corporation has manufactured this system. In addition to this composite one, the Fyfe Company manufactures another FRP system; NYSDOT uses both techniques.

If cracks are located in the aluminium before the application of the FRP composite material, drilling a small hole at the crack tip arrests them. The aluminium surface to receive the FRP composite is prepared generally by washing the area with caustic soda; this area is then washed with water and finally etched in preparation to accept a primer such as Alodine® (Henkel Corp, USA). The surface is then ready for the application of the adhesive and finally the pre-impregnated fibre (prepreg); there are nine steps in the application of the GFRP to each joint. The glass fabric is pre-impregnated with a urethane resin and the polymerisation commences when 0.5–1% water is added to the prepreg. The four different GFRP composite materials are packaged in aluminium foil to enable easy access and a quick application in the field. The GFRP prepreg composite material is packaged into a kit by Air Logistics Corporation specifically designed for on site jointing. A final consolidating veil wrap is added to the resin; this acts as an UV inhibitor and eliminates the need to paint the upgraded joint. Full detail of the repair method for cracked aluminium welded connections between diagonals and chord members using glass fibre composites is given in Pantelides *et al.* (2003).

The application of FRP materials to aluminium is a difficult procedure and it is essential to use the correct design and installation guidelines. A number of such structures have been upgraded by NYSDOT using this method and, with the FRP materials properly selected and applied, the results have been successful.

Further information can be obtained from Professor C Pantelides, University of Utah, Salt Lake City, Utah, USA, and Mr Harry White, NYSDOT – Main Office Structures Division, Albany, New York, USA.

13.7 Case study 5: Metallic structures strengthening with unstressed FRP composites

13.7.1 Steel structural member, Boots Building – Nottingham, UK (2000)

In a situation where it is necessary for the composite to navigate a corner or non-planar section of the structural system during the upgrading operation, the pultrusion manufacturing technique could not normally be used. Although pultruded angles or other more complicated sections are produced, to manufacture a special section is expensive as a new die would need to be manufactured and the engineered angles would not necessarily fit the shape of the member. In this case a pre-impregnated composite could be manufactured and fabricated around the member on site (Section 13.2) or wet lay-up procedures, such as Replark TM (Mitsubishi Chemical America Inc., USA) and the resin infusion under flexible tooling (RIFT), could be used to apply the composite to the structural surface. The following case study is an example of the hot melt factory-made pre-impregnated composite fabricated on to the surface of the structure on site.

The on site pre-impregnation fabrication technique discussed in Chapter 2, Section 2.5.4, has been applied to a principal curved steel beam on the Boots Building, Nottingham. The information described here has been discussed in Garden (2001) and Garden and Shahidi (2004). The strengthening system used a low temperature moulding (LTM) advanced polymer composite material, manufactured by ACG, Derbyshire, UK; Taywood Engineering, UK, was the specialist subcontractor for this work. The beam is curved in plan to connect two straight members around the corner of the building. The purpose of the strengthening scheme was to restore the flexural and torsional capacity of the beam to above its original uncorroded level so that an anticipated increase in floor loading could be accommodated. The thickness of the composite strengthening layer, which comprised unidirectional, 0°/90° carbon fibres and ±45° glass fibre orientation, was 2 mm. The unidirectional fibres were aligned along the direction of the length of the beam for flexural strengthening and the ±45° fibre directions were used to resist shear and torsional loading created due to the curvature of the beam; the matrix used was a low temperature curing epoxy resin, (curing temperature 65 °C). The surface preparation and installation procedure was undertaken as described in Chapters 2 and 3, respectively but, in addition, the steel beam was kept in a dry condition by placing silica packs on to it with the beam enclosed in polythene. To act as a bonding aid, an ambient-cured epoxy adhesive was painted on to the clean and dry steel surfaces. Figure 13.8a shows a vacuum bag used to apply pressure for compaction of the composite material and Fig. 13.8b illustrates the final placement of the carbon fibre prepreg layers around the flanges and web of the steel beam.

Further information may be obtained from Garden (2001) and Garden and Shahidi (2004) and from Dr H Garden, Taylor Woodrow, UK, and ACG, Derbyshire, UK.

Case studies 371

13.8 Boots Store Nottingham, upgrading a curved-in-plan steel beam with high modulus, site cured pre-impregnated CFRP composite (by kind permission of Taylor Woodrow Technical Centre, UK and ACG Derbyshire UK, and Elsevier for Fig. 13.7a from Hollaway and Head, 2001).

13.9 Strengthening Slattocks Steel Bridge with unstressed pultruded CFRP plates. Location: Rochdale, UK (by kind permission of Mouchel Parkman, West Byfleet, UK).

13.7.2 Early-steel beam strengthening, Slattocks Canal Bridge – Rochdale, UK

Slattocks Canal Bridge, shown in Fig. 13.9, was the first early steel bridge to be strengthened in Europe using laminated CFRP plates, where the traffic was allowed to use the bridge during the strengthening operation. The bridge is an historic steel bridge at a 44-degree skew which was strengthened using CFRP plates in 1997. This early riveted steel beam bridge consists of square-spanning RSJs 510 mm deep with 191 mm wide flanges supporting a reinforced concrete deck slab below the top flange; the clear span between masonry-faced concrete flanges is 7.64 m^2 and that between masonry-faced concrete abutments on spread foundations is 7.62 m^2. An assessment of the structure revealed that the bridge required strengthening to allow its 17 t vehicle load capacity to be raised to 40 t. After various strengthening options such as closing the bridge to traffic and adding a central support to the bridge were excluded, 8 mm thick lengths of 100 mm wide CFRP plates were bonded to the soffits of the 12 internal RSJs; the head was thus reduced by less than 10 mm. This design preserved the original historic nature of the bridge. The design of the strengthening system used an elasto-plastic analysis, whereby the plastic capacity of the RSJs was mobilised; the CFRP plates were well within their elastic range. In addition, the bond stresses in the adhesive were limited to values such that no peak stress effects were developed.

During the upgrading procedure this heavily trafficked bridge was required to be kept open. The upgrading of the bridge demonstrated the ease and simplicity of on site fabrication in using unstressed CFRP composites as a cost-effective alternative to steel plate bonding or other traditional methods of strengthening. The cost of the upgrading scheme was £20 000 in 1997.

Further details may be obtained from MouchelPackman, West Byfleet, Surrey, UK.

13.7.3 Cast iron bridge strengthening (scheduled ancient monument), Tickford bridge – Newport Pagnell, Buckinghamshire, UK

Tickford bridge is believed to be the oldest operational cast iron bridge in the world. It was designed by Thomas Wilson and built over the River Ouzel, Buckinghamshire, UK in 1810 and is scheduled as an ancient monument. The span of the bridge is 18.5 m and comprises six cast iron arch ribs supporting 10.8 m wide deck slab, which carries the two lane B526 over the river through the centre of Newport Pagnell. As an ancient monument, any strengthening work was required to have respect for the character and aesthetics of the structure. A weight restriction of 3 t was imposed on the bridge following a structural assessment. To improve the capacity of the bridge, a wet lay-up high modulus carbon fibre composite prepreg strengthening system (Replark™ carbon fibre sheet system) was selected; this had the advantage over other carbon fibre systems which cannot be readily applied to curved surfaces. In addition to the CFRP prepreg, a continuous filament of polyester drape veil was placed between the carbon fibre composite and the cast iron member to prevent any possible galvanic action taking place. In certain areas the curing temperatures of the resins were elevated to between 50 °C and 60 °C by means of ducted hot air. This ensured adequate energy absorption rates in the adhesive layer so as to avoid debonding of the carbon fibre. The upgrading system was applied to the three largest spandrel rings and the lower cord of the bridge. The design employed a fracture mechanics approach to analyse the effects of any stress concentrations at joints or cracks in the cast iron. A total of 120 m^2 of carbon fibre prepreg sheet was applied in 14 layers; the total thickness of the system was 14 mm. FaberMaunsell undertook the design of the upgrading system.

Further details on the upgrading of this bridge may be obtained from FaberMaunsell Ltd, St Albans, Herts, UK. An example of upgrading a wrought iron bridge may be found in Shenton *et al.* (2000).

13.8 Case study 6: Metallic structures strengthening with prestressed FRP composites

13.8.1 Cast iron beam strengthening, Hythe Bridge, Oxford, UK (2001)

During the ROBUST project (1993–1997), structural strengthening of reinforced concrete beams with prestressed FRP plates was investigated. The aim was either to increase the serviceability capacity of the structural strengthening or to extend its ultimate limit state. After the completion of ROBUST, the first known application of the prestressing technique for flexural upgrading of cast iron bridges was utilised on the cast iron Hythe Bridge at Oxford, UK. This bridge was constructed circa 1861 as a two-span simply supported structure with 10 cast iron longitudinal beams at approximately 1.1 m centre and brickwork jack-arches spanning transversely between the cast iron beams. The internal cast iron beams were strengthened with prestressed carbon fibre reinforced polymer plates to the bottom flanges in 1999. The objective of strengthening the bridge was to raise the capacity from Group 2 fire engines to 40 t vehicles. The bridge was weak in mid-span bending, but was able to support the full 40 t assessment in shear. The structure carries a busy city centre road above a tributary of the river Thames, and a means of strengthening was required that did not cause traffic disruption. A feasibility study was carried out and showed theoretically that it was possible to strengthen the structure with either steel plate bonding, unstressed composite plates or stressed composite plates. Each of the three methods involved a degree of uncertainty because they extended past practical previous limits in ways that threatened technical or economical viability. Steel plate bonding and unstressed composite plate bonding both presented significant difficulties, as follows:

- Steel plate bonding:
  - thickness of 135 mm would have resulted in restricted headroom and would have imposed high additional load;
  - such thickness would have required several layers and was beyond previous experience;
  - difficulties in handling and fixing with drilling into cast iron were not advisable;
  - expensive to undertake and final system would require continuing maintenance.
- Unstressed composite plate bonding;
  - multiple layers of laminates (about 70 mm of HM CFRP) would have been required, increasing labour and material costs beyond that normally associated with plate bonding;
  - behaviour of such high build multilayer system is untested;
  - peel forces for such a thick system would require strapping in the absence of representative tests.
- Stressed composite plate bonding:
  - no system for stressing composites in this situation was available;
  - stressing of cast iron could reveal weaknesses.

It was eventually concluded that the stressed composite technique offered the most satisfactory solution.

The proprietary technique developed by MouchelParkman, West Byfleet, UK, requires anchorages at the extremities of the plate to be fixed by bonding, friction or mechanical means or a combination of all methods. End tab plates are bonded to each end of the carbon fibre reinforced composite tendons and provide a means of attaching jacking equipment and anchoring the tendon when extended to the final working strain. The tendons are stressed by hydraulic jack, which reacts against a jacking frame temporarily fixed to the anchorage. The stressed tendon is secured after extension by a shear pin that transfers load from a keyway in the end-tab to the anchorage. The tendons are bonded to the beam by epoxy resin in addition to the end anchorages. The anchorage itself is surrounded by a protective casing and fully grouted. Figure 13.10a is the general view of Hythe Bridge and Fig. 13.10b shows the prestressing plates.

Inspections of bridge post strengthening

The first six years principal inspection of the entire bridge included:

- the brickwork abutments and jack-arches;
- cast iron beams and jack-arches;
- prestressed CFRP plates and anchorages;
- scour protection;
- parapets, verges and carriageway.

(a) (b)

13.10 Hythe bridge, Oxford, UK – prestressed CFRP pultruded composite plates bonded on to cast iron beams and anchored at the free end (by kind permission of Mouchel Parkman, West Byfleet, UK) (from Hollaway and Head (2001), with permission of Elsevier).

The objective of the inspection was to observe, identify and record the conditions of the bridge. The cast iron beams only will be reported here and this will call attention to the environment to which the CFRP composite material was exposed.

The painted cast iron beams were generally in good condition with minor pitting and spots of surface corrosion; however, some of the top surfaces of the bottom flanges had not been painted and these showed extensive surface corrosion on the top surface and along the inner edge. The prestressed CFRP plates were generally in good condition with no evidence of surface damage, indentations, discolouration, splitting or delamination within the plate. A number of voids and other defects were found in the adhesive bond between the CFRP plates and the cast iron beams; however, this had no significant effect on the prestressing as the anchorages fully transferred the prestress force to the cast iron beams. The CFRP plate anchorage was generally in good condition, with no evidence of movement, rotation or slippage. However, a small number of the cover plate bolt fixings were rusted.

The description of the technique applied to Hythe Bridge has been obtained from Luke (2001a, b), and from MouchelParkman UK. Further details may be obtained from MouchelPackman, West Byfleet, Surrey, UK.

A number of cast iron structures in the UK have been strengthened using tapering pultruded and preformed vacuum cured CFRP composites. Luke (2001c) describes applications where CFRP plates have been used to increase the live load capacity of cast iron beams on railway bridges in flexure. Historic cast iron has a relatively low design tensile strength (Luke and Canning, 2005), and generally exhibits linear brittle behaviour. Therefore, to reduce the tensile stresses in the cast iron beams to acceptable levels, ultra-high modulus CFRP plates are bonded to the tensile flange of the cast iron beams (Cadei, *et al.* 2004). Tapered CFRP plates, with a nominal thickness at the ends of the plates, are used to minimise the stress concentrations at the ends of the plates, which are caused by live loading and thermal effects. In these particular cases, environmental control, detailed method statement and site supervision were of particular importance due to the requirement for rail possessions; over-run of a railway possession would cause significant disruption to the rail network, with consequent cost penalties. FRP strengthening of such cast iron railway bridges is often the only feasible and cost-effective strengthening technique, due to access and time constraints, the alternative option being expensive demolition and construction of a new structure (with further disruption to the rail network).

13.9 Case study 7: Preservation of historical structures with FRP composites

In the restoration of historic buildings and structures there will always be conflicts between the notions of structural integrity, permanence of repair, reversal of the repair and the preservation of the historic fabric of the building/structure. Compromises have to be made between these items when considering historic preservation; the approach and final decision will probably be very different to that made on purely structural considerations.

FRP materials can match closely the appearance of the historic building or structure, but their mechanical/in-service properties are likely to be different and these differences need to be clearly understood. Furthermore, the physical properties of the FRP composites need to be considered carefully. Therefore, careful consideration must be given to choosing a replacement material.

13.9.1 Timber beam strengthening of a wooden bridge – Sins, Swtizerland (1992)

In 1807 a covered wooden bridge over the river at Sins, Switzerland, was designed and constructed for horse drawn vehicles and pedestrians. The construction consisted of arches, which were strengthened by suspended truss members. During the civil war of 1847 the bridge was a strategic crossing point and a part of the eastern side was destroyed. This was rebuilt in 1852 with a modified support structure comprising a combination of suspended truss members with interlocking tensioning transoms. Currently the carrying capacity of the bridge is 20 t. At the beginning of the 1990s during an inspection and a load test of the structure several cross beams failed to reach their strength requirements. Rehabilitation work commenced in 1992 by replacing the most highly loaded cross beams with 200 mm thick transverse prestressed bonded wooden planks, and the highly loaded cross beams were strengthened by CFRP plates bonded to their soffits. The strengthening plates were 1 mm thick and 250 mm wide for the top beams and 200 mm wide for the lower beams. The Swiss Federal Laboratories for Material Testing and Research (EMPA) monitored the bonding efficiency by using infrared thermography. Meier (2001) discusses the history of the bridge.

The success of this project has given confidence to researchers and practising engineers to employ the use of FRP for future preservation of historic bridges and building structures. As the FRP plates have high specific strength and stiffness they can be designed and installed at a competitive price; the finished product does not detract visually from the original structure.

13.9.2 Shiriya-Zaki Lighthouse – Honshu Island, Japan

Since the Hyogo-Ken Nambu earthquake of 1995, CFRP jacketing and adhesively bonding FRP composites to historical buildings has been accepted in Japan as a method that can be considered for retrofitting structures. This technique was used for retrofitting FRP to the Shiriya-Zaki Lighthouse, Japan. The lighthouse was constructed in 1876 at the northern part of Honshu Island, Japan, and is currently fully operational. It is approximately 33 m in height and consists of three basic units: (i) the tower which is a double brick/masonry column incorporating eight radial connecting brick walls, with structural brick masonry rooms connected to the tower at ground level; (ii) the stone capital with architectural features at the entablature; and (iii) the light compartment constructed from metallic material. Although this area of Japan is not directly in the earthquake region, the lighthouse has been weakened over the years from the effects of earthquakes, and it required rehabilitation in order to continue to function as a lighthouse.

The retrofitting was undertaken as follows:

- ten vertically prestressed CFRP flat bars and 86 vertical unstressed CFRP flat bars were placed longitudinally;
- to confine the vertical flat bars, transverse CFRP sheets were placed in position;
- rock bolts were inserted into the bedrock and a reinforced concrete jacketing system was constructed at the bottom of the brick tower in order to anchor the CFRP flat bars.

Due to the uncertainties as to the properties of the brick/masonry materials the design for the retrofitting procedure was conservatively estimated.

378   Strengthening and rehabilitation of civil infrastructures

13.11 Strengthening masonry wall (photograph by courtesy of Sika Ltd).

The experience of upgrading the lighthouse against seismic effects has been discussed in Katsumata *et al.* (2001), from which the above discussion has been taken.

13.9.3 Masonry walls

Carbon fibre composites have also been used to strengthen masonry walls for seismic and for wind loading. An office building at Mühlebachstr, Zurich, Switzerland, was converted from apartments and this required structural alterations, including removal of a load-bearing wall. Redistribution of load together with increased seismic and wind load design was required. The final design required that two existing brick walls should be strengthened and stiffened with carbon fibre/epoxy polymer composites. Sika Carbo-Dur® type S1012 plates were applied to the brick wall with Sikadur® 30 (Sika AG, Switzerland); this is shown in Fig. 13.11.

13.9.4 The retrofit of the Elmi-Pandolfi Building – Foligno, Italy

The Elmi-Pandolfi Building is part of a building complex in the historical centre of Foligno, Italy. The building has been modified over centuries, but has changed little from the 17th century when it became a stately home with halls and residential quarters for the noble family and for the house keepers. In addition, it had areas for commerce and storage. During the earthquake of 1997 the building was severely damaged and was subsequently restored structurally by CFRP composite retrofitting techniques.

The peripheral masonry structure adjacent to two main streets was strengthened with three horizontal straps of CFRP composite ribbons placed at various heights, and vertical sheet reinforcement was also provided in one of the main struts. The reinforcement provided substantial tensile strength but will not prevent cracking of the masonry; its main function was to prevent out-of-plane collapse mechanism of the peripheral wall.

The solid brick masonry vault of the drawing room has a length of 10.3 m and a width of 7.4 m with an average thickness of 12 mm; it was retrofitted with CFRP sheets at the extrados. The retrofitting technique was undertaken to prevent some of the probable failure mechanisms; it was dynamically tested before and after applying the retrofit. Cracking is the basic failure mechanism for both the unreinforced and reinforced vault; before cracking the masonry takes tensile load, but after cracking the vault becomes equivalent to a system of arches and supports self-weight and external loads, basically by compression. The major failure criterion under this condition is a mechanism that forms cylindrical hinges, and this reduces the possibility of other types of collapse mechanisms.

The construction stages for the rehabilitation of the vault were (i) removal of the fill material up to the haunches where the solid clay bricks of the arched lintel are joined into the outer wall (six reinforced solid brick arches were discovered during the removal of the fill material), (ii) surface cleaning by sanding and water-based solvents; and (iii) levelling the surface of the outer vault area and surface preparation utilising a suitable epoxy primer for areas where bedding bands were created using suitable epoxy putty. It was required to provide an adequate surface level to prevent any irregularity in the CFRP lay-up, but areas with abrupt variation in curvature did exist and it was shown by experimental tests that a high degree of weakness did exist in the CFRP sheets. Finally, epoxy putty was cast over each lintel onto which a steel plate fitted with a steel wedge was placed. The latter was designed to house four anchoring rods, inserted diagonally and of sufficient length to reach the height of the springer. Sheets of unidirectional carbon fabric 200 mm wide were then bonded to the vault extrados. Figures 13.12 and 13.13 show the detail of the connection between CFRP sheet bonded to the vault and a new hollow brick wall.

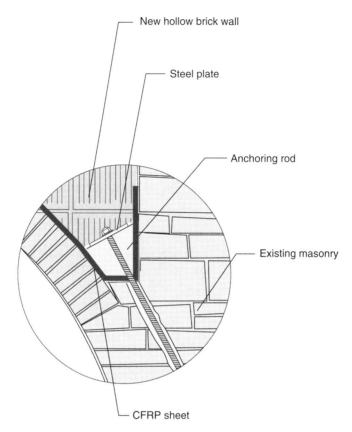

13.12 Detail of the connection between CFRP sheet (glued to the vault) and peripheral wall of drawing room (reprinted from Bastianni (2005) with permission of Elsevier).

Brillouin fibre-optic monitoring systems have been incorporated into most of the vaults and masonry strengthening. On the external walls, fibre circuits have been incorporated to provide strain-sensing monitoring and, in addition, a temperature-sensing section placed in a raceway located near to the bonded fibre. This latter monitoring is to confirm that the glass transition temperature of the polymer will never be reached. Information for this study was derived from Bastianini *et al*. (2005) and Borri, and Giannantoni (2004).

Further details of the retrofitting of Elmi-Pandolfi Building, Foligno, Italy, may be obtained from Professor Marco Corradi of the Department of Civil and Environmental Engineering, University of Perugia, Via Duranti 93, Perugia 06125, Italy.

Case studies 381

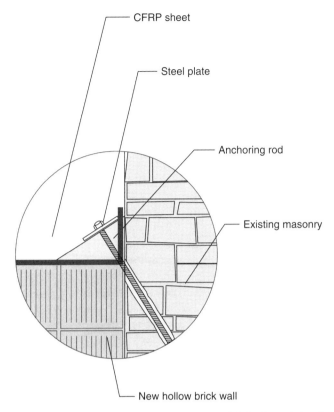

13.13 Detail of the connection between CFRP sheet (glued to the hollow brick wall) and peripheral wall of drawing room (reprinted from Bastianni (2005) with permission of Elsevier).

13.10 Acknowledgements

The author would like to offer his sincere thanks to the engineers and researchers who have assisted and freely given research papers and reports on the upgrading of specific structures mentioned in this chapter and information on subsequent tests undertaken on these rehabilitated structures. Without their help the writing of a chapter of this nature would not have been possible.

13.11 Sources of further information and advice: design codes

In recent years a significant number of design codes and specifications have been published by technical organisations which provide guidance for design with FRP materials for civil engineering. The key publications are listed below.

Europe

Clarke JL (ed.), *Structural Design of Polymer Composites*, Eurocomp Design Code and handbook, London, UK, E and FN Spon (1996)

fib Task Group 9.3, *FRP Reinforcement for Concrete Structures*, Lausanne Switzerland, Federation Internationale du Beton (1999)

(*fib*) Bulletin 14, *Externally Bonded FRP Reinforcement for RC Structures*, Lausanne Switzerland Federation Internationale du Beton (2001)

(*fib*) Bulletin 40, *FRP Reinforcement in RC Structures*, Lausanne, Switzerland, Federation Internationale du Beton (2007)

Concrete Society, *Strengthening Concrete Structures using Fibre Composite Materials: Acceptance, Inspection and Monitoring*, TR57, Camberley, UK, The Concrete Society (2003)

Concrete Society, *Design Guidance for Strengthening Concrete Structures Using Fibre Composite Materials*, TR55, 2nd edn, Camberley, UK, The Concrete Society (2000)

CIRIA Report C595, *Strengthening Metallic Structures Using Externally Bonded Fibre-Reinforced Polymers*, Cadei JM, Stratford TK, Hollaway LC, and Duckett WG, London, UK, CIRIA (2004)

USA

FRP reinforcing rebars and tendons

ACI, *Prestressing Concrete structures with FRP Tendons*, ACI 440.4R-04, Farmington Hills, MI, USA, American Concrete Institute (2004)

ACI, *Guide for the Design and Construction of Structural Concrete Reinforced with FRP Bars*, 440.1R-06, Farmington Hills, MI, USA, American Concrete Institute (2006)

ACI, *Report on Fibre Plastic Reinforcement for Concrete Structures*, 440.R-96 (re-approved 2002), Farmington Hills, MI, USA, American Concrete Institute (2002)

ACI, *Guide Test Methods for Fibre-Reinforced Polymers (FRP) for Reinforcing or Strengthening Concrete Structures*, 440.3R-04, Farmington Hills, MI, USA, American Concrete Institute (2004)

FRP strengthening systems

ACI, *Guide for the Design and Construction of Externally Bonded FRP Systems for Strengthening Concrete Structures*, 440.2R-02, Farmington Hills, MI, USA, American Concrete Institute (2002)

Canada

AC 125, *Acceptance Criteria for Concrete and Reinforced and Unreinforced Masonry Strengthening Using Fibre Reinforced Polymer Composite Systems*, Whittier, British Columbia, Canada, ICC Evaluation Service (1997)

AC 187, *Acceptance Criteria for Inspection and Verification of Concrete and Reinforced and Unreinforced Masonry Strengthening Using Fibre Reinforced Polymer Composite Systems*, Whittier, British Columbia, Canada, ICC Evaluation Service (2001)

CSA, *Canadian Highway Bridge Design Code*, CSA-06–00, Toronto, Ontario, Canada, Canadian Standards Association (2001)

CSA, *Design and Construction of Building Components with Fiber-Reinforced Polymers*, CSA S806-02, Toronto, Ontario, Canada, Canadian Standards Association (2002)

ISIS Canada, *Reinforcing Concrete Structures with Fiber Reinforced Polymers*, Design Manual No. 3, Canadian Network of Centers of Excellence on Intelligent Sensing for Innovative Structures, Winnipeg, Manitoba, Canada, ISIS Canada Corporation (2001)

Japan

JSCE, *Recommendation for Design and Construction of Concrete Structures Using Continuous Fiber Reinforced Materials*, Concrete Engineering Series 23, A Machida (ed.), Tokyo, Japan, Research Committee on Continuous Fiber Reinforcing Materials (1997)

BRI, *Guidelines for Structural Design of FRP Reinforced Concrete Building Structures*, Tsukuba, Japan, Building Research Institute (1995)

JSCE, *Recommendation for Design and Construction of Concrete Structures using Continuous Fiber Reinforcing Materials*, Concrete Engineering Series 23, Tokyo, Japan, Japan Society of Civil Engineers (1997)

JSCE, *Recommendations for Upgrading of Concrete Structures with Use of Continuous Fibre Sheets*, Concrete Engineering Series 41, Tokyo, Japan, Japan Society of Civil Engineers (2001)

13.12 References

BARBONI M, BENEDETTI A and NANNI A (2000) Carbon FRP strengthening of doubly curved precast PC shell, *Journal of Composites for Construction* ASCE, **1**(4), 168–174.

BASTIANINI F, CORRADI M, BORRI A and DI TOMMASO A (2005) Retrofit and monitoring of an historical building using 'Smart' CFRP with embedded fibre optic Brillouin sensors. *Journal of Construction and Building Materials*, **19**, 525–535.

BORRI A and GIANNANTONI A (2004) Esempi di utilizzo dei materiali compositi per il miglioramento sismico degliedifici in muratura, *Proceedings ANIDIS 2004*, Genova, Italy, Jan 25–29 (CD ROM) (in Italian).

CADEI J M, STRATFORD T K, HOLLAWAY L C and DUCKETT W G (2004) *Strengthening Metallic Structures Using Externally Bonded Fibre-Reinforced Polymers*, Report C595, London, UK, CIRIA.

CANNING L (2002) *The Structural Analysis and Optimization of an Advanced Composite/concrete Beam* PhD Thesis University of Surrey, Guildford, UK.

DOYLE P F, HURST J and MADDISON G F (2004) Ring strengthening of West Burton Cooling Tower C1, in Mungan I and Wittek U (eds) *Natural Draught Cooling Towers*, Leiden, The Netherlands, Balkema 1327–1336.

FARMER N (2004) Near surface mounted reinforcement for strengthening – UK experience and development of best practice, in Hollaway L C, Chryssanthopoulos M K, and Moy S S J (eds), *Advanced Polymer Composites For Structural Applications in Construction*, Cambridge, UK, Woodhead, 659–666.

GARDEN H N (2001) Use of composites in civil engineering infrastructure, *Reinforced Plastics*, **45**(7/8), 44–50.

GARDEN H and SHAHIDI E (2002) The use of advanced composite laminates as structural reinforcement in a historic building, in Shenoi R A, Moy S S J and Hollaway L C (eds), *Advanced Polymer Composites For Structural Applications in Construction*, London, UK, Thomas Telford, 457–465.

HEJLL A and NORLING O (2002) Betongbalkar förstärkta med kolfiberkomposit: Dynamisk belasting under limmets härdningsförlopp, Division of Structural Engineering, Luleå University of Technology, Sweden.

HOLLAWAY L C, ZHANG L, PHOTIOU N K, TENG J G and ZHANG S S (2006) Advances in adhesive joining of carbon fibre/polymer composites to steel members for rehabilitation of bridge structures. *Journal of Advances in Structural Engineening*, **9**(6), 101–113.

HULATT J, HOLLAWAY L C and THORNE A M (2004) A novel advanced polymer composite/concrete structural element, *Proceeding Institution of Civil Engineers*, Special Issue: Advanced Polymer Composites for Structural Applications in Construction, 9–17.

JAMES G (2004) *Long Term Monitoring of the Alvik and Gröndal Bridges*, TRITA-BKN Rapport 76, Department of Civil and Architectural Engineering, Royal Institute of Technology (KTH), Stockholm, Sweden.

KATSUMATA H, KIMURA K and MURAHASHI H (2001) Experience of FRP strengthening for Japanese historical structures, in Teng (ed.), *Proceedings International Conference on FRP Composites in Civil Engineering – CICE 2001*, Oxford, UK, Elsevier Science, 1001–1008.

KELLER T (2003) Strengthening of concrete bridges with carbon cables and strips, in Tan K H (ed.), *Proceedings Fibre Reinforced Polymer Reinforcement for Concrete Structures*, 9 FRPRCS-6, Singapore, World Scientific, Vol. 2, 13311340.

LABOSSIÈRE P, NEALE K W, ROCHETTE P, DEMERS M, LAMOTHE P, LAPIERRE P and DESGAGNÉ G (2000) Fibre reinforced polymer strengthening of the Sainte-Émélie-de-l'Énergie bridge: design, instrumentation and field-testing. *Canadian Journal of Civil Engineering*, **27**, 916–927.

LUKE S (2001a) Strengthening structures with carbon fibre plates. Case histories for Hythe Bridge, Oxford and Qafco Prill Tower, *NGCC First Annual conference and*

AGM – *Composites in Construction Through Life Performance*, Watford, UK, Oct 30–31.
LUKE S (2001b) Strengthening of Existing Structures Using Advanced Composite Materials, Network Group for Composites in Construction, Fibre composites – structural materials for the 21st century, NGCC course run through the Institution of Structural Engineers, 18th September 2001.
LUKE S (2001c) The use of carbon fibre plates for the strengthening of two metallic bridges of an historic nature in the UK, in Teng (ed.), *Proceedings International Conference on FRP Composites in Civil Engineering – CICE 2001*, Oxford, UK, Elsevier Science, 975–983.
LUKE S and CANNING L (2005) Strengthening and repair of railway bridges using FRP composites, in Parke G A R and Disney P (eds), *Bridge Management 5*, London, UK, Thomas Telford, 549–556.
LUKE S and CANNING L (2004) Strengthening highway and railway bridges with FRP composites – case studies in Hollaway L C, Chryssanthopoulos M K and Moy S S J (eds), *Advanced Polymer Composites for Structural Applications in Construction: ACIC 2004*, Cambridge, UK, Woodhead, 747–754.
MEIER U (2000) State-of-the-practice of advanced composite materials in structural engineering in Europe with emphasis on transportation, in Bandyopadhyay S, Gowripalan N, Drayton N and Heslehurst R (eds), *Proceedings Composites in the Transportation Industry – ACUN-2*, Sydney, NSW, Australia, University of New South Wales, 44–50.
MEIER U (2001) Strengthening and stiffening of historic wooden structures with CFRP, in Teng (ed.), *Proceedings International Conference on FRP Composites in Civil Engineering – CICE 2001*, Oxford, UK, Elsevier Science, 967–974.
NEALE W and LABOSSIÈRE P (1998) fiber composite sheets in cold climate rehab, *ACI Concrete International*, **20**(6), 22–24.
PANTELIDES C P, NADAULD J and CERCONE L (2003) Repair of cracked aluminium overhead sign structures with glass fibre reinforced polymer composites, *Journal of Composites for Construction*, ASCE, **7**, 118–126.
NADAULD J D and PANTELIDES C P (2007) Rehabilitation of cracked aluminium connections with GFRP composites for fatigue stresses, *Journal of Composites for Construction*, ASCE, **11**(3), 328–335.
RIZKALLA S (1999) Rehabilitation of structures and bridges, in Bandyopadhyay S, Gowripalan N, Rizkalla S, Dutta P and Bhattacharyya D (eds), *Proceedings Composites in the Transportation Industry – ACUN-1*, Sydney, NSW, Australia, University of New South Wales.
ROBUST (1993–1997) *Strengthening of Bridges Using Polymeric Composite Materials*, a UK Government DTI-LINK Structural Composites Programme 1995–1999.
SHENTON H W, CHAJES M J, FINCH W W, HEMPHILL S and CRAIG R (2000) Performance of a historic 19th century wrought iron through truss bridge rehabilitated using advanced composites, in Elgaaly M (ed.), *Proceedings Advanced Technology in Structural Engineering: Structures Congress 2000*, Reston, VA, USA, American Society of Civil Engineers (CD ROM).
TÄLJSTAN T and HEJLL A (2004) CFRP strengthening and monitoring of the Gröndal Bridge in Sweden, in Seracino R (ed.), *Proceedings International Conference on FRP Composites in Civil Engineering – CICE 2004*, London, UK, Taylor and Francis, 961–968.

Index

acoustic emission testing 345
active strengthening 14–16
additives 56
adhesive joints 216–18, 220–1
  crack propagation 223
  debonding 223
  durability 218
  fatigue failure 218, 229–30
  linear-elastic stress analysis 224–7
  see also bonding
adhesives 66, 196, 268, 354
  acceptance criteria 107–8, 333–4
  air voids 269, 336–7
  application 269
  compatibility with substrate 104
  curing 106–7, 271, 284, 336
  elasto-plastic analysis 225
  glass transition temperature 37, 84
  grab 84
  interfacial contact 85–6
  mixing 269
  quality assurance 107–8, 333–4
  rheological characteristics 86
  surface tension 85
  for timber 96–7
  see also surface preparation
advanced polymer composite (APC) materials 2, 35–7, 45–78
  additives 56
  age of the composite 67, 72–3
  chemical properties 46, 55–6, 76–7
  component parts 47
  cure cycle 66

degradation 67, 72–3
delamination 70–1
durability 46, 54–5, 71–6
  testing and monitoring 77
fatigue resistance 70–1
fibre and matrix proportions 65
fibre orientation 65, 142–3, 336
fibre types 56–63, 64
fillers 56, 296
in-service properties 51–6, 71–7
manufacturing methods 45–6, 65–7, 353–5
matrix properties 47–56, 70–1
mechanical properties 45, 49–51, 64, 67–71
prepeg process 45–6, 65–6, 353–4
pultrusion process 45, 65, 67, 331–2, 353
see also prestressed systems; unstressed systems
age of the composite 67, 72–3
air voids 269, 336–7
alkaline environments 72
aluminium structures 367–9
aluminium trihydrate 74
analysis-oriented stress-strain models 172–6
anchoring techniques 6, 263–4, 276
  design guidelines 202–3
  fibre anchors 136–7
  gradually anchored prestressed plates 282–3

387

mechanically anchored prestressed plates 274, 275, 282–3
U-anchor system 8, 10
anodising surface preparations 90
antimony oxide 74
aqueous solutions 72
AR-glass 63
aramid fibres 12, 22, 57, 61, 220, 274
  bare fibres 297
arches 19–20, 29–30, 255–7, 306–7
as installed condition 326–7
axial compression 159
axial modulus 11–12

beams 5–7, 19–21, 230
  cantilever beams 138
  column joints 13–14, 18
  critical sections 134–5
  diagonal compression failure 152
  ductility 203–4
  flexural strengthening 5–9, 14–15, 120–38
  Hythe bridge 373–6
  indeterminate beams 138, 153
  load-deflection response 115–16
  shear strengthening 9–11, 15, 141–54
  Slattocks Canal bridge 371–2
  timber beams 32–4, 376–7
  see also flexural failure modes; shear failure modes
beton plaqué 5
bimetallic corrosion 218
blast strengthening 30–1
blistering 314
bolting 6
bonding
  analysis 223–30
  behaviour 127–31
  bond-critical concrete/FRP systems 301–3
  cementitious bonding agent 271
  concrete structures 5–6, 14, 299
  creating satisfactory bonds 46
  defects 229
  durability 229
  environmental controls 284

fracture mechanics 223–4
masonry structures 239–43, 309
metallic structures 19–21, 223–30, 313–14
operation 105–7
record keeping 286
requirements 83–4
slip models 130–1
steel-epoxy bonds 314–15
strength 129–30, 281
stress analysis 224–7
timber structures 304
see also debonding; surface preparation
Boots Building 369–70
brick masonry 307–8
see also masonry structures
bridges 19–20
  Gröndals 366–7
  Highway 10 overpass 365–6
  Hythe 373–6
  Louisa–Fort Gay 357–9
  prestressed concrete 366–7
  Rivière Noire 363–5
  Sins 376–7
  Slattocks Canal 371–2
  Tickford 372–3
  Wynantskill Creek 256–7
brittle shear failure 26, 27
buckling 231

cantilever beams 138
cantilever slabs 7
carbon fibres 7, 9, 12, 57–61, 220
  bare fibres 297
  King's College Hospital project 355–6
  M1 slip road project 360–1
  prestressed systems 274
  unstressed systems 268
carbon steel see steel
cast iron 16–17, 18, 19–20, 372–6
  surface preparation 95
CDC (critical diagonal crack-induced) debonding 118–19, 201
cementitious bonding agent 271
change in use 3, 23–4, 31–2

Index 389

characteristic strength 197
chemical properties 46, 55–6, 76–7
chemical surface treatments 90
circular columns 159–61, 165, 166–9, 184–7, 209
clay brick/tile masonry 307–8
cold climate structures 362–6
collapse modes 244–7
columns 8, 13, 158–90
  axial compression 159
  beam joints 13–14, 18
  circular 159–61, 165, 166–9, 184–7, 209
  compressive strength 209–10
  concentric compression 176, 177, 184
  design guidelines 180–90, 207–11
  ductility 210–11
  eccentric compression 159, 176, 184–90
  elliptical 163–4, 165, 171–2, 190
  hoop rupture strain 160, 182
  masonry 29, 259–61, 309
  moment-curvature analysis 179–80
  rectangular 158, 162, 165, 169–71, 188–90, 209–10
  section analysis 176–80
  seismic retrofit 210–11
  serviceability requirements 210
  size effects 164–6
  slenderness limits 180, 183–4
  stress-strain curves 161, 166, 173
  stress-strain models
    analysis-oriented 172–6
    design guidelines 208
    design-oriented 166–72, 181–2
  see also confinement
compatibility of FRP 24–5
compression field theory 145
compressive strength 68, 115, 216
  columns 176, 177, 184, 209–10
  failure 247–8, 310
concentric compression 176, 177, 184
concrete cover separation 117, 202
concrete masonry 307, 308

Concrete Society guidelines *see* design
concrete structures 4–16
  active strengthening 14–16
  bond performance 5–6, 14, 299
  bond-critical systems 301–3
  confinement 11–12, 15–16, 158, 159–66
    analysis-oriented models 172–6
    design-oriented models 166–72, 181–2
  contact-critical systems 300–1
  corrosion 14
  definition 1
  design guidelines 195–211
  durability 298–303
  fatigue resistance 16, 301–2
  Louisa-Fort Gay bridge 357–9
  prestressed concrete 5, 9, 366–7
  seismic retrofit 12–14
  shear strengthening 9–11, 15, 141–54
  slabs 7, 359–61
  stress-strain curve 120–5
  structural deficiencies 4–5
  surface preparation 90–2, 354
  water degradation 300
  West Burton Power Station 361–2
  Wynantskill Creek bridge 256–7
  see also beams; columns; flexural strengthening
condition assessment 196
conductivity 53, 76
confinement 11–12, 15–16, 29, 158, 259–61
  analysis-oriented models 172–6
  behaviour of FRP-confined columns 159–66
  design-oriented models 166–72, 181–2
construction errors 4
contact-critical concrete/FRP systems 300–1
contamination-free surfaces 104
contractor experience 267, 278, 281
control samples 285, 331, 337–9
cooling towers 361–2

corona discharge surface preparation 90
corrosion 4, 18, 22, 313, 314
  bimetallic corrosion 218
  steel reinforcements 14, 300
costs 36, 358
  of fibres 56–7
coupling agents 86–8
cracking 71, 204
creep characteristics 50–1, 68–9, 205, 274
critical sections of beams 134–5
cross-linking 48
crystalline polymers 52–3
curing 66, 106–7, 271, 284, 336
curvature of substrate 269, 270

data reliability 292–3
dead load 3, 23, 219, 276–7
debonding 5–6, 10, 113–20
  CDC (critical diagonal crack-induced) 118–19, 201
  concrete cover separation 117, 202
  crack propagation 223
  design guidelines 199–203
  IC (intermediate crack-induced) 116–17, 133–4, 199–201
  masonry structures 28, 242–3, 248, 253, 310–11
  out-of-plane loading 248
  peak adhesive stress reduction 227–8
  plate end interfacial 117–18, 131–3, 136, 202–3
  repairs 347–8
  shear strengthening 149–51
  suppression of debonding failures 136–7
  testing 336–7, 344
defect limits 285–6
deflections 204
degradation 3–4, 24, 32, 67, 72–3
  environmental 324
  water 300
delamination 70–1, 311
demountable mechanical gauges 340

design 195–211
  anchoring techniques 202–3
  characteristic strength 197
  columns 207–11
  concrete structures 195–211
  condition assessment 196
  confined concrete columns 180–90
  debonding 199–203
  elastic modulus 198
  equations 120–5
  errors 4
  fatigue 196
  flexural strengthening 120–5, 198–205
  in-plane loading 253–4
  limit state philosophy 196–7
  masonry structures 259
  material safety factor 197
  out-of-plane loading 248–9
  partial safety factors 197
  prestressed systems 273
  serviceability guidelines 204–5
  shear strengthening 144–5, 146, 205–7
  stress-strain models 166–72, 181–2, 208
detailed inspections 328, 344–6
diagonal compression failure 152
displacement transducers 340–1
domes 258–9, 306–7
ductility 6, 203–4, 210–11
durability 46, 54–5, 71–6, 229, 292–315, 353
  adhesive joints 218
  concrete-FRP systems 298–303
  data reliability 292–3
  environmental degradation 324
  factors affecting 292
  fibre sizings 294
  filled resin systems 296
  freeze/thaw conditions 301, 302–3, 305
  masonry-FRP systems 306–12
  metallic-FRP systems 312–15
  polymer resins 294–5
  temperature effects 302

testing and monitoring 77, 293
timber-FRP systems 303–6
water degradation 300
Dyne pens 105

E-CR-glass 63
E-glass 63, 300
earthquakes *see* seismic performance
eccentric compression 159, 176, 184–90
elastic modulus 198
elastic shortening 274
elasto-plastic adhesive analysis 225
electrical resistance strain gauges 340
electrochemical surface treatments 90
elliptical columns 163–4, 165, 171–2, 190
Elmi-Pandolfi building 378–80
end zones *see* plate end
environmental controls 284
environmental degradation 324
epoxy resins *see* resins
exterior walls 309
extrusion processes 61

failure
  compressive failure 152, 247–8, 310
  delamination 70–1, 311
  fibre-matrix interface 70
  flexural failure 113–20, 141, 198–9
  in-plane loading 252–3
  masonry structures 311
  metallic structures 215–16
  out-of-plane loading 247–8
  rupture failure 149, 310
  shear failure 26, 27, 141, 143–4, 248, 311
  see also debonding
fatigue 4, 18, 70–1, 205, 229–30
  adhesive joints 218
  concrete structures 16, 301–2
  design guidelines 196
  life extension 22, 230–1
fib guidelines *see* design
fibre-optic sensors 340, 379–80

fibres
  costs 56–7
  fibre anchors 136–7
  matrix interface failure 70
  matrix proportions 65
  orientation 65, 142–3, 336
  sizings 294
  types 56–63, 64
filament winding 158
fillers 56, 296
fire resistance 37, 56, 73–6, 90, 296, 325–6
flexural failure modes 113–20, 141, 198–9
flexural modulus tests 338
flexural strengthening 5–9, 14–15, 21, 120–38, 267–8
  bond behaviour 127–31
  contractor qualifications 267
  design equations 120–5
  design guidelines 198–205
  ductility 203–4
  installation process 278, 281, 284–6
  interfacial stresses 126–7
  metallic structures 221–30
  moment capacity 134–5, 198–9
  partial safety factors 137–8
  plate end anchorage 134–5
  preloading effects 125–6
  site activities 267, 286–90
  strength checks 135–6
  surface irregularity 203
  timber structures 32–4
  see also prestressed systems; unstressed systems
floor belts 254–5
forced curing 271
foundation piles 32
fracture mechanics 223–4
freeze/thaw behaviour 301, 302–3, 305
frequency of inspections 328–9

galvanic coupling 300
glass fibres 7, 12, 56–7, 61–3, 220, 274
  bare fibres 297
  durability testing 77

glass transition temperature 37, 51–3, 84, 295, 338–9, 352
grab 84
grit-blasting 91, 92, 94–5
Gröndals bridge 366–7

halogenated fillers 296
halogens 74
hazard events 4, 5, 18, 324–6
heat distortion temperature (HDT) 75
high-pressure washing 91
Highway 10 overpass bridge 365–6
historic structures 23–4, 25, 376–80
Honshu island lighthouse 377–8
hoop rupture strain 160, 182
horizontal bending 247
Hythe bridge 373–6

IC (intermediate crack-induced) debonding 116–17, 133–4, 199–201
identification labels 331, 333
impact damage 4, 5, 18, 324–6
imposed displacements 24
in situ laminated FRP fabrics 269–70, 279
in situ resin infusion 271–2, 280
in situ vacuum cured FRP fabric 271, 279
in-plane loading 27–9, 251–4
  design guidelines 253–4
  failure mechanisms 252–3
in-service properties 51–6, 71–7
indeterminate beams 138, 153
inspection 288–9, 326–39, 374–5
  as installed condition 326–7
  control samples 331, 337–9
  detailed 328, 344–6
  fibre orientation 336
  frequency 328–9
  FRP plates 331–3
  identification labels 331, 333
  material conformity 331–3
  qualifications of inspectors 326
  record keeping 329, 330
  site trials 329–31
  sonic methods 336–7
  special inspections 329, 346
  staff training 329
  surface condition 334
  surface preparation 334, 336
  visual 327, 328, 336–7, 342–4
installation process 278, 281, 284–6
  contractor experience 267, 278, 281
  defect limits 285–6
  environmental controls 284
  flexural strengthening 278, 281, 284–6
  material specification 278, 281, 284
  quality assurance 285
  quality of substrate 281
  sequence of installation 284–5
  workmanship 278
  see also surface preparation; testing
installed condition inspection 326–7
instrumentation 339–42
interfacial adhesive contact 85–6
interfacial stresses 126–7
intumescent coatings 74, 296

jacketing 14, 207–8
  prefabricated FRP jackets 158
  U-jackets 136, 137, 153
jacking 276–7
joints
  beam-column joints 13–14, 18
  metallic structures 18, 216–18, 220–1
  testing 104–5, 107–8
  see also adhesive joints; bonding

King's College Hospital 355–6
Kobe earthquake 1

lamination
  delamination 70–1, 311
  in situ laminated FRP fabrics 269–70, 279
  laminating resin 270
lighthouse 377–8
limit state philosophy 196–7

linear-elastic stress analysis 224–7
lintels 254–5
live load 3, 23
load buckling 216
load path 3
load-deflection response 115–16
Louisa-Fort Gay bridge 357–9

M1 slip road 360, 360–1
maintenance 288–9, 346–8
manufacturing methods 45–6, 65–7, 353–5
masonry structures 23–31, 235–64
　anchorage techniques 263–4
　arches 29–30, 255–7, 306–7
　area of reinforcements 262–3
　blast strengthening 30–1
　bonding mechanics 239–43, 309
　brittle shear failure 26, 27
　clay brick/tile masonry 307–8
　columns 29, 259–61, 309
　compatibility of FRP 24–5
　compression failure 310
　concrete masonry 307, 308
　confinement 29, 259–61
　debonding 28, 242–3, 248, 253, 310–11
　delamination 311
　design procedures 259
　domes 258–9, 306–7
　durability 306–12
　Elmi-Pandoffi building 378–80
　floor belts 254–5
　imposed displacements 24
　in-plane loading 27–9, 251–4
　lintels 254–5
　near-surface mounted (NSM) bars 27, 242
　out-of-plane loading 25–7, 243–51
　portal frames 257–8
　rupture of reinforcement 310
　settlement repair 31
　shear failure 311
　Shiriya-Zaki lighthouse 377–8
　single-side strengthening 28
　spacing of reinforcements 262

stone masonry 307, 308
structural deficiencies 23–4
structural modelling 238
structural repointing 28, 243–4
ties 30, 261
vaults 23, 29–30, 258–9, 306–7
walls
　exterior walls 309
　moisture ingress 308–9
　seismic performance 378
　slenderness ratio 26, 249
material conformity 331–3
material specification 218–21, 278, 281, 284
matrix properties 47–56, 70–1
　additives 56
　chemical properties 46, 55–6, 76–7
　cracking 71
　cross-linking 48
　durability 46, 54–5, 71–6, 77
　fibre-matrix interface failure 70
　fibre-matrix proportions 65
　fillers 56
　mechanical properties 45, 49–51, 64, 67–71
　physical properties 45, 53–4
　thermal properties 51–3, 69–70, 76
　thermoplastic binders 47–8
　thermosetting binders 47, 48, 49–56
mechanical damage 324–6
mechanical properties 45, 49–51, 64, 67–71
mechanical surface treatments 89–90
mechanically fixed FRP plates 8, 272, 280
metallic structures 16–22, 215–32
　aluminium 367–9
　arches 19–20
　beams 18, 19–21, 230
　blistering 314
　bond performance 19–21, 223–30, 313–14
　Boots Building 369–70
　buckling 231
　cast iron 16–17, 18, 19–20, 372–6

compression members 21
connections between members 21–2, 231–2
corrosion 18, 22, 313, 314
durability 312–15
failure modes 215–16
fatigue 18, 22, 230–1
flexural members 19–21
flexural strengthening 221–30
Hythe bridge 373–6
joints 18, 216–18, 220–1
material selection 218–21
prestressed systems 276
safety factors 221
sectional analysis 221–3
Slattocks Canal bridge 371–2
steel 17, 20–1
structural deficiencies 17–18, 215
structural integrity 22, 232
substrate yield 219–20
surface preparation 92–6, 355
Tickford bridge 372–3
wrought iron 17, 20–1
method statements 287
micro-roughness of surfaces 104
mixing adhesives 269
modular ratio 218–19
moisture ingress 72, 303–4, 308–9
moment capacity 134–5, 198–9
moment-curvature analysis 179–80
monitoring 288–9

N-Varg 63
nanoclays 296
nanoparticles 56
near-surface mounted (NSM) bars 8–9, 10–11, 270–1, 277, 279, 282–3
  beam-column joints 13–14
  M1 slip road project 360–1
  masonry structures 27, 242

oligomers 48
one-way slabs 7
out-of-plane loading 25–7, 243–51
  collapse modes 244–7

compression failure 247–8
debonding 248
design of systems 248–9
horizontal bending 247
limitations due to arching 249–51
loss of rigid-body equilibrium 244–5
repointing 28, 243–4
shear failure 248
vertical bending 245–6

PAN precursors 59
partial safety factors 137–8, 197
peak adhesive stress reduction 227–8
peel-ply layers 101
peeling failure see IC (intermediate crack-induced) debonding
permeability 55–6
phosphorus compounds 74
physical properties 45, 53–4, 90
pitch precursors 59
plate end 15
  anchorage 134–5
  interfacial debonding 117–18, 131–3, 136, 202–3
polymer matrices see matrix properties
portal frames 257–8
post-tensioning 273, 274
predictive models 144–6
prefabricated FRP jackets 158
prefabricated FRP plates 268, 353
preloading effects 125–6
prepeg process 45–6, 65–6, 353–4
prestressed concrete 5, 9, 366–7
prestressed systems 15, 22, 67, 230, 272–83, 274, 289
  adhesively bonded plates 277
  anchorage 274, 275, 276, 282–3
  concrete structures 275–6
  creep effects 274
  dead load reduction 276–7
  design and application 273
  elastic shortening 274
  gradually anchored plates 282–3
  Gröndals bridge 366–7
  Hythe bridge 373–6
  jacking 276–7

Index 395

level of prestress force 275
losses in prestress force 274
mechanically anchored plates 274,
    275, 282–3
metallic structures 276
near surface mounted (NSM) bars
    277, 282–3
post-tensioning 273, 274
reasons for using 272–3
tendons and cables 277, 282–3
primers 86–8
pultrusion process 45, 65, 67, 331–2,
    353

qualification for use tests 85
qualifications of inspectors 326
quality assurance 107–8, 196, 285,
    333–4
    instrumentation 339–42
    see also inspection; testing
Québec overpass bridge 365–6

R-glass 63
radiographic testing 345
record keeping 286, 329, 330
rectangular columns 158, 162, 165,
    169–71, 188–90, 209–10
reinforced concrete (RC) structures see
    concrete structures
repairing a structure 1, 347–8
repointing 28, 243–4
resin infusion
    in situ 271–2, 280
    under flexible tooling (RIFT) 46
resin transfer moulding (RTM) 46, 66
resins 48, 74, 83
    durability 294–5
    fillers 56, 296
    freeze/thaw behaviour 301, 302–3
    glass transition temperature 295
    laminating resin 270
    vapor-permeable 300
retrofitting 1, 12–14
rheological characteristics of adhesives
    86
rigid-body equilibrium 244–5

ring strengthening 361–2
Rivière Noirs bridge 363–5
Route 378 bridge 256–7
rupture failure 149, 310

S-glass 63
sacrificial test zones 345–6
safety factors 197, 221
section analysis 176–80, 221–3
seismic performance 1, 4, 12–14, 24, 35,
    210–11, 378
    wall ductility 26–7
sensors 340
sequence of installation 284–5
serviceability guidelines 204–5
serviceability requirements 210
setting-up a site 286–7
settlement repair 31
shear failure modes 141, 143–4, 248,
    311
shear strengthening 9–11, 15, 141–54
    debonding failure 149–51
    design code approach 144–5, 146
    design guidelines 205–7
    efficiency factor 205–6
    fibre orientation 142–3
    FRP contribution to strength 146–8
    predictive models 144–6
    rupture failure 149
    shear capacity 152
    side-bonding 142
    spacing for FRP strips 152–3, 262
    timber structures 34–5
    truss models 144, 145
shells 333
Shiriya-Zaki lighthouse 377–8
shrinkage of wood 32
side-bonding 142
sign support structures 367–9
single-lap shear tests 225–6, 338
single-side strengthening 28
Sins bridge 376–7
site activities 267, 286–90
    maintenance 288–9, 349–8
    method statements 287
    monitoring 288–9

setting-up 286–7
supervision 287–8, 289
see also inspection
site trials 329–31
size of columns effects 164–6
sizings 294
slabs 7, 359–61
Slattocks canal bridge 371–2
slenderness limits 26, 180, 183–4, 249
slip models 130–1
smoke inhibitors *see* fire resistance
solubility 55
solvent degreasing 89
sonic inspection methods 336–7
soundness of surfaces 104
spacing of reinforcements 152–3, 262
special inspections 329, 346
spinning processes 61
sprayed FRP composites 272, 280
stability of surfaces 104
staff training 329, 348
standard tests 332–3
steel 17, 20–1
  durablity of steel-FRP systems 312–15
  epoxy bonds 314–15
  reinforcements corrosion 300
  surface preparation 93–4, 95–6
  see also metallic structures
stiffness 34, 49, 61
stone masonry 307, 308
strain gauges 340
strength 49, 129–30, 135–6, 281
stress analysis 204, 224–7, 297
stress rupture 205
stress-strain curves 120–5, 161, 166, 173
stress-strain models
  analysis-oriented 172–6
  design guidelines 208
  design-oriented 166–72, 181–2
structural deficiencies 2–4
  change in use 3, 23–4, 31–2
  concrete structures 4–5
  masonry structures 23–4

metallic structures 17–18, 215
timber structures 31–2
see also degradation
structural modelling 238
structural repointing 28, 243–4
structural tests 341
substrate yield 219–20
supervision 287–8, 289
suppression of debonding failures 136–7
surface preparation 83–108, 203, 268–9, 284, 334, 354–5
  anodising 90
  assessment prior to bonding 102–5
  cast iron 95
  chemical treatments 90
  concrete 90–2, 354
  corona discharge 90
  coupling agents 86–8
  curvature of substrate 269, 270
  Dyne pens 105
  electrochemical treatments 90
  fire resistance 37, 56, 73–6, 90, 296, 325–6
  of FRP materials 98–102
  high-pressure washing 91
  inspection 334, 336
  key surface qualities 104
  mechanical treatments 89–90
  metallic materials 92–6, 355
  physical treatments 90
  primers 86–8
  purpose of 84–5
  solvent degreasing 89
  steel 93–4, 95–6
  timber 96–8, 355
  wrought iron 95
surface tension 85

T-glass 63
tapping the surface tests 336–7, 344
temperature
  and adhesive joints 217
  effects on durability 302

glass transition 37, 51–3, 84, 295, 338–9, 352
heat distortion (HDT) 75
and testing 352
tendons and cables 277, 282–3
tensile dumb-bell tests 338
tensile properties 67, 115, 216
testing
  acoustic emission techniques 345
  debonding 336–7, 344
  durability 77, 293
  flexural modulus tests 338
  interpretation of results 341–2
  joints 104–5, 107–8
  qualification for use 85
  radiographic techniques 345
  sacrificial test zones 345–6
  sampling 285, 331, 337–9
  single-lap shear tests 225–6, 338
  standard tests 332–3
  structural tests 341
  tapping the surface 336–7, 344
  and temperature 352
  tensile dumb-bell tests 338
  thermography techniques 344–5
  ultrasonic technology 345
thermal properties 51–3, 69–70, 76
thermography testing 344–5
thermoplastic binders 47–8
thermosetting binders 47, 48, 49–56
Theydon Bois viaduct 359
Tickford bridge 372–3
ties 30, 261
timber structures 31–5
  adhesives 96–7
  beams 32–4, 376–7
  bond performance 304
  durability 303–6
  flexural strengthening 32–4
  foundation piles 32
  moisture content 303–4
  seismic upgrade 35
  shear strengthening 34–5
  shrinkage 32
  Sins bridge 376–7

stiffness enhancement 34
structural deficiencies 31–2
surface preparation 96–8, 355
toughness of polymers 50
training 329, 348
truss models 10, 11, 144, 145
two-way slabs 7

U-anchor system 8, 10
U-jackets 136, 137, 153
ultrasonic technology 345
ultraviolet radiation 56, 76
unstressed systems 67, 230–1, 268–72
  adhesively bonded plates 268–9, 279
  Boots Building project 369–70
  in situ laminated FRP fabrics 269–70, 279
  in situ resin infusion 271–2, 280
  in situ vacuum cured FRP fabric 271, 279
  mechanically fixed FRP plates 272, 280
  near surface mounted (NSM) bars 270–1, 279
  pre-formed FRP plates 268
  Slattocks canal bridge 371–2
  sprayed FRP composites 272, 280
  Tickford bridge 372–3

vacuum cured FRP fabric 271, 279
vapor-permeable epoxy resins 300
vaults 23, 29–30, 258–9, 306–7
vertical bending 245–6
vibrating wire strain gauges 340
vinylesters 48, 49, 73
viscosity modifiers 56
visual inspections 327, 328, 336–7, 342–4

walls
  ductility 26–7
  exterior walls 309
  moisture ingress 308–9

seismic performance 378
slenderness ratio 26, 249
see also masonry structures
water degradation 300
West Burton Power Station 361–2
wet lay-up system 46, 65, 354
wet-blasting 92, 95
wettability of a surface 104

wood *see* timber structures
workmanship 278, 281
wraps *see* jacketing
wrought iron 17, 20–1
   surface preparation 95
Wynantskill Creek bridge 256–7

XXsys Technologies 46

B/S. # 515414   624.18923
                         STR